Arithmetic Duality Theorems

PERSPECTIVES IN MATHEMATICS, Vol. 1

J. Coates and S. Helgason, editors

Arithmetic Duality Theorems

J. S. Milne
Department of Mathematics
University of Michigan
Ann Arbor, Michigan

ACADEMIC PRESS, INC.
Harcourt Brace Jovanovich, Publishers

Boston Orlando San Diego
New York Austin London Sydney
Tokyo Toronto

Academic Press, Inc.
Orlando, Florida 32887

Library of Congress Cataloging in Publication Data

Milne, J. S., Date
Arithmetic duality theorems.

(Perspectives in mathematics)
Bibliography: p.
Includes index.
1. Fields, Algebraic. 2. Homology theory.
3. Duality theory (Mathematics) I. Title. II. Series.
QA247.M554 1986 512'.3 86-26451

ISBN 0-12-498040-6 (alk. paper)

9 8 7 6 5 4 3 2 1
Printed in USA

Contents

PREFACE

In the late fifties and early sixties, Tate (and Poitou) found some important duality theorems concerning the Galois cohomology of finite modules and abelian varieties over local and global fields.

About 1964, Artin and Verdier extended some of the results to étale cohomology groups over rings of integers in local and global fields.

Since then many people (Artin, Bester, Bégueri, Mazur, McCallum, the author, Roberts, Shatz, Vvedens'kii) have generalized these results to flat cohomology groups.

Much of the best of this work has not been fully published. My initial purpose in preparing these notes was simply to write down a complete set of proofs before they were forgotten, but I have also tried to give an organized account of the whole subject. Only a few of the theorems in these notes are new, but many results have been sharpened, and a significant proportion of the proofs have not been published before.

The first chapter proves the theorems on Galois cohomology announced by Tate in his talk at the International Congress at Stockholm in 1962, and describes later work in the same area. The second chapter proves the theorem of Artin and Verdier on étale cohomology and also various generalizations of it. In the final chapter improvements using flat cohomology are described.

As far as possible, theorems are proved in the context in which they are stated: thus theorems on Galois cohomology are proved using only Galois cohomology, and theorems on étale cohomology are proved using only étale cohomology.

Each chapter begins with a summary of its contents; each section ends with a list of its sources.

It is a pleasure to thank all those with whom I have discussed these questions over the years, but especially M. Artin, P. Berthelot, L. Breen, S. Bloch, K. Kato, S. Lichtenbaum, W. McCallum, B. Mazur, W. Messing, L. Roberts, and J. Tate.

Parts of the author's research contained in this volume has been supported by N.S.F.

Finally I mention that, thanks to the computer, it has been possible to produce this volume without recourse to typist, copy editor, or type-setter.

NOTATIONS AND CONVENTIONS

We list our usual notations and conventions. When they are not followed in a particular section, this is noted at the start of the section.

A *global field* is a finite extension of $\mathbb{Q}$ or is finitely generated and of transcendence degree one over a finite field. A *local field* is $\mathbb{R}$, $\mathbb{C}$, or a field that is locally compact relative to a discrete valuation. Thus it is a finite extension of $\mathbb{Q}_p$, $\mathbb{F}_p((T))$, or $\mathbb{R}$. If v is a prime of a global field, then $|\ |_v$ denotes the valuation at v normalized in the usual way so that the product formula holds, and $\mathcal{O}_v = \{a \in K \mid |a|_v \leq 1\}$. The completions of K and $\mathcal{O}_v$ relative to $|\ |_v$ are denoted by K_v and $\hat{\mathcal{O}}_v$.

For a field K, K_a and K_s denote the algebraic and separable algebraic closures of K, and K_{ab} denotes the maximal abelian extension of K. For a local field K, K_{un} is the maximal unramified extension of K. We sometimes write G_K for the absolute Galois group $Gal(K_s/K)$ of K and $G_{F/K}$ for $Gal(F/K)$. By char(K) we mean the characteristic exponent of K, that is, char(K) is p if K has characteristic $p \neq 0$ and is 1 otherwise. For a Hausdorff topological group G, G^{ab} is the quotient of G by the closure of its commutator subgroup. Thus G^{ab} is the maximal abelian Hausdorff quotient group of G, and $G_K^{ab} = Gal(K_{ab}/K)$.

If M is an abelian group (or, more generally, an object in an abelian category) and m is an integer, then M_m and $M^{(m)}$ are the kernel and cokernel of multiplication by m on M. Moreover M(m) is the m-primary component $\bigcup_n M_{m^n}$ and $M_{m\text{-div}}$ is the m-divisible subgroup $\bigcap_n \mathrm{Im}(m^n\colon M \longrightarrow M)$. The divisible subgroup M_{div} of M is $\bigcap_m M_{m\text{-div}}$. We write T_mM for $\varprojlim M_{m^n}$, and $\hat{M}$ for the completion of M with respect to the topology defined by the subgroups of finite index (sometimes the subgroups are restricted to those of finite index a power of a fixed integer m, and sometimes to those that are open with respect to some topology on M). If M is finite [M] denotes its order. A group is of *cofinite-type* if it is torsion and M_m is finite for all integers m.

As befits a work with the title of this one, we shall need to consider a many different types of duals. In general, M^* will denote $\mathrm{Hom}_{\mathrm{cts}}(M,\mathbb{Q}/\mathbb{Z})$, the group of continuous characters of finite order of M. Thus if M is a discrete torsion abelian group, then M^* is its compact Pontryagin dual, and if M is an profinite abelian group, then M^* is its discrete torsion Pontryagin dual. If M is a module over G_K for some field K, then M^D denotes the dual $\mathrm{Hom}(M,K_s^{\times})$; when M is a finite group scheme, M^D denotes the Cartier dual $\mathcal{H}om(M,\mathbb{G}_m)$. The dual (Picard variety) of an abelian variety is denoted by A^t. For a vector space M, $M^{\vee}$ denotes the linear dual of M.

All algebraic groups and group schemes will be commutative (unless stated otherwise). If T is a torus over a field k, then $X^*(T)$ is the group $\mathrm{Hom}_{k_s}(T_{k_s},\mathbb{G}_m)$ of characters of T and $X_*(T)$ is the group $\mathrm{Hom}_{k_s}(\mathbb{G}_m,T_{k_s})$ of cocharacters (also called multiplicative one-parameter subgroups).

There seems to be no general agreement on what signs should be used in homological algebra. Fortunately, the signs of the maps in these notes will not be important, but the reader should be aware that when a diagram is said to commute, it may only anticommute. I have generally followed the sign conventions in [Berthelot, Breen, and Messing (1982), Chapter 0].

We sometimes use = to denote a canonical isomorphism, and the symbol $X \overset{df}{=} Y$ means that X is defined to be Y, or that X equals Y by definition.

In Chapters II and III, we shall need to consider several different topologies on a scheme X (always assumed to be locally Noetherian or the perfection of a locally Noetherian scheme). These are denoted as follows:

X_{et} (small étale site) is the category of schemes étale over X endowed with the étale topology;

X_{Et} (big étale site) is the category of schemes locally of finite-type over X endowed with the étale topology;

X_{sm} (smooth site) is the category of schemes smooth over X endowed with the smooth topology (covering families are surjective families of smooth maps);

X_{qf} (small fpqf site) is the category of schemes flat and quasi-finite over X endowed with the flat topology;

X_{fl} (big flat site) is the category of schemes locally of finite-type over X endowed with the flat topology;

X_{pf} (perfect site) see (III.0).

The category of sheaves of abelian groups on a site X_* is denoted by $S(X_*)$.

CHAPTER I

GALOIS COHOMOLOGY

In §1 we prove a very general duality theorem that applies whenever one has a class formation. The theorem is used in §2 to prove a duality theorem for modules over the Galois group of a local field. This section also contains an expression for the Euler-Poincaré characteristic of such a module. In §3, these results are used to prove Tate's duality theorem for abelian varieties over a local field.

The next four sections concern global fields. Tate's duality theorem on modules over the Galois group of a global field is obtained in §4 by applying the general result in §1 to the class formation of the global field and combining the resulting theorem with the local results in §2. Section 5 derives a formula for the Euler-Poincaré characteristic of such a module. Tate's duality theorems for abelian varieties over global fields are proved in §6, and in the following section it is shown that the validity of the conjecture of Birch and Swinnerton-Dyer for an abelian variety over a number field depends only on the isogeny class of the variety.

The final three sections treat rather diverse topics. In §8 a duality theorem is proved for tori that implies the abelian case of Langlands's conjectures for a nonabelian class field theory. The next section briefly describes some of the applications that have been made of the duality theorems: to the Hasse principle for finite modules and algebraic groups, to the existence of forms of algebraic groups, to Tamagawa numbers of algebraic tori over global fields, and to the central embedding problem for Galois groups. In the appendix, a class field theory is developed for Henselian local fields whose residue fields are quasi-finite and for function fields in one variable over quasi-finite fields.

In this chapter, the reader is assumed to be familiar with basic Galois cohomology (the first two chapters of [Serre (1964)] or the first four chapters of [Shatz (1972)]), class field theory ([Serre (1967a)] and [Tate (1967a)]), and, in a few sections, abelian varieties ([Milne (1986b)]).

Throughout the chapter, when G is a profinite group, "G-module" will mean "discrete G-module", and the cohomology group $H^r(G,M)$ will be defined using continuous cochains. The category of discrete G-modules is denoted by $\mathbf{Mod}_G$.

§0 Preliminaries

Throughout this section, G will be a profinite group. By a torsion-free G-module, we mean a G-module that is torsion-free as an abelian group.

Tate (modified) cohomology groups

([Serre (1962), VII], [Weiss (1969)].)

When G is finite, there are *Tate cohomology groups* $H_T^r(G,M)$, $r \in \mathbb{Z}$, M a G-module, such that

$$H_T^r(G,M) = H^r(G,M),\quad r > 0,$$
$$H_T^0(G,M) = M^G/N_GM,\quad \text{where } N_G = \sum_{\sigma\in G}\sigma,$$
$$H_T^{-1}(G,M) = \mathrm{Ker}(N_G)/I_GM,\quad \text{where } I_G = \{\,\sum n_\sigma\sigma \mid \sum n_\sigma = 0\},$$
$$H_T^{-r}(G,M) = H_{r-1}(G,M),\quad -r < -1.$$

A short exact sequence of G-modules gives rise to a long exact sequence of Tate cohomology groups (infinite in both directions).

A *complete resolution* for G is an exact sequence

$$L. = \cdots \longrightarrow L_2 \xrightarrow{d_2} L_1 \xrightarrow{d_1} L_0 \xrightarrow{d_0} L_{-1} \xrightarrow{d_{-1}} L_{-2} \longrightarrow \cdots$$

of finitely generated free $\mathbb{Z}[G]$-modules, together with an element $e \in L_{-1}^G$ that generates the image of d_0. For any complete resolution of G, $H_T^r(G,M)$ is the rth cohomology group of the complex $\mathrm{Hom}_G(L.,M)$. The map d_0 factors as

$$L_0 \xrightarrow{\epsilon} \mathbb{Z} \xrightarrow{\iota} L_{-1}$$

with $\epsilon(x)e = d_0(x)$ and $\iota(m) = me$. If we let

$$L.^+ = \cdots \longrightarrow L_2 \xrightarrow{d_2} L_1 \xrightarrow{d_1} L_0$$

$$L.^- = L_{-1} \xrightarrow{d_{-1}} L_{-2} \longrightarrow L_{-3} \longrightarrow \cdots,$$

then $H^r(G,M) = H^r(\mathrm{Hom}_G(L.^+,M))$, $r \geq 0$, and $H_r(G,M) = H^{-r-1}(\mathrm{Hom}_G(L.^-,M))$, $r \geq 0$.

By the *standard resolution* $L.^+$ for G we mean the complex with $L_r^+ = \mathbb{Z}[G^r]$ and the usual boundary map, so that $\mathrm{Hom}(L.^+,M)$ is the complex of nonhomogeneous cochains of M (see [Serre (1962), VII.3]). By the *standard complete resolution* for G, we mean the complete resolution obtained by splicing together $L.^+$ with its dual (see [Weiss (1969), I-4-1]).

Except for Tate cohomology groups, we always set $H^r(G,M) = 0$ for $r < 0$.

For any bilinear G-equivariant pairing of G-modules

$$M \times N \longrightarrow P$$

there is a family of cup-product pairings

$$(x,y) \mapsto x \cup y\colon H_T^r(G,M) \times H_T^s(G,N) \longrightarrow H_T^{r+s}(G,M)$$

with the following properties:

(0.1.1) $dx \cup y = d(x \cup y)$;

(0.1.2) $x \cup dy = (-1)^{\deg(x)} d(x \cup y)$;

(0.1.3) $x \cup (y \cup z) = (x \cup y) \cup z$;

(0.1.4) $x \cup y = (-1)^{\deg(x)\deg(y)} y \cup x$;

(0.1.5) $\mathrm{Res}(x \cup y) = \mathrm{Res}(x) \cup \mathrm{Res}(y)$;

(0.1.6) $\mathrm{Inf}(x \cup y) = \mathrm{Inf}(x) \cup \mathrm{Inf}(y)$;

(d = boundary map, Res = restriction map; Inf = inflation map).

Theorem 0.2. (Tate-Nakayama) *Let G be a finite group, C a G-module, and u an element of $H^2(G,C)$. Suppose that for all subgroups H of G*

(a) $H^1(H,C) = 0$, *and*

(b) $H^2(H,C)$ *has order equal to that of H and is generated by* $\mathrm{Res}(u)$.

Then, for any G-module M such that $\mathrm{Tor}_1^{\mathbb{Z}}(M,C) = 0$, *cup-product with u defines an isomorphism*

$$x \mapsto x \cup u\colon H_T^r(G,M) \longrightarrow H_T^{r+2}(G,M\otimes C)$$

for all integers r.

Proof: [Serre (1962), IX.8].

Extensions of G-modules

For G-modules M and N, define $\mathrm{Ext}_G^r(M,N)$ to be the set of homotopy classes of morphisms $M^{\cdot} \longrightarrow N^{\cdot}$ of degree r, where $M^{\cdot}$ is any resolution of M by G-modules and $N^{\cdot}$ is any resolution of N by injective G-modules. One sees readily that different resolutions of M and N give rise to canonically isomorphic groups $\mathrm{Ext}_G^r(M,N)$. On taking $M^{\cdot}$ to be M itself, we see that $\mathrm{Ext}_G^r(M,N) = H^r(\mathrm{Hom}_G(M,N^{\cdot}))$, and so $\mathrm{Ext}_G^r(M,-)$ is the rth right derived functor of $N \mapsto \mathrm{Hom}_G(M,N)\colon \mathbf{Mod}_G \longrightarrow \mathbf{Ab}$. In particular, $\mathrm{Ext}_G^r(\mathbb{Z},N) = H^r(G,N)$.

There is a canonical product

$$(f,g) \mapsto f.g\colon \operatorname{Ext}^r_G(N,P) \times \operatorname{Ext}^s_G(M,N) \longrightarrow \operatorname{Ext}^{r+s}_G(M,P)$$

such that f.g is obtained from $f\colon N^\cdot \to P^\cdot$ and $g\colon M \to N^\cdot$ by composition (here $N^\cdot$ and $P^\cdot$ are injective resolutions of N and P). For $r = s = 0$, the product can be identified with composition

$$(f,g) \mapsto f\circ g\colon \operatorname{Hom}_G(N,P) \times \operatorname{Hom}_G(M,N) \longrightarrow \operatorname{Hom}_G(M,P).$$

When we take $M = \mathbb{Z}$, and replace N and P with M and N, the pairing becomes

$$\operatorname{Ext}^r_G(M,N) \times H^s(G,M) \longrightarrow H^{r+s}(G,N).$$

An r-fold extension of M by N defines in a natural way a class in $\operatorname{Ext}^r_G(M,N)$ (see [Bourbaki Alg. X.7.3] for one correct choice of signs). Two such extensions define the same class if and only if they are equivalent in the usual sense, and for $r \geq 1$, every element of $\operatorname{Ext}^r_G(M,N)$ arises from such an extension (ibid. X.7.5). Therefore $\operatorname{Ext}^r_G(M,N)$ can be identified with the set of equivalence classes of r-fold extensions of M by N. With this identification, products are obtained by splicing extensions (ibid. X.7.6). Let $f \in \operatorname{Ext}^r_G(N,P)$; then the map $g \mapsto f.g\colon \operatorname{Ext}^r_G(M,N) \longrightarrow \operatorname{Ext}^{r+s}_G(M,P)$ is the r-fold boundary map defined by any r-fold extension of N by P representing f.

A spectral sequence for Exts

Let M and N be G-modules, and write Hom(M,N) for the set of homomorphisms from M to N as abelian groups. For $f \in \operatorname{Hom}(M,N)$ and $\sigma \in G$, define σf to be $m \mapsto \sigma(f(\sigma^{-1}m))$. Then Hom(M,N) is a G-module, but it is not in general a discrete G-module. For a closed normal subgroup H of G, set

$$\mathcal{H}om_H(M,N) = \bigcup_U \operatorname{Hom}(M,N)^U \quad \text{(union over the open subgroups } U,\ H \subset U \subset G)$$

$= \{f \in \mathrm{Hom}(M,N) \mid \sigma f = f$ for all σ in some $U\}$.

Then $\mathcal{H}om_H(M,N)$ is a discrete G/H-module, and we define $\mathcal{E}xt^r_H(M,N)$ to be the rth right derived functor of the left exact functor

$$N \mapsto \mathcal{H}om_H(M,N): \mathbf{Mod}_G \longrightarrow \mathbf{Mod}_{G/H}.$$

In the case that $H = \{1\}$, we drop it from the notation; in particular, $\mathcal{H}om(M,N) = \cup \mathrm{Hom}(M,N)^U$ with U running over all the open subgroups of G. If M is finitely generated, then $\mathcal{H}om_H(M,N) = \mathrm{Hom}_H(M,N)$, and so $\mathcal{E}xt^r_H(M,N) = \mathrm{Ext}^r_H(M,N)$; in particular, $\mathcal{H}om(M,N) = \mathrm{Hom}(M,N)$ (homomorphisms as abelian groups).

Theorem 0.3. *Let H be a closed normal subgroup of G, and let N and P be G-modules. Then, for any G/H-module M such that* $\mathrm{Tor}^{\mathbb{Z}}_1(M,N) = 0$, *there is a spectral sequence*

$$\mathrm{Ext}^r_{G/H}(M, \mathcal{E}xt^s_H(N,P)) \Rightarrow \mathrm{Ext}^{r+s}_G(M \otimes_{\mathbb{Z}} N, P).$$

Proof: This will be shown to be the spectral sequence of a composite of functors, but first we need some lemmas.

Lemma 0.4. *For any G-modules N and P and G/H-module M, there is a canonical isomorphism*

$$\mathrm{Hom}_{G/H}(M, \mathcal{H}om_H(N,P)) \xrightarrow{\approx} \mathrm{Hom}_G(M \otimes_{\mathbb{Z}} N, P).$$

Proof: There is a standard isomorphism

$$\mathrm{Hom}(M, \mathrm{Hom}(N,P)) \xrightarrow{\approx} \mathrm{Hom}(M \otimes_{\mathbb{Z}} N, P).$$

Take G-invariants. On the left we get $\mathrm{Hom}_G(M, \mathrm{Hom}(N,P))$, which equals $\mathrm{Hom}_G(M, \mathrm{Hom}_H(N,P))$ because M is a G/H-module, and equals $\mathrm{Hom}_G(M, \mathcal{H}om_H(N,P))$ because M is a *discrete* G/H-module. On the right we get $\mathrm{Hom}_G(M \otimes_{\mathbb{Z}} N, P)$.

Lemma 0.5. *If I is an injective G-module and N is a torsion-free G-module, then* $\mathcal{H}om_H(N,I)$ *is an injective G/H-module.*

Proof: We have to check that

$$\mathrm{Hom}_{G/H}(-,\mathcal{H}om_H(N,I))\colon \mathbf{Mod}_{G/H} \longrightarrow \mathbf{Ab}$$

is an exact functor, but (0.4) expresses it as the composite of the two exact functors $-\otimes_{\mathbb{Z}}N$ and $\mathrm{Hom}_G(-,I)$.

Lemma 0.6. *Let N and I be G-modules with I injective, and let M be a G/H-module. Then there is a canonical isomorphism*

$$\mathrm{Ext}^r_{G/H}(M,\mathcal{H}om_H(N,I)) \xrightarrow{\approx} \mathrm{Hom}_G(\mathrm{Tor}^{\mathbb{Z}}_r(M,N),I).$$

Proof: We use a resolution of N

$$0 \longrightarrow N_1 \longrightarrow N_0 \longrightarrow N \longrightarrow 0$$

by torsion-free G-modules to compute $\mathrm{Tor}^{\mathbb{Z}}_r(M,N)$. Thus $\mathrm{Tor}^{\mathbb{Z}}_1(M,N)$ and $\mathrm{Tor}^{\mathbb{Z}}_0(M,N) = M\otimes_{\mathbb{Z}}N$ fit into an exact sequence

$$0 \longrightarrow \mathrm{Tor}^{\mathbb{Z}}_1(M,N) \longrightarrow M\otimes_{\mathbb{Z}}N_1 \longrightarrow M\otimes_{\mathbb{Z}}N_0 \longrightarrow \mathrm{Tor}^{\mathbb{Z}}_0(M,N) \longrightarrow 0,$$

and $\mathrm{Tor}^{\mathbb{Z}}_r(M,N) = 0$ for $r \geq 2$. For each open subgroup U of G containing H, there is a short exact sequence

$$\begin{array}{ccccccccc} 0 \to & \mathrm{Hom}_G(\mathbb{Z}[G/U]\otimes_{\mathbb{Z}}N,I) & \to & \mathrm{Hom}_G(\mathbb{Z}[G/U]\otimes_{\mathbb{Z}}N_0,I) & \to & \mathrm{Hom}_G(\mathbb{Z}[G/U]\otimes_{\mathbb{Z}}N_1,I) & \to 0. \\ & \| & & \| & & \| & \\ & \mathrm{Hom}_U(N,I) & & \mathrm{Hom}_U(N_0,I) & & \mathrm{Hom}_U(N_1,I) & \end{array}$$

The direct limit of these sequences is an injective resolution

$$0 \longrightarrow \mathcal{H}om_H(N,I) \longrightarrow \mathcal{H}om_H(N_0,I) \longrightarrow \mathcal{H}om_H(N_1,I) \longrightarrow 0$$

of $\mathcal{H}om_H(N,I)$, which we use to compute $\mathrm{Ext}^r_{G/H}(M,\mathcal{H}om_H(N,I))$. In the diagram

$$\begin{array}{ccc} \mathrm{Hom}_{G/H}(M,\mathcal{H}om_H(N_0,I)) & \xrightarrow{\alpha} & \mathrm{Hom}_{G/H}(M,\mathcal{H}om_H(N_1,I)) \\ \downarrow\approx & & \downarrow\approx \\ \mathrm{Hom}_G(M\otimes_{\mathbb{Z}}N_0,I) & \xrightarrow{\beta} & \mathrm{Hom}_G(M\otimes_{\mathbb{Z}}N_1,I), \end{array}$$

we have

$$\mathrm{Ker}(\alpha) = \mathrm{Hom}_{G/H}(M,\mathcal{H}om_H(N,I)), \quad \mathrm{Coker}(\alpha) = \mathrm{Ext}^1_{G/H}(M,\mathcal{H}om_H(N,I)),$$

$$\mathrm{Ker}(\beta) = \mathrm{Hom}_G(\mathrm{Tor}^{\mathbb{Z}}_0(M,N),I), \quad \mathrm{Coker}(\beta) = \mathrm{Hom}_G(\mathrm{Tor}^{\mathbb{Z}}_1(M,N),I).$$

Thus the required isomorphisms are induced by the vertical maps in the diagram.

We now prove the theorem. Lemma 0.4 shows that $\mathrm{Hom}_G(M\otimes_{\mathbb{Z}}N,-)$ is the composite of the functors $\mathcal{H}om_H(N,-)$ and $\mathrm{Hom}_{G/H}(M,-)$, and Lemma 0.6 shows that the first of these maps injective objects I to objects that are acyclic for the second functor. Thus the spectral sequence arises in the standard way from a composite of functors [Hilton and Stammbach (1970)].

Example 0.7. Let M = N = $\mathbb{Z}$, and replace P with M. The spectral sequence then becomes the Hochschild-Serre spectral sequence

$$H^r(G/H,H^s(H,M)) \Rightarrow H^{r+s}(G,M).$$

Example 0.8. Let M = $\mathbb{Z}$ and H = {1}, and replace N and P with M and N. The spectral sequence then becomes

$$H^r(G,\mathcal{E}xt^s(M,N)) \Rightarrow \mathrm{Ext}^{r+s}_G(M,N).$$

When M is finitely generated, this is simply a long exact sequence

$$0 \to H^1(G,\mathrm{Hom}(M,N)) \to \mathrm{Ext}^1_G(M,N) \to H^0(G,\mathrm{Ext}^1(M,N)) \to H^2(G,\mathrm{Hom}(M,N)) \to.$$

In particular, when we also have that N is divisible by all primes occurring as the order of an element of M, then $\mathrm{Ext}^1(M,N) = 0$, and so

$$H^r(G,\mathrm{Hom}(M,N)) = \mathrm{Ext}^r_G(M,N).$$

Example 0.9. In the case that N = $\mathbb{Z}$, the spectral sequence becomes

$$\mathrm{Ext}^r_{G/H}(M, H^s(H,P)) \Rightarrow \mathrm{Ext}^{r+s}_G(M,P).$$

The map $\mathrm{Ext}^r_{G/H}(M,P^H) \longrightarrow \mathrm{Ext}^r_G(M,P)$ is obviously an isomorphism for $r = 0$; the spectral sequence shows that it is an isomorphism for $r = 1$ if $H^1(H,P) = 0$, and that it is an isomorphism for all r if $H^r(H,P) = 0$ for all $r > 0$.

Remark 0.10. Assume that M is finitely generated. It follows from the long exact sequence in (0.8) that $\mathrm{Ext}^r_G(M,N)$ is torsion for $r \geq 1$. Moreover, if G and N are written compatibly as $G = \varprojlim G_i$ and $N = \varinjlim N_i$ (N_i is a G_i-module) and the action of G on M factors through each G_i, then

$$\mathrm{Ext}^r_G(M,N) = \varinjlim \mathrm{Ext}^r_{G_i}(M,N_i).$$

Remark 0.11. Let H be an closed subgroup of G, and let M be an H-module. The corresponding *induced* G-module M_* is the set of continuous maps $a\colon G \longrightarrow M$ such that $a(hx) = h.a(x)$ all $h \in H$, $x \in G$. The group G acts on M_* by the rule: $(ga)(x) = a(xg)$. The functor $M \mapsto M_*\colon \mathbf{Mod}_H \longrightarrow \mathbf{Mod}_G$ is right adjoint to the functor $\mathbf{Mod}_G \longrightarrow \mathbf{Mod}_H$ "regard a G-module as an H-module"; in other words,

$$\mathrm{Hom}_G(N,M_*) \xrightarrow{\approx} \mathrm{Hom}_H(N,M), \text{ N a G-module, M an H-module.}$$

Both functors are exact, and therefore $M \mapsto M_*$ preserves injectives and the isomorphism extends to isomorphisms $\mathrm{Ext}^r_G(N,M_*) \xrightarrow{\approx} \mathrm{Ext}^r_H(N,M)$ all r. In particular, there are canonical isomorphisms $H^r(G,M_*) \xrightarrow{\approx} H^r(H,M)$ for all r. (Cf. [Serre (1964), I.2.5].)

Augmented cup-products

Certain pairs of pairings give rise to cup-products with a dimension shift.

Proposition 0.12. *Let*

$$0 \longrightarrow M' \longrightarrow M \longrightarrow M'' \longrightarrow 0$$

$$0 \longrightarrow N' \longrightarrow N \longrightarrow N'' \longrightarrow 0$$

be exact sequences of G*-modules. Then a pair of pairings*

$$M' \times N \longrightarrow P$$

$$M \times N' \longrightarrow P$$

coinciding on $M' \times N'$ *defines a canonical family of (augmented cup-product) pairings*

$$H^r(G,M'') \times H^s(G,N'') \longrightarrow H^{r+s+1}(G,P).$$

Proof: See [Lang (1966), Chapter V].

Remark 0.13. (a) The augmented cup-products have properties similar to those listed in (0.1) for the usual cup-product.

(b) Augmented cup-products have a very natural definition in terms of hypercohomology. The tensor product of two complexes

$$(M^0 \xrightarrow{d_M} M^1) \otimes (N^0 \xrightarrow{d_N} N^1)$$

is defined to be the complex

$$M^0 \otimes N^0 \xrightarrow{d^0} M^1 \otimes N^0 \oplus M^0 \otimes N^1 \xrightarrow{d^1} M^1 \otimes N^1$$

with

$$d^0(x \otimes y) = d_M(x) \otimes y + x \otimes d_N(y),$$

$$d^1(x \otimes y + x' \otimes y') = x \otimes d_N(y) - d_M(x') \otimes y'.$$

With the notations in the proposition, let $M^{\cdot} = (M' \longrightarrow M)$ and $N^{\cdot} = (N' \longrightarrow N)$. Also write $P[-1]$ for the complex with P in the degree one and zero elsewhere. Then the hypercohomology groups $\mathbb{H}^r(G,M^{\cdot})$, $\mathbb{H}^r(G,N^{\cdot})$, and $\mathbb{H}^r(G,P[-1])$ equal $H^{r-1}(G,M'')$, $H^{r-1}(G,N'')$, and $H^{r-1}(G,P)$ respectively, and to give a pair of pairings as in the proposition is the same as to give a map of complexes $M^{\cdot} \otimes N^{\cdot} \longrightarrow P[-1]$. Such a pair therefore defines a cup-product pairing

$$\mathbb{H}^r(G,M^{\cdot}) \times \mathbb{H}^s(G,N^{\cdot}) \longrightarrow \mathbb{H}^{r+s}(G,P[-1]),$$

and this is the augmented cup-product.

Compatibility of pairings

We shall need to know how the Ext and cup-product pairings compare.

Proposition 0.14. (a) *Let* $M \times N \longrightarrow P$ *be a pairing of G-modules, and consider the maps* $M \longrightarrow \mathcal{H}om(N,P)$ *and*

$$H^r(G,M) \longrightarrow H^r(G,\mathcal{H}om(N,P)) \longrightarrow \mathrm{Ext}^r_G(N,P)$$

induced by the pairing and the spectral sequence in (0.3). *Then the diagram*

$$\begin{array}{ccccc} H^r(G,M) & \times & H^s(G,N) & \longrightarrow & H^{r+s}(G,P) \quad (\textit{cup-product}) \\ \downarrow & & \| & & \| \\ \mathrm{Ext}^r_G(N,P) & \times & H^s(G,N) & \longrightarrow & H^{r+s}(G,P) \quad (\textit{Ext pairing}) \end{array}$$

commutes (up to sign).

(b) *Consider a pair of exact sequences*

$$0 \longrightarrow M' \longrightarrow M \longrightarrow M'' \longrightarrow 0$$

$$0 \longrightarrow N' \longrightarrow N \longrightarrow N'' \longrightarrow 0$$

and a pair of pairings

$$M' \times N \longrightarrow P$$

$$M \times N' \longrightarrow P$$

coinciding on $M' \times N'$. *This data gives rise to canonical maps* $H^r(G,M'') \longrightarrow \mathrm{Ext}^{r+1}_G(N'',P)$, *and the diagram*

$$\begin{array}{ccccc} H^r(G,M'') & \times H^s(G,N'') & \longrightarrow H^{r+s+1}(G,P) & & (\textit{augmented cup-product}) \\ \downarrow & \| & \| & & \\ \mathrm{Ext}_G^{r+1}(N'',P) & \times H^s(G,N'') & \longrightarrow H^{r+s+1}(G,P) & & (\textit{Ext pairing}) \end{array}$$

commutes (up to sign).

Proof: (a) This is standard, at least in the sense that everyone assumes it to be true. There is a proof in a slightly more general context in [Milne (1980), V.1.20], and [Gamst and Hoechsmann, (1970)] contains a very full discussion of such things. (See also the discussion of pairings in the derived category in III.0.)

(b) The statement in (a) holds also if M, N, and P are complexes. If we regard the pair of pairings in (b) as a pairing of complexes $M^{\cdot} \times N^{\cdot} \longrightarrow P[-1]$ (notations as (0.13b)) and replace M, N, and P in (a) with $M^{\cdot}$, $N^{\cdot}$, and $P[-1]$, then the diagram in (a) becomes that in (b). (Explicity, the map $H^r(G,M'') \longrightarrow \mathrm{Ext}_G^{r+1}(N'',P)$ is obtained as follows: the pair of pairings defines a map of complexes $M^{\cdot} \longrightarrow \mathcal{H}om(N^{\cdot},P[-1])$, and hence a map $\mathbb{H}^r(G,M^{\cdot}) \longrightarrow \mathbb{H}^r(G,\mathcal{H}om(N^{\cdot},P[-1]))$; but $\mathbb{H}^r(G,M^{\cdot}) = H^{r-1}(G,M)$, and there is an edge morphism $\mathbb{H}^r(G,\mathcal{H}om(N^{\cdot},P[-1])) \longrightarrow \mathrm{Ext}_G^r(N^{\cdot},P[-1]) = \mathrm{Ext}_G^r(N'',P)$.)

Conjugation of cohomology groups

Consider two profinite groups G and G', a G-module M, and a G'-module M'. A homomorphism $f\colon G' \longrightarrow G$ and an additive map $h\colon M \longrightarrow M'$ are said to be *compatible* if $h(f(g').m) = g'.h(m)$ for $g' \in G'$ and $m \in M$. Such a pair induces homomorphisms $(f,h)_*^r\colon H^r(G,M) \longrightarrow H^r(G',M')$ for all r.

Proposition 0.15. *Let M be a G-module, and let* $\sigma \in G$. *The maps* $\mathrm{ad}(\sigma) = (g \mapsto \sigma g \sigma^{-1})\colon G \longrightarrow G$ *and* $\sigma^{-1} = (m \mapsto \sigma^{-1}m)\colon M \longrightarrow M$ *are comp-*

atible, and $(\mathrm{ad}(\sigma),\sigma^{-1})^r_* : H^r(G,M) \longrightarrow H^r(G,M)$ *is the identity map for all* r.

Proof: The first assertion is obvious, and the second needs only to be checked for r = 0, where it is also obvious (see [Serre (1962), VII.5]).

The proposition is useful in the following situation. Let K be a global field and v a prime of K. The choice of an embedding $K_s \hookrightarrow K_{v,s}$ over K amounts to choosing an extension w of v to K_s, and the embedding identifies G_{K_v} with the decomposition group D_w of w in G_K. A second embedding is the composite of the first with ad(σ) for some σ ∈ G (because G_K acts transitively on the extensions of v to K_s). Let M be a G_K-module. An embedding $K_s \hookrightarrow K_{v,s}$ defines a map $H^r(G_K,M) \longrightarrow H^r(G_{K_v},M)$, and the proposition shows that the map is independent of the choice of the embedding.

Extensions of algebraic groups

Let k be a field, and let $G = \mathrm{Gal}(k_s/k)$. The category of algebraic group schemes over k is an abelian category $\mathbf{Gp}_k$ (recall that all group schemes are assumed to be commutative), and therefore it is possible to define $\mathrm{Ext}^r_k(A,B)$ for objects A and B of $\mathbf{Gp}_k$ to be the set of equivalence classes of r-fold extensions of A by B (see [Mitchell (1965), VII]). Alternatively, one can chose a projective resolution $A^{\cdot}$ of A in the pro-category $\mathbf{Pro\text{-}Gp}_k$, and define $\mathrm{Ext}^r_k(A,B)$ to be the set of homotopy classes of maps $A^{\cdot} \longrightarrow B$ of degree r (see [Oort (1966)] or [Demazure and Gabriel (1970), V.2]). For any object A of $\mathbf{Gp}_k$, $A(k_s)$ is discrete G-module, and we often write $H^r(k,A)$ for $H^r(G,A(k_s))$.

Proposition 0.16. *Assume that* k *is perfect.*

(a) *The functor* $A \mapsto A(k_s)$: $\mathbf{Gp}_k \longrightarrow \mathbf{Mod}_G$ *is exact.*

(b) *For all objects* A *and* B *in* $\mathbf{Gp}_k$, *there exists a canonical pairing*

$$\mathrm{Ext}^r_k(A,B) \times H^s(k,A) \to H^{r+s}(k,B).$$

Proof: (a) This is obvious since k_s is algebraically closed.

(b) The functor in (a) sends an r-fold exact sequence in $\mathbf{Gp}_k$ to an r-fold exact sequence in $\mathbf{Mod}_G$, and it therefore defines a canonical map $\mathrm{Ext}^r_k(A,B) \longrightarrow \mathrm{Ext}^r_G(A(k_s),B(k_s))$. We define the pairing to be that making

$$\begin{array}{ccccc} \mathrm{Ext}^r_G(A,B) & \times & H^s(k,A) & \longrightarrow & H^{r+s}(k,A) \\ \downarrow & & \| & & \| \\ \mathrm{Ext}^r_G(A(k_s),B(k_s)) & \times & H^s(G,A(k_s)) & \longrightarrow & H^{r+s}(G,B(k_s)) \end{array}$$

commute.

Proposition 0.17. *Assume that* k *is perfect, and let* A *and* B *be algebraic group schemes over* k. *Then there is a spectral sequence*

$$H^r(G,\mathrm{Ext}^s_{k_s}(A,B)) \Rightarrow \mathrm{Ext}^{r+s}_k(A,B).$$

Proof: See [Milne (1970a)].

Corollary 0.18. *If* k *is perfect and* N *is a finite group scheme over* k *of order prime to* char(k), *then* $\mathrm{Ext}^r_k(N,\mathbb{G}_m) = \mathrm{Ext}^r_G(N(k_s),k_s^\times)$ *all* r.

Proof: Clearly $\mathrm{Hom}_{k_s}(N,\mathbb{G}_m) = \mathrm{Hom}_G(N(k_s),k_s^\times)$, and the table [Oort (1966), p II.14-2] shows that $\mathrm{Ext}^s_{k_s}(N,\mathbb{G}_m) = 0$ for $s > 0$. Therefore the proposition implies that $\mathrm{Ext}^r_k(N,\mathbb{G}_m) = H^r(G,\mathrm{Hom}_G(N(k_s),k_s^\times)$, which equals $\mathrm{Ext}^r_G(N(k_s),k_s^\times)$ by (0.8).

Topological abelian groups

Let M be an abelian group. In the next proposition we write M^ for the m-adic completion $\varprojlim M/m^nM$ of M, and we let $\mathbb{Z}_m = \prod_{\ell | m} \mathbb{Z}_\ell = \mathbb{Z}\hat{}$ and $\mathbb{Q}_m = \prod_{\ell | m} \mathbb{Q}_\ell = \mathbb{Z}_m \otimes_{\mathbb{Z}} \mathbb{Q}$.

Proposition 0.19. (a) *For any abelian group* M, $M\hat{} = (M/M_{m\text{-div}})\hat{}$; *if* M *is finite, then* $M\hat{} = M(m)$, *and if* M *is finitely generated, then* $M = M \otimes_{\mathbb{Z}} \mathbb{Z}_m$.

(b) *For any abelian group* M, $\varinjlim M^{(m^n)} = (M \otimes_{\mathbb{Z}} \mathbb{Q}/\mathbb{Z})(m)$, *which is zero if* M *is torsion and is isomorphic to* $(\mathbb{Q}_m/\mathbb{Z}_m)^r$ *if* M *is finitely generated of rank* r.

(c) *For any abelian group*, $T_m M = \operatorname{Hom}(\mathbb{Q}_m/\mathbb{Z}_m, M) = T_m(M_{m\text{-div}})$; *it is torsion-free.*

(d) *Write* $M^* = \operatorname{Hom}_{cts}(M, \mathbb{Q}_m/\mathbb{Z}_m)$; *then for any finitely generated abelian group* M, $M^* = (M\hat{})^*$ *and* $M^{**} = M\hat{}$.

(e) *Let* M *be a discrete torsion abelian group and* N *a totally disconnected compact abelian group, and let*

$$M \times N \longrightarrow \mathbb{Q}/\mathbb{Z}$$

be a continuous pairing that identifies each group with the Pontryagin dual of the other. Then the exact annihilator of N_{tors} *is* M_{div}, *and so there is a nondegenerate pairing*

$$M/M_{div} \times N_{tors} \longrightarrow \mathbb{Q}/\mathbb{Z}.$$

Proof: Easy.

Note that the proposition continues to hold if we take $m = "\prod p"$, that is, we take M^ be the profinite completion of M, $M_{m\text{-div}}$ to be M_{div}, M(m) to be M_{tor}, and so on.

We shall be concerned with the exactness of completions and duals of exact sequences. Note that the completion of the exact sequence

$$0 \longrightarrow \mathbb{Z} \longrightarrow \mathbb{Q} \longrightarrow \mathbb{Q}/\mathbb{Z} \longrightarrow 0$$

for the profinite topology is

$$0 \longrightarrow \hat{\mathbb{Z}} \longrightarrow 0 \longrightarrow 0 \longrightarrow 0,$$

which is far from being exact. To be able to state a good result, we need the notion of a strict morphism. Recall [Bourbaki Tpgy, III.2.8] that a continuous homomorphism $f\colon G \longrightarrow H$ of topological groups is said to be a *strict morphism* if the induced map $G/\mathrm{Ker}(f) \longrightarrow f(G)$ is an isomorphism of topological groups. Equivalently, f is strict if the image of every open subset of G is open in $f(G)$. Every continuous homomorphism of a compact group to a Hausdorff group is strict, and obviously every continuous homomorphism from a topological group to a discrete group is strict. The Baire category theorem implies that a continuous homomorphism from a locally compact σ-compact group to a locally compact Hausdoff group is a strict morphism [Hewitt and Ross (1963), 5.29]. (A space is *σ-compact* if it is a countable union of compact subspaces.)

Recall also that it is possible to define the completion $\hat{G}$ of a topological group when the group has a basis of neighbourhoods (G_i) for the identity element consisting of normal subgroups; in fact, $\hat{G} = \varprojlim G/G_i$. In the next proposition, we write G^* for the full Pontryagin dual of a topological group G.

Proposition 0.20. *Let*

$$G' \xrightarrow{f} G \xrightarrow{g} G''$$

be an exact sequence of abelian topological groups and strict morphisms.

(a) *Assume that the topologies on* G', G, *and* G'' *are defined by neighbourhood bases consisting of subgroups; then the sequence of completions*

$$\hat{G}' \longrightarrow \hat{G} \longrightarrow \hat{G}''$$

is also exact.

(b) *Assume that the groups are locally compact and Hausdorff and that the image of* G *is closed in* G''; *then the dual sequence*

$$G''^{*} \longrightarrow G^{*} \longrightarrow G'^{*}$$

is also exact.

Proof: By assumption, we have a diagram

$$\begin{array}{ccccc} & & G/\mathrm{Im}(f) & \xrightarrow{\approx} & \mathrm{Im}(g) \\ & & \uparrow & & \downarrow b \\ G' & \xrightarrow{f} & G & \xrightarrow{g} & G'' \\ \downarrow a & & \uparrow & & \\ G'/\mathrm{Ker}(f) & \xrightarrow{\approx} & \mathrm{Im}(f). & & \end{array}$$

When we complete, the map a remains surjective, the middle column remains a short exact sequence, and b remains injective because in each case a subgroup has the subspace topology and a quotient group the quotient topology (see [Atiyah and MacDonald (1969), 10.3]). Since the isomorphisms obviously remain isomorphisms, (a) is now clear.

The proof of (b) is similar, except that it makes use of the fact that for any closed subgroup K of a locally compact abelian group G, the exact sequence

$$0 \longrightarrow K \longrightarrow G \longrightarrow G/K \longrightarrow 0$$

gives rise to an exact dual sequence

$$0 \longrightarrow (G/K)^* \longrightarrow G^* \longrightarrow K^* \longrightarrow 0.$$

Note that in (b) of the theorem, the image of G in G" will be closed if it is the kernel of a homomorphism from G" into a Hausdorff group.

The right derived functors of $\varprojlim$

The category of abelian groups satisfies the condition Ab5: the direct limit of an exact sequence of abelian groups is again exact. Unfortunately, the corresponding statement for inverse limits is false, although the formation of inverse limits is always a left exact operation (and the product of a family of exact sequences is exact).

Proposition 0.21. *Let* **A** *be an abelian category satisfying the condition* Ab5 *and having enough injectives, and let* I *be a filtered ordered set. Then for any object* B *of* **A** *and any direct system* (A_i) *of objects of* **A** *indexed by* I, *there is a spectral sequence*

$$\varprojlim^{(r)} \mathrm{Ext}^s_{\mathbf{A}}(A_i, B) \Rightarrow \mathrm{Ext}^{r+s}_{\mathbf{A}}(\varinjlim A_i, B),$$

where $\varprojlim^{(r)}$ *denotes the* r^{th} *right derived functor of* $\varprojlim$.

Proof: [Roos (1961)].

Proposition 0.22. *Let* (A_i) *be an inverse system of abelian groups indexed by* $\mathbb{N}$ *with its natural order.*

(a) *For* $r \geq 2$, $\varprojlim^{(r)} A_i = 0$.

(b) *If each* A_i *is finitely generated, then* $\varprojlim^{(1)} A_i$ *is divisible, and it is uncountable when nonzero.*

(c) *If each* A_i *is finite, then* $\varprojlim^{(1)} A_i = 0$.

Proof: (a) See [Roos (1961)].

(b) See [Jensen (1972), 2.5].

(c) See [Jensen (1972), 2.3].

Corollary 0.23. *Let* $\mathbf{A}$ *be an abelian category satisfying* Ab5 *and having enough injectives, and let* (A_i) *be a direct system of objects of* $\mathbf{A}$ *indexed by* $\mathbb{N}$. *If* B *is such that* $\mathrm{Ext}^s_{\mathbf{A}}(A_i, B)$ *is finite for all* s *and* i, *then*

$$\varprojlim \mathrm{Ext}^s_{\mathbf{A}}(A_i, B) = \mathrm{Ext}^s_{\mathbf{A}}(\varinjlim A_i, B).$$

The kernel-cokernel exact sequence of a pair of maps

The following simple result will find great application in these notes.

Proposition 0.24. *For any pair of maps* $A \xrightarrow{f} B \xrightarrow{g} C$ *of abelian groups, there is an exact sequence*

$$0 \longrightarrow \mathrm{Ker}(f) \longrightarrow \mathrm{Ker}(g \circ f) \longrightarrow \mathrm{Ker}(g) \longrightarrow$$
$$\mathrm{Coker}(f) \longrightarrow \mathrm{Coker}(g \circ f) \longrightarrow \mathrm{Coker}(g) \longrightarrow 0.$$

Proof: An easy exercise.

Notes: The subsection "A spectral sequence for Exts" is based on [Tate (1966)]. The rest of the material is fairly standard.

§1 Duality relative to a class formation

Class formations

Consider a profinite group G, a G-module C, and a family of

isomorphisms

$$\mathrm{inv}_U\colon H^2(U,C) \xrightarrow{\approx} \mathbb{Q}/\mathbb{Z}$$

indexed by the open subgroups U of G. Such a system is said to be a *class formation* if

(1.1a) for all open subgroups $U \subset G$, $H^1(U,C) = 0$, and

(1.1b) for all pairs of open subgroups $V \subset U \subset G$, the diagram

$$\begin{array}{ccc} H^2(U,C) & \xrightarrow{\mathrm{Res}_{V,U}} & H^2(V,C) \\ \downarrow \mathrm{inv}_U & & \downarrow \mathrm{inv}_V \\ \mathbb{Q}/\mathbb{Z} & \xrightarrow{n} & \mathbb{Q}/\mathbb{Z} \end{array}$$

commutes with $n = (U:V)$. The map inv_U is called the *invariant map relative to* U.

When V is a normal subgroup of U of index n, the conditions imply that there is an exact commutative diagram

$$\begin{array}{ccccccccc} 0 & \longrightarrow & H^2(U/V,C^V) & \longrightarrow & H^2(U,C) & \xrightarrow{\mathrm{res}_{U,V}} & H^2(V,C) & \longrightarrow & 0 \\ & & \approx\downarrow \mathrm{inv}_{U/V} & & \approx\downarrow \mathrm{inv}_U & & \approx\downarrow \mathrm{inv}_V & & \\ 0 & \longrightarrow & \frac{1}{n}\mathbb{Z}/\mathbb{Z} & \longrightarrow & \mathbb{Q}/\mathbb{Z} & \xrightarrow{n} & \mathbb{Q}/\mathbb{Z} & \longrightarrow & 0 \end{array}$$

in which $\mathrm{inv}_{U/V}$ is defined to be the restriction of inv_U. In particular, for a normal open subgroup U of G of index n, there is an isomorphism

$$\mathrm{inv}_{G/U}\colon H^2(G/U,C^U) \xrightarrow{\approx} \tfrac{1}{n}\mathbb{Z}/\mathbb{Z},$$

and we write $u_{G/U}$ for the element of $H^2(G/U,C^U)$ mapping to $1/n$. Thus $u_{G/U}$ is the unique element of $H^2(G/U,C^U)$ such that $\mathrm{inv}_G(\mathrm{Inf}(u_{G/U})) = 1/n$.

Lemma 1.2. *Let* M *be a* G*-module such that* $\mathrm{Tor}_1^{\mathbb{Z}}(M,C) = 0$. *Then the map*

$$a \mapsto a \cup u_{G/U}: H_T^r(G/U,M) \longrightarrow H_T^{r+2}(G/U,M\otimes_{\mathbb{Z}} C^U)$$

is an isomorphism for all open normal subgroups U *of* G *and integers* r.

Proof: Apply (0.2) to G/U, C^U, and $u_{G/U}$.

Theorem 1.3. *Let* (G,C) *is a class formation; then there is a canonical map* $rec_G: C^G \longrightarrow G^{ab}$ *whose image in* G^{ab} *is dense and whose kernel is the group* $\cap N_{G/U}C^U$ *of universal norms.*

Proof: Take $M = \mathbb{Z}$ and $r = -2$ in the lemma. As $H_T^{-2}(G/U,\mathbb{Z}) = (G/U)^{ab}$ and $H_T^0(G/U,C^U) = C^G/N_{G/U}C^U$, the lemma gives an isomorphism $(G/U)^{ab} \xrightarrow{\approx} C^G/N_{G/U}C^U$. On passing to the projective limit over the inverses of these maps, we obtain an injective map $C^G/\cap N_{G/U}U \longrightarrow G^{ab}$. The map rec_G is the composite of this with the projection of C^G onto $C^G/\cap N_{G/U}U$. It has dense image because, for all open normal subgroups U of G, its composite with $G^{ab} \longrightarrow (G/U)^{ab}$ is surjective.

The map rec_G is called the *reciprocity map*.

Question 1.4. Is there a derivation of (1.3), no more difficult than the above one, that avoids the use of homology groups?

Remark 1.5. (a) The following description of rec_G will be useful. The cup-product pairing

$$H^0(G,C) \times H^2(G,\mathbb{Z}) \longrightarrow H^2(G,C)$$

can be identified with a pairing

$$\langle\ ,\ \rangle: C^G \times \mathrm{Hom}_{cts}(G,\mathbb{Q}/\mathbb{Z}) \longrightarrow \mathbb{Q}/\mathbb{Z},$$

and the reciprocity map is uniquely determined by the equation

$$\langle c,\chi\rangle = \chi(\mathrm{rec}_G(c)), \text{ all } c \in C^G,\ \chi \in \mathrm{Hom}_{cts}(G^{ab},\mathbb{Q}/\mathbb{Z}).$$

See [Serre (1962), XI.3, Pptn 2].

(b) The definition of a class formation that we have adopted is slightly stronger than the usual definition (see [Artin and Tate (1961), XIV]) in that we require inv_U to be an isomorphism rather than an injection inducing isomorphisms $H^2(U/V,C^V) \xrightarrow{\approx} (U:V)^{-1}\mathbb{Z}/\mathbb{Z}$ for all open subgroups $V \subset U$ with V normal in U. It is equivalent to the usual definition plus the condition that the order of G (as a profinite group) is divisible by all integers n.

Example 1.6. (a) Let G be a profinite group isomorphic to $\hat{\mathbb{Z}}$ (completion of $\mathbb{Z}$ for the topology of subgroups of finite index), and let $C = \mathbb{Z}$ with G acting trivially. Choose a topological generator σ of G. For each m, G has a unique open subgroup U of index m, and σ^m generates U. The boundary map in the cohomology sequence of

$$0 \longrightarrow \mathbb{Z} \longrightarrow \mathbb{Q} \longrightarrow \mathbb{Q}/\mathbb{Z} \longrightarrow 0$$

is an isomorphism $H^1(U,\mathbb{Q}/\mathbb{Z}) \longrightarrow H^2(U,\mathbb{Z})$, and we define inv_U to be the composite of the inverse of this isomorphism with

$$\begin{array}{rcl} H^1(U,\mathbb{Q}/\mathbb{Z}) = \mathrm{Hom}_{cts}(U,\mathbb{Q}/\mathbb{Z}) & \longrightarrow & \mathbb{Q}/\mathbb{Z} \\ f & \mapsto & f(\sigma^m) \end{array}.$$

Note that inv_U depends on the choice of σ. Clearly $(G,\mathbb{Z})$ with these maps is a class formation. The reciprocity map is injective but not surjective.

(b) Let G be the Galois group $\mathrm{Gal}(K_s/K)$ of a nonarchimedean local field K, and let $C = K_s^\times$. If $I = \mathrm{Gal}(K_s/K_{un})$, then the inflation map $H^2(G/I,K_{un}^\times) \longrightarrow H^2(G,K_s^\times)$ is an isomorphism, and we define inv_G to be the composite of its inverse with the isomorphisms

$$H^2(G/I,K_{un}^\times) \xrightarrow{\mathrm{ord}} H^2(G/I,\mathbb{Z}) \xrightarrow{\mathrm{inv}_{G/I}} \mathbb{Q}/\mathbb{Z},$$

where $\mathrm{inv}_{G/I}$ is the map in defined in (a) (with the choice of the Frobenius automorphism for σ). Define inv_U analogously. Then $(G, K_s^\times)$ is a class formation (see [Serre (1967a), §1] or the appendix to this chapter). The reciprocity map is injective but not surjective.

(c) Let G be the Galois group $\mathrm{Gal}(K_s/K)$ of a global field K, and let $C = \varinjlim C_L$ where L runs through the finite extensions of K in K_s and C_L is the idèle class group of L. For each prime v of K, choose an embedding of K_s into $K_{v,s}$ over K. Then there is a unique isomorphism $\mathrm{inv}_G\colon H^2(G,C) \longrightarrow \mathbb{Q}/\mathbb{Z}$ making the diagram

$$\begin{array}{ccc} \mathrm{inv}_G\colon H^2(G,C) & \longrightarrow & \mathbb{Q}/\mathbb{Z} \\ \uparrow & & \| \\ \mathrm{inv}_v\colon H^2(G_v, K_{v,s}^\times) & \longrightarrow & \mathbb{Q}/\mathbb{Z} \end{array}$$

commute for all v (including the real primes) with inv_v the map defined in (b) unless v is real, in which case it is the unique injection. Define inv_U analogously. Then (G,C) is a class formaion (see [Tate (1967a), §11]). In the number field case, the reciprocity map is surjective with divisible kernel, and in the function field case it is injective but not surjective.

(d) Let K be a field complete with respect to a discrete valuation having an algebraically closed residue field k, and let G = $\mathrm{Gal}(K_s/K)$. For a finite separable extension L of K, let R_L be the ring of integers in L. There is a pro-algebraic group U_L over k such that $U_L(k) = R_L^\times$. Let $\pi_1(U_L)$ be the pro-algebraic étale fundamental group of U_L, and let $\pi_1(U) = \varinjlim \pi_1(U_L)$, $K \subset L \subset K_s$, $[L:K] < \infty$. Then $\pi_1(U)$ is a discrete G-module and $(G, \pi_1(U))$ is a class formation. In this case the reciprocity map is an isomorphism. See [Serre (1961), 2.5 Pptn 11, 4.1 Thm 1].

(e) Let K be an algebraic function field in one variable over an algebraically closed field k of characteristic zero. For each finite extension L of K, let $C_L = \mathrm{Hom}(\mathrm{Pic}(X_L), \mu(k))$, where X_L is the smooth complete algebraic curve over k with function field L and $\mu(k)$ is the group of roots on unity in k. Then the duals of the norm maps $\mathrm{Pic}(X_{L'}) \longrightarrow \mathrm{Pic}(X_L)$, $L' \supset L$, make the family (C_L) into a direct system, and we let C be the limit of the system. The pair (G,C) is a class formation for which the reciprocity map is surjective but not injective. See [Kawada and Tate (1955)] and [Kawada (1960)].

(f) For numerous other examples of class formations, see [Kawada (1971)].

The main theorem

For each G-module M, the pairings of §0

$$\mathrm{Ext}^r_G(M,C) \times H^{2-r}(G,M) \longrightarrow H^2(G,C) \xrightarrow{\mathrm{inv}} \mathbb{Q}/\mathbb{Z}$$

induce maps

$$\alpha^r(G,M)\colon \mathrm{Ext}^r_G(M,C) \longrightarrow H^{2-r}(G,M)^*.$$

In particular, for r = 0 and $M = \mathbb{Z}$, we obtain a map

$$\alpha^0(G,\mathbb{Z})\colon C^G \longrightarrow H^2(G,\mathbb{Z})^* = \mathrm{Hom}_{cts}(G,\mathbb{Q}/\mathbb{Z})^* = G^{ab}.$$

Lemma 1.7. *In the case that* $M = \mathbb{Z}$, *the maps* $\alpha^r(G,M)$ *have the following description:*

$\alpha^0(G,\mathbb{Z})\colon C^G \longrightarrow G^{ab}$ *is equal to* rec_G;

$\alpha^1(G,\mathbb{Z})\colon 0 \longrightarrow 0$;

$\alpha^2(G,\mathbb{Z})\colon H^2(G,C) \xrightarrow{\approx} \mathbb{Q}/\mathbb{Z}$ *is equal to* inv_G.

In the case that $M = \mathbb{Z}/m\mathbb{Z}$, *the maps* $\alpha^r(G,M)$ *have the following description:*

$\alpha^0(G,\mathbb{Z}/m\mathbb{Z})\colon (C^G)_m \longrightarrow (G^{ab})_m$ *is induced by* rec_G;

$\alpha^1(G,\mathbb{Z}/m\mathbb{Z})\colon (C^G)^{(m)} \longrightarrow (G^{ab})^{(m)}$ *is induced by* rec_G;

$\alpha^2(G,\mathbb{Z}/m\mathbb{Z})\colon H^2(G,C)_m \longrightarrow \frac{1}{m}\mathbb{Z}/\mathbb{Z}$ *is the isomorphism induced by* inv_G.

Proof: Only the assertion about $\alpha^0(G,\mathbb{Z})$ requires proof. As we observed in (1.5a), $\mathrm{rec}_G\colon H^0(G,C) \longrightarrow H^2(G,\mathbb{Z})^*$ is the map induced by the cup-product pairing

$$H^0(G,C) \times H^2(G,\mathbb{Z}) \longrightarrow H^2(G,C) \approx \mathbb{Q}/\mathbb{Z},$$

and we know (0.14) that this agrees with the Ext pairing.

Theorem 1.8. *Let* (G,C) *be a class formation, and let* M *be a finitely generated* G*-module.*

(a) *The map* $\alpha^r(G,M)$ *is bijective for all* $r \geq 2$, *and* $\alpha^1(G,M)$ *is bijective for all torsion-free* M. *In particular,* $\mathrm{Ext}^r_G(M,C) = 0$ *for* $r \geq 3$.

(b) *The map* $\alpha^1(G,M)$ *is bijective for all* M *if* $\alpha^1(U,\mathbb{Z}/m\mathbb{Z})$ *is bijective for all open subgroups* U *of* G *and all* m.

(c) *The map* $\alpha^0(G,M)$ *is surjective (respectively bijective) for all finite* M *if in addition* $\alpha^0(U,\mathbb{Z}/m\mathbb{Z})$ *is surjective (respectively bijective) for all* U *and* m.

Proof: The first step is to show that the domain and target of $\alpha^r(G,M)$ are both zero for large r.

Lemma 1.9. *For* $r \geq 4$, $\mathrm{Ext}^r_G(M,C) = 0$; *when* M *is torsion-free,* $\mathrm{Ext}^3_G(M,C)$ *is also zero.*

Proof: Every finitely generated G-module M can be resolved

$$0 \longrightarrow M_1 \longrightarrow M_0 \longrightarrow M \longrightarrow 0$$

by finitely generated torsion-free G-modules M_i. It therefore suffices to prove that for any torsion-free module M, $\mathrm{Ext}^r_G(M,C) = 0$ for $r \geq 3$. Let $N = \mathrm{Hom}(M,\mathbb{Z})$. Then $N\otimes_{\mathbb{Z}}C \approx \mathrm{Hom}(M,C)$ as G-modules, and so

(0.8) provides an isomorphism $\mathrm{Ext}^r_G(M,C) \approx H^r(G,N\otimes_{\mathbb{Z}}C)$. Note that this last group is equal to $\varinjlim H^r(G/U,N\otimes_{\mathbb{Z}}C^U)$ where the limit is over the open normal subgroups of G for which $N^U = N$. The theorem of Tate and Nakayama (0.2) shows that

$$a \mapsto a\cup u_{G/U}\colon H^{r-2}(G/U,N) \longrightarrow H^r(G/U,N\otimes_{\mathbb{Z}}C^U)$$

is an isomorphism for all $r \geq 3$. The diagram

$$\begin{array}{ccc} H^{r-2}(G/U,N) & \xrightarrow{\approx} & H^r(G/U,N\otimes_{\mathbb{Z}}C^U) \\ \downarrow (U:V)\mathrm{Inf} & & \downarrow \mathrm{Inf} \\ H^{r-2}(G/V,N) & \xrightarrow{\approx} & H^r(G/V,N\otimes_{\mathbb{Z}}C^V) \end{array}$$

commutes because $\mathrm{Inf}(u_{G/U}) = (U:V)u_{G/V}$ and $\mathrm{Inf}(a\cup b) = \mathrm{Inf}(a)\cup\mathrm{Inf}(b)$. As $H^{r-2}(G/U,N)$ is torsion for $r - 2 \geq 1$, and the order of U is divisible by all integers n, the limit $\varinjlim H^{r-2}(G/U,N)$ (taken relative to the maps $(U:V)\mathrm{Inf}$) is zero for $r - 2 \geq 1$, and this shows that $H^r(G,N\otimes_{\mathbb{Z}}C) = 0$ for $r \geq 3$.

Lemma 1.9 shows that the statements of the theorem are true for $r \geq 4$, and (1.7) shows that they are true for $r \leq 2$ whenever the action of G on M is trivial. Moreover, (1.9) shows that $\mathrm{Ext}^3_G(\mathbb{Z},C) = 0$, and it follows that $\mathrm{Ext}^3_G(\mathbb{Z}/m\mathbb{Z},C) = 0$ because $\mathrm{Ext}^2_G(\mathbb{Z},C)$ is divisible. Thus the theorem is true whenever the action of G on M is trivial. We embed a general M into an exact sequence

$$0 \longrightarrow M \longrightarrow M_* \longrightarrow M_1 \longrightarrow 0$$

with U an open normal subgroup of G such that $M^U = M$ and $M_* = \mathrm{Hom}(\mathbb{Z}[G/U],M) = \mathbb{Z}[G/U]\otimes_{\mathbb{Z}}M$. As $H^r(G,M_*) = H^r(U,M)$ and $\mathrm{Ext}^r_G(M_*,C) = \mathrm{Ext}^r_U(M,C)$ (apply (0.3) to $\mathbb{Z}[G/U]$, M, and C), there is an exact commutative diagram

$$\begin{array}{ccccccccc}
\to \mathrm{Ext}^r_G(M_1,C) & \to & \mathrm{Ext}^r_U(M,C) & \to & \mathrm{Ext}^r_G(M,C) & \to & \mathrm{Ext}^{r+1}_G(M_1,C) & \to & \dots \\
\downarrow \alpha^r(G,M_1) & & \downarrow \alpha^r(U,M) & & \downarrow \alpha^r(G,M) & & \downarrow \alpha^{r+1}(G,M_1) & & \\
\to H^{2-r}(G,M_1)^* & \to & H^{2-r}(U,M)^* & \to & H^{2-r}(G,M)^* & \to & H^{1-r}(G,M_1)^* & \to & \dots .
\end{array} \quad (1.9.1)$$

The maps $\alpha^3(U,M)$, $\alpha^4(G,M_1)$, and $\alpha^4(U,M)$ are all isomorphisms, and so the five-lemma shows that $\alpha^3(G,M)$ is surjective. Since this holds for all M, $\alpha^3(G,M_1)$ is also surjective, and now the five-lemma shows that $\alpha^3(G,M)$ is an isomorphism. The same argument shows that $\alpha^2(G,M)$ is an isomorphism. If M is torsion-free, so also are M_* and M_1, and so the same argument shows that $\alpha^1(G,M)$ is an isomorphism when M is torsion-free. The rest of the proof proceeds similarly.

Example 1.10. Let $(G,\mathbb{Z})$ be the class formation defined by a group $G \approx \hat{\mathbb{Z}}$ and a generator σ of G. The reciprocity map is the inclusion $n \mapsto \sigma^n\colon \mathbb{Z} \to G$. As $\hat{\mathbb{Z}}/\mathbb{Z}$ is uniquely divisible, we see that both $\alpha^0(U,\mathbb{Z}/m\mathbb{Z})$ and $\alpha^1(U,\mathbb{Z}/m\mathbb{Z})$ are isomorphisms for all m, and so the theorem implies that $\alpha^r(G,M)$ is an isomorphism for all finitely generated M, $r \geq 1$, and $\alpha^0(G,M)$ is an isomorphism for all finite M.

In fact, $\alpha^0(G,M)$ defines an isomorphism $\mathrm{Hom}_G(M,\mathbb{Z})^\wedge \to H^2(G,M)$ for all finitely generated M. To see this, note that $\mathrm{Hom}_G(M,\mathbb{Z})$ is finitely generated and $\mathrm{Ext}^1(M,\mathbb{Z})$ is finite (because $H^1(G,M)$ is) for all finitely generated M. Therefore, on tensoring the first four terms of the long exact sequence of Exts with $\hat{\mathbb{Z}}$, we obtain an exact sequence

$$0 \to \mathrm{Hom}_G(M_1,\mathbb{Z})^\wedge \to \mathrm{Hom}_U(M,\mathbb{Z})^\wedge \to \mathrm{Hom}_G(M,\mathbb{Z})^\wedge \to \mathrm{Ext}^1_G(M_1,\mathbb{Z}) \to \dots .$$

When we replace the top row of (1.9.1) with this sequence, the argument proving the theorem descends all the way to $r = 0$.

When M is finite, $\mathrm{Ext}^r(M,\mathbb{Z}) = 0$ for $r \neq 1$ and $\mathrm{Ext}^1(M,\mathbb{Z}) = \mathrm{Hom}(M,\mathbb{Q}/\mathbb{Z}) = M^*$. Therefore $\mathrm{Ext}^r_G(M,\mathbb{Z}) = H^{r-1}(G,M^*)$ (by (0.3)), and so we have a non-degenerate cup-product pairing

$$H^r(G,M) \times H^{1-r}(G,M^*) \longrightarrow H^1(G,\mathbb{Q}/\mathbb{Z}) \approx \mathbb{Q}/\mathbb{Z}.$$

When M is torsion-free, $\mathrm{Ext}^r(M,\mathbb{Z}) = 0$ for $r \neq 0$ and $\mathrm{Hom}(M,\mathbb{Z})$ is the linear dual $M^\vee$ of M. Therefore $\mathrm{Ext}^r(M,\mathbb{Z}) = H^r(G,M^\vee)$, and so the map $H^r(G,M^\vee) \longrightarrow H^{2-r}(G,M)^*$ defined by cup-product is bijective for $r \geq 1$, and induces a bijection $H^0(G,M^\vee)^\wedge \longrightarrow H^2(G,M)^*$ in the case $r = 0$.

Example 1.11. Let K be a field for which there exists a class formation (G,C) with $G = \mathrm{Gal}(K_s/K)$, and let T be a torus over K. The character group $X^*(T)$ of T is a finitely generated torsion-free G-module with $\mathbb{Z}$-linear dual the cocharacter group $X_*(T)$, and so the pairing

$$\mathrm{Ext}^r_G(X^*(T),C) \times H^{2-r}(G,X^*(T)) \longrightarrow H^2(G,C) \approx \mathbb{Q}/\mathbb{Z}$$

defines an isomorphism $\mathrm{Ext}^r_G(X^*(T),C) \longrightarrow H^{2-r}(G,X^*(T))^*$ for $r \geq 1$. According to (0.8), $\mathrm{Ext}^r_G(X^*(T),C) = H^r(G,\mathrm{Hom}(X^*(T),C))$, and $\mathrm{Hom}(X^*(T),C) = X_*(T)\otimes C$. Therefore the cup-product pairing

$$H^r(G,X_*(T)\otimes C) \times H^{2-r}(G,X^*(T)) \longrightarrow H^2(G,C) \approx \mathbb{Q}/\mathbb{Z}$$

induced by the natural pairing between $X_*(T)$ and $X^*(T)$ defines an isomorphism $H^r(G,X_*(T)\otimes C) \longrightarrow H^{2-r}(G,X^*(T))^*$ for $r \geq 1$.

Remark 1.12. Let (G,C) be a class formation. In [Brumer (1966)] there is a very useful criterion for C to have strict cohomological dimension 2. Let $V \subset U \subset G$ be open subgroups with V normal in U. We get an exact sequence

$$0 \longrightarrow \mathrm{Ker}(\mathrm{rec}_V) \longrightarrow C^V \xrightarrow{\mathrm{rec}_V} V^{ab} \longrightarrow \mathrm{Coker}(\mathrm{rec}_V) \longrightarrow 0,$$

of U/V-modules which induces a double connecting homomorphism

$$d\colon H^{r-2}_T(U/V,\mathrm{Coker}(\mathrm{rec}_V)) \longrightarrow H^r_T(U/V,\mathrm{Ker}(\mathrm{rec}_U)).$$

The theorem states that $scd_p(G) = 2$ if and only if, for all such pairs $V \subset U$, d induces an isomorphism on the p-primary components for all r. In each of the examples (1.6a,b,d) and in the function field case of (c), the kernel of rec_V is zero and the cokernel is uniquely divisible and hence has trivial cohomology. In the number field case of (c) the cohomology groups of the kernel are elementary 2-groups, which are zero if and only if the field is totally imaginary [Artin and Tate (1961), IX.2]. Consequently $scd_p(G) = 2$ in examples (1.6a,b,c,d) except when p = 2 and K is a number field having a real prime.

On the other hand, let K be a number field and let G_S be the Galois group over K of the maximal extension of K unramified outside a set of primes S. The statement in [Tate (1962), p292] that $scd_p(G_S) = 2$ for all primes p that are units at all v in S (except for p = 2 when K is not totally complex) is still unproven in general. As was pointed out by A. Brumer, it is equivalent to the non-vanishing of certain p-adic regulators.

A generalization

We shall need a generalization of Theorem 1.8. For any set P of rational prime numbers, we define a *P-class formation* to be a system $(G,C,(inv_U)_U)$ as at the start of this section except that, instead of requiring the maps inv_U to be isomorphisms, we require them to be injections satisfying the following two conditions:

(a) for all open subgroups V and U of G with V a normal subgroup of U, the map $inv_{U/V}$: $H^2(U/V,C^V) \longrightarrow (U:V)^{-1}\mathbb{Z}/\mathbb{Z}$ is an isomorphism, and

(b) for all open subgroups U of G and all primes ℓ in P, the map on ℓ-primary components $H^2(U,C)(\ell) \longrightarrow (\mathbb{Q}/\mathbb{Z})(\ell)$ induced by inv_U is an isomorphism.

Thus when P contains all prime numbers, a P-class formation is a class formation in the sense of the first paragraph of this section, and when P is the empty set, a P-class formation is a class formation in the sense of [Artin and Tate (1961)]. Note that, in the presence of the other conditions, (b) is equivalent to the order of G being divisible by ℓ^{∞} for all ℓ in P. If (G,C) is a class formation and H is a normal closed subgroup of G, then $(G/H, C^H)$ is a P-class formation with P equal to the set primes ℓ such that ℓ^{∞} divides (G:H).

If (G,C) is a P-class formation, then everything said above continues to hold provided that, at certain points, one restricts attention to the ℓ-primary components for ℓ in P. (Recall (0.10) that $\mathrm{Ext}^r_G(M,N)$ is torsion for $r \geq 1$.) In particular, the following theorem holds.

Theorem 1.13. *Let (G,C) be a P-class formation, let ℓ be a prime in P, and let M be a finitely generated G-module.*

(a) *The map* $\alpha^r(G,M)(\ell)\colon \mathrm{Ext}^r_G(M,C)(\ell) \longrightarrow H^{2-r}(G,M)^*(\ell)$ *is bijective for all $r \geq 2$, and $\alpha^1(G,M)(\ell)$ is bijective for all torsion-free M.*

(b) *The map $\alpha^1(G,M)(\ell)$ is bijective for all M if $\alpha^1(U,\mathbb{Z}/\ell^m\mathbb{Z})$ is bijective for all open subgroups U of G and all m.*

(c) *The map $\alpha^0(G,M)$ is surjective (respectively bijective) for all finite ℓ-primary M if in addition $\alpha^0(U,\mathbb{Z}/\ell^m\mathbb{Z})$ is surjective (respectively bijective) for all U and m.*

Exercise 1.14. Let $K = \mathbb{Q}(d^{1/2})$ where d is chosen so that the 2-class field tower of K is infinite. Let K_{un} be the maximal unramified extension of K, and let $H = \mathrm{Gal}(K_s/K_{un})$. Then $(G_K/H, C^H)$ is a P-class formation with $P = \{2\}$. Investigate the maps $\alpha^r(G_K/H,M)$ in this case.

Notes: Theorem 1.8 and its proof are taken from [Tate (1966)]

§2 Local fields

Unless stated otherwise, K will be a nonarchimedean local field, complete with respect to the discrete valuation ord: $K^\times \twoheadrightarrow \mathbb{Z}$, and with finite residue field k. Let R be the ring of integers in K, and let K_{un} be the maximal unramified extension of K. Write G = $Gal(K_s/K)$ and $I = Gal(K_s/K_{un})$. As we noted in (1.6b), $(G, K_s^\times)$ has a natural structure of a class formation. The reciprocity map $rec_G \colon K^\times \longrightarrow G^{ab}$ is known to be injective with dense image. More precisely, there is an exact commutative diagram

$$\begin{array}{ccccccccc} 0 & \longrightarrow & R^\times & \longrightarrow & K^\times & \xrightarrow{ord} & \mathbb{Z} & \longrightarrow & 0 \\ & & \downarrow \approx & & \downarrow & & \downarrow & & \\ 0 & \longrightarrow & I^{ab} & \longrightarrow & G^{ab} & \longrightarrow & \hat{\mathbb{Z}} & \longrightarrow & 0 \end{array}$$

in which all the vertical arrows are injective and I^{ab} is the inertia subgroup of G^{ab}. The norm groups in $K^\times$ are the open subgroups of finite index. See [Serre (1962), XIII.4, XIV.6].

In this section N^ will denote the completion of a group N relative to the topology defined by the subgroups of N of finite index unless N has a topology induced in a natural way from that on K, in which case we allow only subgroups of finite index that are open relative to the topology. With this definition, $(R^\times)\hat{} = R^\times$, and the reciprocity map defines an isomorphism $(K^\times)\hat{} \longrightarrow G_K^{ab}$. When M is a discrete G-module, the group $Hom_G(M, K_s^\times)$ inherits a topology from that on K_s, and in the next theorem $\mathrm{Hom}_G(M,K_s^\times)\hat{}$ denotes its completion for the topology defined by the open subgroups of finite index[1].

As $\hat{\mathbb{Z}}/\mathbb{Z}$ is uniquely divisible, $\alpha^0(G,\mathbb{Z}/m\mathbb{Z})$ and $\alpha^1(G,\mathbb{Z}/m\mathbb{Z})$ are isomorphisms for all m. Thus most of the following theorem is an immediate consequence of Theorem 1.8.

Theorem 2.1. *Let M be a finitely generated G-module, and consider*

$$\alpha^r(G,M)\colon \mathrm{Ext}^r_G(M,K_s^\times) \longrightarrow H^{2-r}(G,M)^*.$$

Then $\alpha^r(G,M)$ *is an isomorphism for all* $r \geq 1$, *and* $\alpha^0(G,M)$ *defines an isomorphism (of profinite groups)*

$$\mathrm{Hom}_G(M,K_s^\times)\hat{} \longrightarrow H^2(G,M)^*.$$

The ^ *can be omitted if* M *is finite. The groups* $\mathrm{Ext}^r_G(M,K_s^\times)$ *and* $H^r(G,M)$ *are finite for all* r *if* M *is of finite order prime to* char(K), *and the groups* $\mathrm{Ext}^1_G(M,K_s^\times)$ *and* $H^1(G,M)$ *are finite for all finitely generated* M *whose torsion subgroup is of order prime to* char(K).

Proof: We begin with the finiteness statements. For n prime to char(K), the cohomology sequence of the Kummer sequence

$$0 \longrightarrow \mu_n(K_s) \longrightarrow K_s^\times \xrightarrow{\ n\ } K_s^\times \longrightarrow 0$$

---

[1] If n is prime to the characteristic of K, then $K^{\times n}$ is an open subgroup of finite index in $K^\times$. It follows that every subgroup of $K^\times$ (hence of $\mathrm{Hom}_G(M,K_s^\times)$) of finite index prime to char(K) is open. In contrast, when the characteristic of K is $p \neq 0$, there are many subgroups of finite index in $K^\times$ that are not closed. In fact (see [Weil (1967), II.3, Pptn 10]), $1 + \mathfrak{m} \approx \prod \mathbb{Z}_p$ (product of countably many copies of $\mathbb{Z}_p$), and a proper subgroup of $\prod \mathbb{Z}_p$ containing $\oplus\, \mathbb{Z}_p$ can not be closed.

shows that the cohomology groups $H\ \mu_n(K_s)$ are $\mu_n(K)$, $K^{\times}/K^{\times n}$, $\frac{1}{n}\mathbb{Z}/\mathbb{Z}$, 0 respectively for $r = 0, 1, 2, \geq 3$. In particular, they are all finite.

Let M be a finite G-module of order prime to char(K), and choose a finite Galois extension L of K containing all m^{th} roots of 1 for m dividing the order of M and such that $\mathrm{Gal}(K_s/L)$ acts trivially on M. Then M is isomorphic as a $\mathrm{Gal}(K_s/L)$-module to a direct sum of copies of modules of the form μ_m, and so the groups $H^s(\mathrm{Gal}(K_s/L),M)$ are finite for all s, and zero for $s \geq 3$. The Hochschild-Serre spectral sequence

$$H^r(\mathrm{Gal}(L/K),H^s(\mathrm{Gal}(K_s/L),M)) \Rightarrow H^{r+s}(G,M)$$

now shows that the groups $H^r(G,M)$ are all finite because the cohomology groups of a finite group with values in a finite (even finitely generated for $r \geq 1$) module are finite. This proves that $H^r(G,M)$ is finite for all r and all M of finite order prime to char(K), and Theorem 1.8 shows that all the $\alpha^r(G,M)$ are isomorphisms for finite M, and so the groups $\mathrm{Ext}^r_G(M,K_s^{\times})$ are also finite.

Let M be a finitely generated G-module whose torsion subgroup has order prime to char(K). In proving that $H^1(G,M)$ is finite, we may assume that M is torsion-free. Let L be a finite Galois extension of K such that $\mathrm{Gal}(K_s/L)$ acts trivially on M. The exact sequence

$$0 \longrightarrow H^1(\mathrm{Gal}(L/K),M) \longrightarrow H^1(\mathrm{Gal}(K_s/K),M) \longrightarrow H^1(\mathrm{Gal}(K_s/L),M)$$

shows that $H^1(G,M)$ is finite because the last group in the sequence is zero and the first is finite. Theorem 1.8 implies that $\alpha^r(G,M)$ is an isomorphism for $r \geq 1$ and all finitely generated M, and so $\mathrm{Ext}^1_G(M,K_s^{\times})$ is also finite.

It remains to prove the assertion about $\alpha^0(G,M)$. Note that

$\alpha^0(G,\mathbb{Z})$ defines an isomorphism $(K^\times)^\wedge \longrightarrow G^{ab}$, and so the statement is true if G acts trivially on M. Let L be a finite Galois extension of K such $Gal(K_s/L)$ acts trivially on M. Then $Hom_G(M,K_s^\times) = Hom_G(M,L^\times)$, and $Hom_G(M,L^\times)$ contains an open compact group $Hom_G(M,\mathcal{O}_w^\times)$, where $\mathcal{O}_w$ is the ring of integers in L. Using this, it is easy to prove that the maps

$$0 \longrightarrow Hom_G(M_1,K_s^\times) \longrightarrow Hom_G(M_*,K_s^\times) \longrightarrow Hom_G(M,K_s^\times) \longrightarrow$$

in the top row of (1.9.1) are strict morphisms. Therefore the sequence remains exact when we complete the first three terms (see (0.20)), and so the same argument as in (1.8) completes the proof.

Corollary 2.2. *If M is a countable G-module whose torsion is prime to* char(K), *then*

$$\alpha^1(G,M)\colon Ext^1_G(M,K_s^\times) \longrightarrow H^1(G,M)^*$$

is an isomorphism.

Proof: Write M as a countable union of finitely generated G-modules M_i and note that $Ext^1_G(M,K_s^\times) = \varprojlim Ext^1_G(M_i,K_s^\times)$ by (0.23).

For any finitely generated G-module M, write $M^D = Hom(M,K_s^\times)$. It is again a discrete G-module, and it acquires a topology from that on $K_s^\times$.

Corollary 2.3. *Let M be a finitely generated G-module whose torsion subgroup has order prime to* char(K). *Then cup-product defines an isomorphism*

$$H^r(G,M^D) \longrightarrow H^{2-r}(G,M)^*$$

for all $r \geq 1$, *and an isomorphism (of compact groups)*

$$H^0(G,M^D)^\wedge \longrightarrow H^2(G,M)^*.$$

The groups $H^1(G,M)$ *and* $H^1(G,M^D)$ *are finite.*

Proof: As $K_s^\times$ is divisible by all primes other than char(K), $\mathrm{Ext}^r(M,K_s^\times) = 0$ for all $r > 0$, and so $\mathrm{Ext}_G^r(M,K_s^\times) = H^r(G,M^D)$ for all r (see (0.8)).

Corollary 2.4. *Let* T *be a commutative algebraic group over* K *whose identity component* T^o *is a torus. Assume that the order of* T/T^o *is not divisible by the characteristic of* K, *and let* $X^*(T)$ *be the group of characters of* T. *Then cup-product defines a dualities between:*

the compact group $H^0(K,T)^\wedge$ *(completion relative to the topology of open subgroups of finite index) and the discrete group* $H^2(G,X^*(T))$;

the finite groups $H^1(K,T)$ *and* $H^1(G,X^*(T))$;

the discrete group $H^2(K,T)$ *and the compact group* $H^0(G,X^*(T))^\wedge$ *(completion relative to the topology of subgroups of finite index). In particular,* $H^2(K,T) = 0$ *if and only if* $X^*(T)^G = 0$ *(when* T *is connected, this last condition is equivalent to* T(K) *being compact).*

Proof: The G-module $X^*(T)$ is finitely generated without char(K)-torsion, and $X^*(T)^D = T(K_s)$, and so this follows from the preceding corollary (except for the parenthetical statement, which we leave as an exercise — cf. [Serre (1964), pII-26]).

Remark 2.5. (a) If the characteristic of K is $p \neq 0$ and M has elements of order p, the $\mathrm{Ext}_G^1(M,K_s^\times)$ and $H^1(G,M)$ are usually infinite. For example $\mathrm{Ext}_G^1(\mathbb{Z}/p\mathbb{Z},K_s^\times) = K^\times/K^{\times p}$ and $H^1(G,\mathbb{Z}/p\mathbb{Z}) = K/\wp K$, $\wp(x) = x^p - x$, which are both infinite.

(b) If n is prime to the characteristic of K and K contains a primitive n^{th} root of unity, then $\mathbb{Z}/n\mathbb{Z} \approx \mu_n$ noncanonically and

$(\mathbb{Z}/n\mathbb{Z})^D \approx \mu_n$ canonically. The pairing

$$H^1(K,\mathbb{Z}/n\mathbb{Z}) \times H^1(K,\mu_n) \longrightarrow H^2(K,\mu_n) \approx \mathbb{Z}/n\mathbb{Z}$$

in (2.3) gives rise to a canonical pairing

$$H^1(K,\mu_n) \times H^1(K,\mu_n) \longrightarrow H^2(K,\mu_n \otimes \mu_n) \approx \mu_n.$$

The group $H^1(K,\mu_n) = K^{\times}/K^{\times n}$, and the pairing can be identified with

$$(f,g) \mapsto (-1)^{v(f)v(g)} f^{v(g)}/g^{v(f)}\colon K^{\times}/K^{\times n} \times K^{\times}/K^{\times n} \longrightarrow \mu_n$$

(see [Serre (1962), XIV.3]).

If K has characteristic $p \neq 0$, then the pairing

$$\mathrm{Ext}^1_G(\mathbb{Z}/p\mathbb{Z},K_s^{\times}) \times H^1(G,\mathbb{Z}/p\mathbb{Z}) \longrightarrow H^2(G,K_s^{\times}) \approx \mathbb{Q}/\mathbb{Z}$$

can be identified with

$$(f,g) \mapsto p^{-1}\mathrm{Tr}_{k/\mathbb{F}_p}(\mathrm{Res}(f\tfrac{dg}{g}))\colon K^{\times}/K^{\times p} \times K/\wp K \longrightarrow \mathbb{Q}/\mathbb{Z}$$

(see [Serre (1962), XIV.5] or (III.6) below).

Unramified cohomology

A G-module M is said to be *unramified* if $M^I = M$. For a finitely generated G-module, we write M^d for the submodule $\mathrm{Hom}(M,R_{un}^{\times})$ of $M^D = \mathrm{Hom}(M,K_s^{\times})$. Note that if M is unramified, then $H^1(G/I,M)$ makes sense and is a subgroup of $H^1(G,M)$. Moreover, when M is finite, $H^1(G/I,M)$ is dual to $\mathrm{Ext}^1_{G/I}(M,\mathbb{Z})$ (see (1.10)).

Theorem 2.6. *If M is a finitely generated unramified G-module whose torsion is prime to* char(k), *then the groups* $H^1(G/I,M)$ *and* $H^1(G/I,M^d)$ *are the exact annihilators of each other in the cup-product pairing*

$$H^1(G,M) \times H^1(G,M^D) \longrightarrow H^2(G,K_s^{\times}) = \mathbb{Q}/\mathbb{Z}.$$

Proof: From the spectral sequence (0.3)

$$\mathrm{Ext}^r_{G/I}(M,\mathrm{Ext}^s_I(\mathbb{Z},K_s^\times)) \Rightarrow \mathrm{Ext}^{r+s}_G(M,K_s^\times)$$

and the vanishing of $\mathrm{Ext}^1_I(\mathbb{Z},K_s^\times) = H^1(I,K_s^\times)$, we find that

$$\mathrm{Ext}^1_{G/I}(M,K_{un}^\times) \xrightarrow{\approx} \mathrm{Ext}^1_G(M,K_s^\times).$$

From the split-exact sequence of G-modules

$$0 \longrightarrow R_{un}^\times \longrightarrow K_{un}^\times \longrightarrow \mathbb{Z} \longrightarrow 0,$$

we obtain an exact sequence

$$0 \longrightarrow \mathrm{Ext}^1_{G/I}(M,R_{un}^\times) \longrightarrow \mathrm{Ext}^1_{G/I}(M,K_{un}^\times) \longrightarrow \mathrm{Ext}^1_{G/I}(M,\mathbb{Z}) \longrightarrow 0,$$

and so the kernel of $\mathrm{Ext}^1_G(M,K_s^\times) \longrightarrow \mathrm{Ext}^1_{G/I}(M,\mathbb{Z})$ is $\mathrm{Ext}^1_{G/I}(M,R_{un}^\times)$. It is easy to see from the various definitions (especially the definition of inv_G in (1.6b)) that

$$\begin{array}{lccc} \alpha^1(G,M): & \mathrm{Ext}^1_G(M,K_s^\times) & \xrightarrow{\approx} & H^1(G,M)^* \\ & \downarrow & & \downarrow \mathrm{Inf}^* \\ \alpha^1(G/I,M): & \mathrm{Ext}^1_{G/I}(M,\mathbb{Z}) & \xrightarrow{\approx} & H^1(G/I,M)^* \end{array}$$

commutes. Therefore the kernel of $\mathrm{Ext}^1_G(M,K_s^\times) \longrightarrow H^1(G/I,M)^*$ is $\mathrm{Ext}^1_{G/I}(M,R_{un}^\times)$. Example (0.8) allows us to identify $\mathrm{Ext}^1_G(M,K_s^\times)$ with $H^1(G,M^D)$ and $\mathrm{Ext}^1_{G/I}(M,R_{un}^\times)$ with $H^1(G/I,M^d)$, and so the last statement says that the kernel of $H^1(G,M^D) \longrightarrow H^1(G/I,M)^*$ is $H^1(G/I,M^d)$. (When M is finite, this result can also be proved by a counting argument; see [Serre (1964), II.5.5].)

Remark 2.7. A finite G-module M is unramified if and only if it extends to a finite étale group scheme over spec(R). In Chapter III below, we shall see that flat cohomology allows us to prove a similar result to (2.6) under the much weaker hypothesis that M extends to a finite flat group scheme over Spec(R) (see III.1 and III.7).

Euler-Poincaré characteristics

If M is a finite G-module, then the groups $H^r(G,M)$ are finite for all r and zero for $r \geq 2$. We define

$$\chi(G,M) = \frac{[H^0(G,M)]\,[H^2(G,M)]}{[H^1(G,M)]}.$$

Theorem 2.8. *Let* M *be a finite* G-module *of order* m *relatively prime to* char(K). *Then* $\chi(G,M) = (R:mR)^{-1}$.

Proof: We first dispose of a simple case.

Lemma 2.9. *If the order of* M *is prime to* char(k), *then* $\chi(G,M) = 1$.

Proof: Let p = char(k). The Sylow p-subgroup I_p of I is normal in I, and the quotient I/I_p is isomorphic to $\hat{\mathbb{Z}}/\mathbb{Z}_p$ (see [Serre (1962), IV.2, Ex 2]). As $H^r(I_p,M) = 0$ for $r > 0$, the Hochschild-Serre spectral sequence for $I \supset I_p$ shows that $H^r(I,M) = H^r(I/I_p, M^{I_p})$, and this is finite for all r and zero for $r > 1$ (cf. [Serre (1962), XIII.1]). The Hochschild-Serre sequence for $G \supset I$ now shows that $H^0(G,M) = H^0(G/I,M^I)$, that $H^1(G,M)$ fits into an exact sequence

$$0 \longrightarrow H^1(G/I,M^I) \longrightarrow H^1(G,M) \longrightarrow H^0(G/I,H^1(I,M)) \longrightarrow 0,$$

and that $H^2(G,M) = H^1(G/I,H^1(I,M))$. But $G/I = \hat{\mathbb{Z}}$, and the exact sequence

$$0 \longrightarrow H^0(\hat{\mathbb{Z}},N) \longrightarrow N \xrightarrow{\sigma-1} N \longrightarrow H^1(\hat{\mathbb{Z}},N) \longrightarrow 0$$

(with σ a generator of $\hat{\mathbb{Z}}$; see [Serre (1962), XIII.1]) shows that $[H^0(\hat{\mathbb{Z}},N)] = [H^1(\hat{\mathbb{Z}},N)]$ for any finite $\hat{\mathbb{Z}}$-module. Therefore

$$\chi(G,M) = \frac{[H^0(G/I,M^I)]}{[H^1(G/I,M^I)]}\,\frac{[H^0(G/I,H^1(I,M))]}{[H^1(G/I,H^1(I,M))]} = 1.$$

Since both sides of equation in (2.8) are additive in M, the lemma allows us to assume that M is killed by p = char(k) and that K is of characteristic zero. We shall prove the theorem for all G-modules M such that $M = M^{G_L}$, where L is some fixed finite Galois extension of K contained in K_s. Let $\bar{G} = Gal(L/K)$. Our modules can be regarded as $\mathbb{F}_p[\bar{G}]$-modules, and we let $R_{\mathbb{F}_p}(\bar{G})$, or simply $R(\bar{G})$, be the Grothendieck group of the category of such modules. Then the left and right hand sides of the equation in (2.8) define homomorphisms $\chi_\ell, \chi_r\colon R(\bar{G}) \longrightarrow \mathbb{Q}_{>0}$. As $\mathbb{Q}_{>0}$ is a torsion-free group, it suffices to show that χ_ℓ and χ_r agree on a set of generators for $R_{\mathbb{F}_p}(\bar{G})\otimes_{\mathbb{Z}}\mathbb{Q}$. The next lemma describes one such set.

Lemma 2.10. *Let* G *be a finite group and, for any subgroup* H *of* G, *let* Ind_H^G *be the homomorphism* $R_{\mathbb{F}_p}(H)\otimes\mathbb{Q} \longrightarrow R_{\mathbb{F}_p}(G)\otimes\mathbb{Q}$ *taking the class of an* H*-module to the class of the corresponding induced* G*-module. Then* $R_{\mathbb{F}_p}(G)\otimes\mathbb{Q}$ *is generated by the images of the* Ind_H^G *as* H *runs over the set of cyclic subgroups of* G *of order prime to* p.

Proof: Write $R_F(G)$ for the Grothendieck group of finitely generated F[G]-modules, F any field. Then [Serre (1967b), 12.5, Thm 26] shows that, in the case that F has characteristic zero, $R_F(G)\otimes\mathbb{Q}$ is generated by the images of the maps Ind_H^G with H cyclic. It follow from [Serre (1967b), 16.1, Thm 33] that the same statement is then true for any field F. Finally [Serre (1967b), 8.3, Pptn 26] shows that, in the case that F has characteristic $p \neq 0$, the cyclic groups of p-power order make no contribution.

It suffices therefore to prove the theorem for a module M of the form $\mathrm{Ind}_H^{\bar{G}} N$. Let $K' = L^H$, let R' be the ring of integers in K', and let n be the order of N. Then $\chi(G,M) = \chi(\mathrm{Gal}(K_s/K'),N)$ and $(R:mR) = (R:nR)^{[K':K]} = (R':nR')$, and so it suffices to prove the theorem for N. This means that *we can assume that* $\bar{G}$ *is a cyclic group of order prime to* p. Therefore $H^r(\bar{G},M) = 0$ for $r > 0$, and so $H^r(G,M) = H^r(\mathrm{Gal}(K_s/L),M)^{\bar{G}}$.

Let χ' be the homomorphism $R(\bar{G}) \longrightarrow R(\bar{G})$ sending M to $\sum (-1)^i[H^i(\mathrm{Gal}(K_s/L),M)]$, where $[*]$ now denotes the class of $*$ in $R(\bar{G})$.

Lemma 2.11. *The following formula holds:*

$$\chi'(M) = -\dim(M).[K:\mathbb{Q}_p].[\mathbb{F}_p[\bar{G}]].$$

Before proving the lemma, we show that it implies the theorem. Let $\theta\colon R_{\mathbb{F}_p}(\bar{G}) \longrightarrow \mathbb{Q}_{>0}$ be the homomorphism sending the class of a module N to the order of $N^{\bar{G}}$. Then $\theta\chi' = \chi$ and $\theta([\mathbb{F}_p[\bar{G}]]) = p$, and so (2.11) shows that $\chi(M) = \theta\chi'(M) = p^{-[K:\mathbb{Q}_p].\dim(M)} = 1/(R:mR)$.

It therefore remains to prove (2.11). On tensoring M with a resolution of $\mathbb{Z}/p\mathbb{Z}$ by injective $\mathbb{Z}/p\mathbb{Z}[\bar{G}]$-modules, we find that cup-product defines isomorphisms of $\bar{G}$-modules

$$H^r(\mathrm{Gal}(K_s/L),\mathbb{Z}/p\mathbb{Z})\otimes M \longrightarrow H^r(\mathrm{Gal}(K_s/L),M),$$

and so

$$\chi'(M) = \chi'(\mathbb{Z}/p\mathbb{Z}).[M].$$

Let M_0 be the G-module with the same underlying abelian group as M but with the trivial G-action. The map $\sigma\otimes m \mapsto \sigma\otimes\sigma m$ extends to an isomorphism $\mathbb{F}_p[\bar{G}]\otimes M_0 \xrightarrow{\approx} \mathbb{F}_p[\bar{G}]\otimes M$, and so

$$\dim(M)[\mathbb{F}_p[\bar{G}]] = [\mathbb{F}_p[\bar{G}]].[M].$$

The two displayed equalities show that the general case of (2.11) is a consequence of the special case $M = \mathbb{Z}/p\mathbb{Z}$.

Note that

$$H^0(\mathrm{Gal}(K_s/L),\mathbb{Z}/p\mathbb{Z}) = \mathbb{Z}/p\mathbb{Z},$$
$$H^1(\mathrm{Gal}(K_s/L),\mathbb{Z}/p\mathbb{Z}) = H^1(\mathrm{Gal}(K_s/L),\mu_p(K_s))^* = (L^\times/L^{\times p})^*,$$
$$H^2(\mathrm{Gal}(K_s/L),\mathbb{Z}/p\mathbb{Z}) = (\mu_p(L))^*,$$

where N^* denotes $\mathrm{Hom}(N,\mathbb{F}_p)$ (still regarded as a $\bar{G}$-module; as $\mathrm{Hom}(-,\mathbb{F}_p)$ is exact, it is defined for objects in $R(\bar{G})$). Therefore

$$\chi'(\mathbb{Z}/p\mathbb{Z})^* = [\mathbb{Z}/p\mathbb{Z}] - [L^\times/L^{\times p}] + [\mu_p(L)].$$

Let U be the group of units $R_L^\times$ in R_L. From the exact sequence

$$0 \longrightarrow U/U^p \longrightarrow L^\times/L^{\times p} \longrightarrow \mathbb{Z}/p\mathbb{Z} \longrightarrow 0,$$

we find that

$$[\mathbb{Z}/p\mathbb{Z}] - [L^\times/L^{\times p}] = [U^{(p)}].$$

and so

$$\chi'(\mathbb{Z}/p\mathbb{Z})^* = - [U^{(p)}] + [\mu_p(L)].$$
$$= - [U^{(p)}] + [U_p].$$

We need one last lemma.

Lemma 2.12. *Let* W *and* W' *be finitely generated* $\mathbb{Z}_p[H]$*-modules for some finite group* H. *If* $W\otimes\mathbb{Q}_p \approx W'\otimes\mathbb{Q}_p$ *as* $\mathbb{Q}_p[H]$*-modules, then*

$$[W^{(p)}] - [W_p] = [W'^{(p)}] - [W'_p]$$

in $\mathbb{F}_p[H]$.

Proof: One reduces the question easily to the case that $W \supset W' \supset pW$, and for such a module the lemma follow immediately from the exact sequence

$$0 \longrightarrow W'_p \longrightarrow W_p \longrightarrow W/W' \longrightarrow W'^{(p)} \longrightarrow W^{(p)} \longrightarrow W/W' \longrightarrow 0$$

given by the snake lemma.

The exponential map sends an open subgroup of U onto an open subgroup of the ring of integers R_L of L, and so (2.12) shows that

$$[U^{(p)}] - [U_p] = [R_L^{(p)}] - [(R_L)_p] = [R_L^{(p)}].$$

The normal basis theorem shows that $L \approx \mathbb{Q}_p[\bar{G}]^{[K:\mathbb{Q}_p]}$ (as $\bar{G}$-modules), and so (2.12) implies that

$$[R_L^{(p)}] = [K:\mathbb{Q}_p].[\mathbb{F}_p[\bar{G}]].$$

As $[\mathbb{F}_p[\bar{G}]]^* = [\mathbb{F}_p[\bar{G}]]$, this completes the proof of (2.11).

Archimedean local fields

Corollaries 2.3, 2.4 and Theorem 2.8 all have analogues for $\mathbb{R}$ and $\mathbb{C}$.

Theorem 2.13. (a) *Let* $G = \mathrm{Gal}(\mathbb{C}/\mathbb{R})$. *For any finite generated G-module M with dual* $M^D = \mathrm{Hom}(M, \mathbb{C}^\times)$, *cup-product defines a non-degenerate pairing*

$$H_T^r(G, M^D) \times H_T^{2-r}(G, M) \longrightarrow H^2(G, \mathbb{C}^\times) \xrightarrow{\approx} \tfrac{1}{2}\mathbb{Z}/\mathbb{Z}$$

of finite groups for all r.

(b) *Let* $G = \mathrm{Gal}(\mathbb{C}/\mathbb{R})$. *For any commutative algebraic group* T *over* $\mathbb{R}$ *whose identity component is a torus, cup-product defines dualities between* $H_T^r(G, X^*(T))$ *and* $H_T^{2-r}(G, T(\mathbb{C}))$ *for all* r.

(c) *Let* $K = \mathbb{R}$ *or* $\mathbb{C}$, *and let* $G = \mathrm{Gal}(\mathbb{C}/K)$. *For any finite G-module* M

$$\frac{[H^0(G,M)][H^0(G,M^D)]}{[H^1(G,M)]} = |m|_v.$$

Proof: (a) Suppose first that M is finite. As G has order 2, the ℓ-primary components for ℓ odd do not contribute to the cohomology groups. We can therefore assume that M is 2-primary, and furthermore that it is simple. Then $M = \mathbb{Z}/2\mathbb{Z}$ with the trivial action of G, and the theorem can be proved in this case by direct calculation.

When $M = \mathbb{Z}$ the result can again be proved by direct calculation, and when $M = \mathbb{Z}[G]$ all groups are zero. Since every torsion-free G-module contains a submodule of finite index that is a direct sum of copies of $\mathbb{Z}$ or $\mathbb{Z}[G]$, this proves the result for such modules, and the general case follows by combining the two cases.

(b) Take $M = X^*(T)$ in (a).

(c) The complex case is obvious because $H^0(G,M) = M$ and $H^0(G,M^D) = M^D$ both have order m, $H^1(G,M) = 0$, and $|m|_v = m^2$. In the real case, let σ generate G, and note that for $m \in M$ and $f \in M^D$

$$((1-\sigma)f)(m) = f(m)/\sigma(f(\sigma m)) = f(m).(f(\sigma m)) \quad (\text{because } \bar{\zeta} = \zeta^{-1})$$
$$= f((1+\sigma)m).$$

Therefore $1 - \sigma\colon M^D \longrightarrow M^D$ is adjoint to $1 + \sigma\colon M \longrightarrow M$, and so, in the pairing $M^D \times M \longrightarrow \mathbb{C}^\times$, $(M^D)^G$ and $N_{\mathbb{C}/\mathbb{R}}M$ are exact annihilators. Consequently

$$[M] = [(M^D)^G][N_{\mathbb{C}/\mathbb{R}}M] = [H^0(G,M^D)][H^0(G,M)]/[H^0_T(G,M)],$$

and the periodicity of the cohomology of cyclic groups shows that $[H^0_T(G,M)] = [H^1(G,M)]$. As $[M] = m = |m|_v$, this proves the formula.

Henselian local fields

Let K be the field of fractions of an excellent Henselian discrete valuation ring R with finite residue field k. (See Appendix A for definitions.) It is shown in the Appendix that the pair $(G_K, K_s^\times)$ is a class formation, and that the norm groups are precisely the open subgroups of finite index. The following theorem generalizes some of the preceding results.

Theorem 2.14. *Let M be a finitely generated G-module whose torsion subgroup is prime to* char(K).

(a) *The map* $\alpha^r(G,M)\colon \mathrm{Ext}^r_G(M,K_s^\times) \longrightarrow H^{2-r}(G,M)^*$ *is an isomorphism for all* $r \geq 1$, *and* $\alpha^0(G,M)$ *defines an isomorphism (of compact groups)* $\mathrm{Hom}_G(M,K_s^\times)^\wedge \longrightarrow H^2(G,M)^*$. *The* ^ *can be omitted if M is finite. The groups* $\mathrm{Ext}^r_G(M,K_s^\times)$ *and* $H^r(G,M)$ *are finite for all r if M is finite, and the groups* $\mathrm{Ext}^1_G(M,K_s^\times)$ *and* $H^1(G,M)$ *are finite for all finitely generated M.*

(b) *If K is countable, then for any algebraic group A over K,*

$$\alpha^1(G,A(K_s))\colon \mathrm{Ext}^1_G(A(K_s),K_s^\times) \longrightarrow H^1(G,A(K_s))^*$$

is an isomorphism, except possibly on the p-primary component when $\mathrm{char}(K) = p \neq 1$.

(c) *Cup-product defines isomorphisms* $H^r(G,M^D) \longrightarrow H^r(G,M)^*$ *for all* $r \geq 1$, *and an isomorphism* $H^0(G,M^D)^\wedge \longrightarrow H^2(G,M)^*$ *of compact groups. The groups* $H^1(G,M^D)$ *and* $H^1(G,M)$ *are both finite.*

Proof: (a) Let $\hat{R}$ be the completion of R. There is a commutative diagram

$$\begin{array}{ccccccccc} 0 & \longrightarrow & R^\times & \longrightarrow & K^\times & \longrightarrow & \mathbb{Z} & \longrightarrow & 0 \\ & & \downarrow & & \downarrow \text{rec} & & \downarrow & & \\ 0 & \longrightarrow & \hat{R}^\times & \longrightarrow & G & \longrightarrow & \hat{\mathbb{Z}} & \longrightarrow & 0. \end{array}$$

All the vertical maps are injective, and the two outside vertical maps have cokernels that are uniquely divisible by all primes $\ell \neq$ char(K). Therefore the reciprocity map $K^\times \longrightarrow G$ is injective and has a cokernel that is uniquely divisible prime to char(K). The first two assertions now follow easily from (1.8). The finiteness statements follow from the fact that $\mathrm{Gal}(K_s/K) = \mathrm{Gal}(\hat{K}_s/\hat{K})$.

(b) The group $A(K_s)$ is countable, and therefore it is a countable union of finitely generated submodules. The statement can therefore be proved the same way as (2.2).

(c) The proof is the same as that of (2.3).

Remark 2.15. (a) Part (a) of the theorem also holds for modules M with p-torsion, except that it is necessary to complete $\mathrm{Ext}^1_G(M, K_s^{\times})$. For example, when $M = \mathbb{Z}/p\mathbb{Z}$, the map

$$K^{\times}/K^{\times p} \longrightarrow \mathrm{Hom}(G_K, \mathbb{Z}/p\mathbb{Z}).$$

Because K is excellent, the map $K^{\times}/K^{\times p} \longrightarrow \hat{K}^{\times}/\hat{K}^{\times p}$ is injective and induces an isomorphism $(K^{\times}/K^{\times p})^{\wedge} \xrightarrow{\approx} \hat{K}^{\times}/\hat{K}^{\times p}$. We know $\mathrm{Gal}(K_s/K) = \mathrm{Gal}(\hat{K}_s/\hat{K})$, and so in this case the assertion follows from the corresponding statement for $\hat{K}$.

(b) As was pointed out to the author by M. Hochster, it is easy to construct nonexcellent Henselian discrete valuation rings. Let k be a field of characteristic p, and choose an element $u \in k[[t]]$ that is transcendental over k(t). Let R be the discrete valuation ring $k(t, u^p) \cap k[[t]]$, and consider the Henselization R^h of R. Then the elements of R^h are separable over R (R^h is a union of étale R-subalgebras), and so $u \notin R^h$, but $u \in (R^h)^{\wedge} = k[[t]]$.

Complete fields with quasi-finite residue fields

Exercise 2.16. Let K be complete with respect to a discrete valuation, but assume that its residue field is quasi-finite rather than finite. (See Appendix A for definitions.) Investigate to what extent the results of this section continue to hold for K. References: [Serre (1962), XIII] and Appendix A for the basic class field theory of such fields; [Serre (1964), pII-24, pII-29] for statements of what is true; [Vvedens'kii and Krupjak (1976)] and [Litvak (1980)] for a proof of (2.3) for a finite module in the case the field has characteristic zero.)

d-local fields

A 0-*local field* is a finite field, and a d-*local field* for $d \geq 1$ is a field that is complete with respect to a discrete valuation and has a (d-1)-local field as residue field. If K is d-local, we shall write K_i, $0 \leq i \leq d$, for the i-local field in the inductive definition of K. We write μ_{ℓ^n} for the G_K-module $\{ \zeta \in K_s \mid \zeta^{\ell^n} = 1\}$, $\mu_{\ell^\infty}(r)$ for $\varinjlim \mu_{\ell^n}^{\otimes r}$, and $\mathbb{Z}_\ell(r)$ for $\varprojlim \mu_{\ell^n}^{\otimes r}$. If M is an ℓ-primary G_K-module, we set $M(r) = M \otimes \mathbb{Z}_\ell(r)$ and $M^*(r) = \mathrm{Hom}(M, \mu_{\ell^\infty}(r))$.

Theorem 2.17. *Let K be a d-local field with $d \geq 1$, and let ℓ be a prime $\neq$ char(K_1).*

(a) *There is a canonical trace map*

$$H^{d+1}(G_K, \mu_{\ell^\infty}(d)) \xrightarrow{\approx} \mathbb{Q}_\ell/\mathbb{Z}_\ell.$$

(b) *For all G_K-modules M of finite order a power of ℓ, the cup-product pairing*

$$H^r(G_K, M^*(d)) \times H^{d+1-r}(G_K, M) \longrightarrow H^{d+1}(G_K, \mathbb{Q}_\ell/\mathbb{Z}_\ell(d)) = \mathbb{Q}_\ell/\mathbb{Z}_\ell$$

is a nondegenerate pairing of finite groups for all r.

Proof: For $d = 1$, this is a special case of (2.3). For $d > 1$, it follows by an easy induction argument from the next lemma.

Lemma 2.18. *Let K be any field complete with respect to a discrete valuation, and let k be the residue field of K. For any finite G_K-module of order prime to char(k), there is a long exact sequence*

$$\cdots \longrightarrow H^r(G_k,M^I) \longrightarrow H^r(G_K,M) \longrightarrow H^{r-1}(G_k,M(-1)_I) \longrightarrow H^{r+1}(G_k,M^I) \longrightarrow \cdots$$

where I *is the inertia group of* G_K.

Proof: Let char(k) = p, and let I_p be a p-Sylow subgroup of I (so $I_p = 1$ if p = 1). Then $I' \overset{df}{=} I/I_p$ is canonically isomorphic to $\prod_{\ell \neq p} \mathbb{Z}_\ell(1)$ (see [Serre (1962), IV.2]). The same argument that shows that $H^r(G,M) = M^G$, M_G, 0 for r = 0, 1, > 2 when $G = \hat{\mathbb{Z}}$ and M is torsion [Serre (1962), XIII.1], shows in our case that

$$H^r(I,M) = H^r(I',M) = \begin{cases} M^I & \text{for } r = 0 \\ M(-1)_I & \text{for } r = 1 \\ 0 & \text{for } r > 1. \end{cases}$$

The lemma therefore follows immediately from the Hochschild-Serre spectral sequence for $G \supset I$.

Write K_rR for the rth Quillen K-group of a ring R.

Corollary 2.19. *Let* K *be a* 2-*local field, and let* m *be an integer prime to* $char(K_1)$ *and such that* K *contains the* m^{th} *roots of* 1. *Then there is a canonical injective homomorphism* $K_2K^{(m)} \longrightarrow Gal(K_{ab}/K)^{(m)}$ *with dense image.*

Proof: On taking $M = \mathbb{Z}/m\mathbb{Z}$ in the theorem, we obtain an isomorphism $H^2(G,\mu_m \otimes \mu_m) \longrightarrow H^1(G,\mathbb{Z}/m\mathbb{Z})^*$. But $H^1(G,\mathbb{Z}/m\mathbb{Z}) = Hom_{cts}(G,\mathbb{Z}/m\mathbb{Z})$, and so this gives us with an injection $H^2(G,\mu_m \otimes \mu_m) \longrightarrow (G^{ab})^{(m)}$ with dense image. Now the theorem of [Merkur'ev and Suslin (1982)] provides us with an isomorphism $(K_2K)^{(m)} \xrightarrow{\approx} H^2(G,\mu_m \otimes \mu_m)$.

Theorem 2.17 is a satisfactory generalization of Theorem 2.3 in the case that the characteristic drops from p to zero at the first step. The general case is not yet understood.

Some exercises

Exercise 2.20. (a) Let G be a profinite group, and let M be a finitely generated G-module. Write $T = \mathrm{Hom}(M,\mathbb{C}^{\times})$, and regard it as an algebraic torus over $\mathbb{C}$. Let G act on T through its action on M. Show that $\mathrm{Ext}^0_G(M,\mathbb{Z}) = X_*(T)^G$, $\mathrm{Ext}^1_G(M,\mathbb{Z}) = \pi_0(T^G)$, and $\mathrm{Ext}^r_G(M,\mathbb{Z}) = H^{r-1}(G,T)$ for $r \geq 2$. If M is torsion-free, show that $\mathrm{Ext}^r_G(M,\mathbb{Z}) = H^r(G,X_*(T))$.

(b) Let K be a local field (archimedean or nonarchimedean), and let T be a torus over K. Let $T^{\vee}$ be the torus such that $X^*(T^{\vee}) = X_*(T)$. Show that the finite group $H^1(K,T)$ is dual to $\pi_0(T^{\vee G})$ and that $H^1(K,T^{\vee})$ is canonically isomorphic to the group $T(K)^*$ of continuous characters of finite order of $T(K)$. (In §8 we shall obtain a similar description of the group of generalized characters of $T(K)$.) [Hint: To prove the first part of (a), use the spectral sequence (0.8)

$$H^r(G,\mathrm{Ext}^s(M,\mathbb{C}^{\times})) \Rightarrow \mathrm{Ext}^{r+s}_G(M,\mathbb{C}^{\times})$$

and the exponential sequence $0 \longrightarrow \mathbb{Z} \longrightarrow \mathbb{C} \longrightarrow \mathbb{C}^{\times} \longrightarrow 0$.]
Reference: [Kottwitz (1984)].

Exercise 2.21. Let K be a 2-local field of characteristic zero such that K_1 has characteristic $p \neq 0$. Assume

(a) K has p-cohomological dimension ≤ 3 and there is a canonical isomorphism $H^3(G,\mu_{p^n}\otimes\mu_{p^n}) \longrightarrow \mathbb{Z}/p^n\mathbb{Z}$ [Kato (1979), §5, Thm 1];

(b) if K contains a primitive p^{th} root of 1, then the cup-product pairing

$$H^1(G,\mu_p) \times H^2(G,\mu_p) \longrightarrow H^3(G,\mu_p\otimes\mu_p) = \mathbb{Z}/p\mathbb{Z}$$

is a nondegenerate pairing of finite groups [ibid. §6].

Prove then that (2.17) holds for K with $\ell = p$.

Exercise 2.22. Let $K = k((t_1, \dots, t_d))$ with k a finite field, and let $p = \mathrm{char}(k)$. Define $\nu(r) = \mathrm{Ker}(\Omega^r_{K/k,d=0} \xrightarrow{C-1} \Omega^r_{K/k})$, where C is the Cartier operator (see [Milne (1976)]). Show that there is a canonical trace map $H^1(G_K, \nu(d)) \xrightarrow{\approx} \mathbb{Z}/p\mathbb{Z}$, and show that the cup-product pairings

$$H^r(G_K, \nu(r)) \times H^{1-r}(G_K, \nu(d-r)) \longrightarrow H^1(G_K, \nu(d)) = \mathbb{Z}/p\mathbb{Z}$$

are nondegenerate in the sense that their left and right kernels are zero. Let d = 2, and assume that there is an exact sequence

$$0 \longrightarrow K_2K \xrightarrow{p} K_2K \longrightarrow \nu(2) \longrightarrow 0$$

with the second map being $\mathrm{dlog}\wedge\mathrm{dlog}\colon K_2K \longrightarrow \nu(2)$. (In fact such a sequence exists: the exactness at the first term is due to [Suslin (1983)]; the exactness at the middle term is a theorem of Bloch [Bloch and Kato (1986)]; and the exactness at the last term has been proved by several people.) Deduce that there is a canonical injective homomorphism $(K_2K)^{(p)} \longrightarrow (G_K^{ab})^{(p)}$. (These results can be extended to groups killed by powers of p rather that p itself by using the sheaves $\nu_n(r)$ of [Milne (1986a)].)

Notes: The main theorems concerning local fields in the classical sense are due to Tate. The proofs are those of Tate except for that of (2.8), which is due to Serre (see [Serre (1964), II.5]). Theorem 2.17 is taken from [Deninger-Wingberg (1986)]

§3 Abelian varieties over local fields

We continue with the notations at the start of the last section.

In particular, K is a local field, complete with respect to a discrete valuation ord, and with finite residue field k. When G and H are algebraic groups over a field f, we write $\mathrm{Ext}^r_F(G,H)$ for the group of formed in the category $\mathbf{Gp}_F$ (see §0).

Let A be an abelian variety over K. The Weil-Barsotti formula [Serre (1959), VII, §3] states that $A^t(K_s) = \mathrm{Ext}^1_{K_s}(A,\mathbb{G}_m)$ where A^t is the dual abelian variety.

Lemma 3.1. *For any abelian variety* A *over a perfect field* F, *there is a canonical isomorphism*

$$H^r(F,A^t) \longrightarrow \mathrm{Ext}^{r+1}_F(A,\mathbb{G}_m),$$

all $r \geq 0$.

Proof: The group $\mathrm{Ext}^r_{F_s}(A,\mathbb{G}_m)$ is shown to be zero for $r \geq 2$ in [Oort (1966), Pptn 12.3], and $\mathrm{Hom}_{F_s}(A,\mathbb{G}_m) = 0$ because all maps from a projective variety to an affine variety are constant. This together with the Weil-Barsotti formula show that the spectral sequence (0.17)

$$H^r(\mathrm{Gal}(F_s/F),\mathrm{Ext}^s_{F_s}(A,\mathbb{G}_m)) \Rightarrow \mathrm{Ext}^{r+s}_F(A,\mathbb{G}_m)$$

degenerates to a family of isomorphisms $H^r(F,A^t) \xrightarrow{\approx} \mathrm{Ext}^{r+1}_F(A,\mathbb{G}_m)$.

In particular $\mathrm{Ext}^1_K(A,\mathbb{G}_m) = A^t(K)$ when K has characteristic zero, and the Ext group therefore acquires a topology from that on K. Recall that there is a canonical pairing (0.16)

$$\mathrm{Ext}^r_K(A,\mathbb{G}_m) \times H^{2-r}(K,A) \longrightarrow H^2(K,\mathbb{G}_m),$$

and an isomorphism $\mathrm{inv}_G\colon H^2(K,\mathbb{G}_m) \xrightarrow{\approx} \mathbb{Q}/\mathbb{Z}$ (1.6b). Therefore there is a canonical map

$$\alpha^r(K,A)\colon \mathrm{Ext}^r_K(A,\mathbb{G}_m) \longrightarrow H^{2-r}(K,A)^*.$$

Theorem 3.2. *If K has characteristic zero, then* $\alpha^1(K,A)$ *is an isomorphism of compact groups*

$$\mathrm{Ext}^1_K(A,\mathbb{G}_m) \xrightarrow{\approx} H^1(K,A)^*$$

and $\alpha^2(K,A)$ *is an isomorphism of torsion groups of cofinite type*

$$\mathrm{Ext}^2_K(A,\mathbb{G}_m) \xrightarrow{\approx} A(K)^*.$$

For $r \neq 1,2$, $\mathrm{Ext}^r_K(A,\mathbb{G}_m)$ *and* $H^{2-r}(K,A)$ *are both zero.*

Proof: We first need a lemma.

Lemma 3.3. *In the situation of the theorem,* A(K) *contains an open subgroup of finite index isomorphic to* $R^{\dim(A)}$; *therefore* $A(K) = A(K)\hat{}$ *(completion for the profinite topology), and*

$$[A(K)^{(n)}]/[A(K)_n] = (R:nR)^{\dim(A)}.$$

Proof: The existence of the subgroup follows from the theory of the logarithm (see [Mattuck (1955)] or [Tate (1967b), p168-169]), and the remaining statements are obvious.

Proof (of 3.2): From

$$0 \longrightarrow A_n \longrightarrow A \xrightarrow{n} A \longrightarrow 0$$

we get the rows of the following diagram

$$\begin{array}{ccccccc}
0 \longrightarrow & \mathrm{Ext}^r_K(A,\mathbb{G}_m)^{(n)} & \longrightarrow & \mathrm{Ext}^r_K(A_n,\mathbb{G}_m) & \longrightarrow & \mathrm{Ext}^{r+1}_K(A,\mathbb{G}_m)_n & \longrightarrow 0 \\
& \downarrow \alpha^r(K,A)^{(n)} & & \downarrow \alpha^r(G,A_n) & & \downarrow \alpha^{r+1}(K,A)_n & \\
0 \longrightarrow & (H^{2-r}(K,A)_n)^* & \longrightarrow & H^{2-r}(K,A_n)^* & \longrightarrow & H^{1-r}(K,A)^{(n)*} & \longrightarrow 0.
\end{array}$$

As is explained in (0.18), $\mathrm{Ext}^r_K(A_n,\mathbb{G}_m) \xrightarrow{\approx} \mathrm{Ext}^r_G(A_n(K_s),K_s^\times)$ for all r, and if we take $\alpha^r(G,A_n)$ to be the map $\alpha^r(G,A_n(K_s))$ of §2, then it is clear that the diagram commutes. As $\alpha^r(G,A_n)$ is an isomorphism of finite groups for all r, we see that

$$\alpha^r(K,A)^{(n)}\colon \operatorname{Ext}^r_K(A,\mathbb{G}_m)^{(n)} \longrightarrow (H^{2-r}(K,A)_n)^*$$

is an injective map of finite groups for all r, and, in the limit,

$$\varprojlim \alpha^r(K,A)^{(n)}\colon \varprojlim \operatorname{Ext}^r_K(A,\mathbb{G}_m)^{(n)} \longrightarrow (H^{2-r}(K,A)_{tors})^*$$

is injective. As $\operatorname{Ext}^1_K(A,\mathbb{G}_m) = A^t(K)$, the lemma shows that $\operatorname{Ext}^1_K(A,\mathbb{G}_m) = \varprojlim \operatorname{Ext}^1_K(A,\mathbb{G}_m)^{(n)}$. Thus we have shown that $\alpha^1(K,A)$ is injective.

We next show that $H^r(K,A) = 0$ for $r \geq 2$. For $r > 2$, this follows from the fact that G has cohomological dimension 2 (see 2.1). On taking r = 0 in the above diagram, we get an exact commutative diagram

$$\begin{array}{ccccccc}
0 \longrightarrow & 0 & \longrightarrow & \operatorname{Hom}_K(A_n,\mathbb{G}_m) & \longrightarrow & \operatorname{Ext}^1_K(A,\mathbb{G}_m) \\
 & \downarrow & & \downarrow \approx & & \downarrow \\
0 \longrightarrow & (H^2(K,A)_n)^* & \longrightarrow & H^2(K,A_n)^* & \longrightarrow & H^1(K,A)^* & .
\end{array}$$

As the right hand vertical arrow is injective, the snake lemma shows that $H^2(K,A)_n = 0$, and therefore that $H^2(K,A) = 0$. Because $H^r(G,A^t) \approx \operatorname{Ext}^{r+1}_K(A,\mathbb{G}_m)$, this also shows that $\operatorname{Ext}^r_K(A,\mathbb{G}_m) = 0$ for $r \neq 1,2$.

We now prove that $\alpha^1(K,A)$ is an isomorphism. We have already seen that it is an injective map $A^t(K) \longrightarrow H^1(K,A)^*$, and it remains to show that the maps $A^t(K)^{(n)} \longrightarrow (H^1(K,A)_n)^*$ are surjective for all integers n. As these maps are injective, this can be accomplished by showing that the groups have the same order. Let $M = A_n(K_s)$ and $M^D = A^t_n(K_s)$. Then (2.8) shows that

$$\chi(G,M) = (R:nR)^{-2d} = \chi(G,M^D),$$

where d is the dimension of A, and (3.3) shows that

$$[A(K)^{(n)}]/[A(K)_n] = (R:nR)^d = [A^t(K)^{(n)}]/[A^t(K)_n].$$

From the cohomology sequence of

$$0 \longrightarrow M \longrightarrow A(K_s) \xrightarrow{n} A(K_s) \longrightarrow 0$$

we find that

$$\chi(G,M) = \frac{[A(K)_n]}{[A(K)^{(n)}]}\frac{[H^2(G,M)]}{[H^1(K,A)_n]}$$

or

$$\frac{1}{(R:nR)^{2d}} = \frac{1}{(R:nR)^d}\frac{[H^0(G,M^D)]}{[H^1(K,A)_n]}.$$

As $H^0(G,M^D) \approx A^t(K)_n$, this can be rewritten as

$$[H^1(K,A)_n] = (R:nR)^d[A^t(K)_n] = [A^t(K)^{(n)}],$$

which completes the proof that $\alpha^1(K,A)$ is an isomorphism.

It remains to show that $\alpha^2(K,A)$ is an isomorphism. The diagram at the start of the proof shows that $\alpha^2(K,A)$ is surjective, and we know that it can be identified with a map $H^1(K,A^t) \longrightarrow A(K)^*$. The above calculation with A and A^t interchanged shows that $[H^1(K,A^t)_n] = [A(K)^{(n)}]$ for all n, which implies that $\alpha^2(K,A)$ is an isomorphism.

Corollary 3.4. *If K has characteristic zero, then there is a canonical pairing*

$$H^r(K,A^t) \times H^{1-r}(K,A) \longrightarrow \mathbb{Q}/\mathbb{Z},$$

which induces an isomorphism of compact groups $A^t(K) \xrightarrow{\approx} H^1(K,A)^*$ *(case* $r = 0$*) and an isomorphism of discrete groups of cofinite-type* $H^1(K,A^t) \xrightarrow{\approx} A(K)^*$ *(case* $r = 1$*). For* $r \neq 0,1$, *the groups* $H^r(K,A)$ *and* $H^r(K,A^t)$ *are zero.*

Proof: Lemma (3.1) allows us to replace $\mathrm{Ext}^r_K(A,\mathbb{G}_m)$ in the statement of the theorem with $H^{r-1}(K,A^t)$.

Remark 3.5. There is an alternative approach to defining the pairings

$$H^r(K,A^t) \times H^{1-r}(K,A) \longrightarrow H^2(K,\mathbb{G}_m)$$

of (3.4). For any abelian variety A over K, write Z(A) for the group of zero cycles on A_{K_s} of degree zero (that is, the set of formal sums $\sum n_i P_i$ with $P_i \in A(K_s)$ and $\sum n_i = 0$). There is a surjective map $S\colon Z(A) \twoheadrightarrow A(K_s)$ sending a formal sum to the corresponding actual sum on A, and we write Y(A) for its kernel. There are exact sequences

$$0 \longrightarrow Y(A^t) \longrightarrow Z(A^t) \longrightarrow A^t(K_s) \longrightarrow 0.$$

$$0 \longrightarrow Y(A) \longrightarrow Z(A) \longrightarrow A(K_s) \longrightarrow 0,$$

Let D be a divisor on $A^t \times A$, and let $\mathfrak{a}$ and $\mathfrak{b}$ be elements of $Y(A^t)$ and Z(A) such that the support of D does not meet the support of $\mathfrak{a}\times\mathfrak{b}$. The projection $D(\mathfrak{a})$ of $D.(\mathfrak{a}\times\mathfrak{b})$ onto A is then defined and, because $\mathfrak{a}$ is in Y(A), it is principal, say $D(\mathfrak{a}) = \operatorname{div}(f)$. It is now possible to define

$$D(\mathfrak{a},\mathfrak{b}) = f(\mathfrak{b}) \overset{\mathrm{df}}{=} \prod_{b\in\operatorname{supp}(\mathfrak{b})} f(b)^{\operatorname{ord}_b(\mathfrak{b})} \in K_s^\times.$$

Now let D be a Poincaré divisor on $A^t \times A$, and let D^t be its transpose. A reciprocity law [Lang (1959), VI.4, Thm 10] shows that the pairings

$$(\mathfrak{a},\mathfrak{b}) \mapsto D(\mathfrak{a},\mathfrak{b})\colon Y(A^t) \times Z(A) \longrightarrow K_s^\times$$

$$(\mathfrak{b},\mathfrak{a}) \mapsto D^t(\mathfrak{b},\mathfrak{a})\colon Y(A) \times Z(A^t) \longrightarrow K_s^\times$$

satisfy the equality $D(\mathfrak{a},\mathfrak{b}) = D^t(\mathfrak{b},\mathfrak{a})$ if $\mathfrak{a} \in Y(A^t)$ and $\mathfrak{b} \in Y(A)$. They therefore give rise to augmented cup-product pairings (0.12)

$$H^r(K,A^t) \times H^{1-r}(K,A) \longrightarrow H^2(K,\mathbb{G}_m).$$

It is possible to show that these pairings agree with those in (3.4) (up to sign) by checking that each is compatible with the pairings

$$H^r(K,A_n^t) \times H^{2-r}(K,A_n) \longrightarrow H^2(K,\mathbb{G}_m)$$

defined by the e_n-pairing $A_n^t \times A_n \longrightarrow \mathbb{G}_m$. Alternatively, one can show directly that the maps $H^r(K,A) \longrightarrow \mathrm{Ext}_G^{r+1}(A^t(K_s),K_s^\times)$ defined by this pair of pairings (see 0.14b) equal those defined by the Weil-Barsotti formula. (In fact the best way of handling these pairings is to make use of biextensions and derived categories, see Chapter III, especially Appendix C.)

Remark 3.6. When K has characteristic $p \neq 0$, (3.4) can still be proved by similarly elementary methods provided one omits the p-parts of the groups. More precisely, write A(K)(non-p) for $\varprojlim A(K)^{(n)}$ where n runs over all integers not divisible by p, and let $H^r(K,A)(\text{non-}p) = \bigoplus_{\ell \neq p} H^r(K,A)(\ell)$ for $r > 0$. Then the pair of pairings in (3.5) defines augmented cup-products

$$H^r(K,A^t) \times H^{1-r}(K,A) \longrightarrow \mathbb{Q}/\mathbb{Z},$$

which induce an isomorphism of compact groups

$$A^t(K)(\text{non-}p) \xrightarrow{\approx} H^1(K,A)(\text{non-}p)^* \quad (\text{case } r = 0)$$

and an isomorphism of discrete groups of cofinite type

$$H^1(K,A^t)(\text{non-}p) \xrightarrow{\approx} A(K)(\text{non-}p)^* \quad (\text{case } r = 1).$$

For $r \neq 0,1$, the groups $H^r(K,A^t)(\text{non-}p)$ and $H^r(K,A)(\text{non-}p)$ are zero.

Probably this can be proved by the same method as above, but I have not checked this. Instead I give a direct proof.

There is a commutative diagram

$$\begin{array}{ccccccccc} 0 \longrightarrow & A^t(K)^{(n)} & \longrightarrow & H^1(K,A^t_n) & \longrightarrow & H^1(K,A^t)_n & \longrightarrow 0 \\ & \downarrow & & \downarrow & & \downarrow & \\ 0 \longrightarrow & H^1(K,A)_n^* & \longrightarrow & H^1(K,A_n)^* & \longrightarrow & A(K)^{(n)*} & \longrightarrow 0 \end{array}$$

for all n prime to p. As the middle vertical arrow is an isomorphism (2.3), we see on passing to the limit that

$$H^1(K,A^t)(\text{non-}p) \longrightarrow A(K)(\text{non-}p)^*$$

is surjective. To show that it is injective, it suffices to show that for any n prime to p, $H^1(K,A^t)_n \longrightarrow A(K)^{(n)*}$ is injective, and this we can do by showing that the two groups have the same order. There is a subgroup of finite index in A(K) that is uniquely divisible by all integers prime to p (namely, the kernel of the specialization map $\mathcal{A}(R) \longrightarrow \mathcal{A}_0(k)$, where $\mathcal{A}$ is the Néron model of $\mathcal{A}$; it follows from Hensel's lemma shows that this is uniquely divisible prime to the characteristic of k). Consequently $[A(K)_n] = [A(K)^{(n)}]$. Now the same argument as in the proof of (3.2) shows that the groups in question have the same order. The rest of the proof is exactly as in (3.2).

In §7 of Chapter III, we shall use flat cohomology to prove that (3.4) is valid even for the p-components of the groups.

Remark 3.7. The duality in (3.4) extends in a rather trivial fashion to archimedean local fields. Let $G = \mathrm{Gal}(\mathbb{C}/\mathbb{R})$ and let A be an abelian variety over $\mathbb{R}$. Then the pair of pairing in (3.5) defines a pairing of finite groups

$$H^r_T(G,A^t(\mathbb{C})) \times H^{1-r}_T(G,A(\mathbb{C})) \longrightarrow H^2_T(G,\mathbb{C}^\times) \approx \tfrac{1}{2}\mathbb{Z}/\mathbb{Z}$$

for all integers r. The pairing can be seen to be nondegenerate from the following diagram

$$\begin{array}{ccccccccc} 0 \longrightarrow & H_T^{r-1}(G,A^t) & \longrightarrow & H_T^r(G,A_2^t) & \longrightarrow & H_T^r(G,A^t) & \longrightarrow 0. \\ & \downarrow & & \downarrow & & \downarrow & \\ 0 \longrightarrow & H_T^{r-1}(G,A)^* & \longrightarrow & H_T^{2-r}(G,A_2)^* & \longrightarrow & H_T^{1-r}(G,A)^* & \longrightarrow 0. \end{array}$$

Part (a) of (2.13) shows that the middle arrow is an isomorphism, and the two ends of the diagram show respectively that $H_T^r(G,A^t) \longrightarrow H_T^{1-r}(G,A)^*$ is injective for all r and surjective for all r.

The group $A(\mathbb{R})^o$ is a connected, commutative, compact real Lie group of dimension dim(A), and therefore it is isomorphic to $(\mathbb{R}/\mathbb{Z})^{\dim(A)}$. The norm map $A(\mathbb{C}) \longrightarrow A(\mathbb{R})$ is continuous and $A(\mathbb{C})$ is compact and connected, and so its image is a closed connected subgroup of $A(\mathbb{R})$. Since it contains the subgroup $2A(\mathbb{R})$ of $A(\mathbb{R})$, which has finite index in $A(\mathbb{R})$, it must also be open, and therefore it equals $A(\mathbb{R})^o$. Consequently $H_T^0(G,A) = \pi_0(A(\mathbb{R}))$. The exact sequence

$$0 \longrightarrow A^o(\mathbb{R})_2 \longrightarrow A(\mathbb{R})_2 \longrightarrow \pi_0(A) \longrightarrow 0$$

shows that $[\pi_0(A)].2^{\dim(A)} = [A(\mathbb{R})_2]$, and so $H^1(\mathbb{R},A^t) \neq 0$ if and only if $[A(\mathbb{R})_2] > 2^{\dim(A)}$. For example, when A is an elliptic curve, $H^1(\mathbb{R},A) \neq 0$ if and only if (in the standard form) the graph of A in $\mathbb{R} \times \mathbb{R}$ intersects the x-axis in three points.

When A is an algebraic group over a field k, we now write $\pi_0(A)$ for the set of connected components (for the Zariski topology) of A over k_s; that is, $\pi_0(A) = A_{k_s}/A^o_{k_s}$ regarded as a G-module.

Proposition 3.8. *Let A be an abelian variety over K, and let $\mathcal{A}$ be its Néron model over R. Then*

$$H^1(G/I,A(K_{un})) = H^1(G/I,\pi_0(\mathcal{A}_0))$$

where $\mathcal{A}_0$ is the closed fibre of $\mathcal{A}/R$. In particular, if A has good *reduction, then* $H^1(G/I, A(K_{un})) = 0$.

Proof: Let $\mathcal{A}^o$ be the open subgroup scheme of $\mathcal{A}$ whose generic fibre is A and whose special fibre is the identity component of $\mathcal{A}_0$. Because $\mathcal{A}$ is smooth over R, Hensel's lemma implies that the reduction map $\mathcal{A}(R_{un}) \to \mathcal{A}_0(k_s)$ is surjective (see for example [Milne (1980), I.4.13]), and it follows that there is an exact sequence

$$0 \longrightarrow \mathcal{A}^o(R_{un}) \longrightarrow \mathcal{A}(R_{un}) \longrightarrow \pi_0(\mathcal{A}_0) \longrightarrow 0.$$

Moreover $\mathcal{A}(R_{un}) = A(K_{un})$ (because $\mathcal{A}\otimes_R R_{un}$ is the Néron model of $A\otimes_K K_{un}$), and so it remains to show that $H^r(G/I, \mathcal{A}^o(R_{un})) = 0$ for $r = 1,2$.

An element α of $H^1(G/I, \mathcal{A}^o(R_{un}))$ can be represented by an $\mathcal{A}^o$-torsor P. As $\mathcal{A}^o_0$ is a connected algebraic group over a finite field, Lang's lemma [Serre (1959), VI.4] shows that the $\mathcal{A}^o_0$-torsor $P\otimes_R k$ is trivial, and so P(k) is nonempty. Hensel's lemma now implies that P(R) is nonempty, and so $\alpha = 0$.

Finally, for each n, $H^2(G/I, \mathcal{A}^o(R_{un}/m^n)) = 0$ because G/I has cohomological dimension 1, and this implies that $H^2(G/I, \mathcal{A}^o(R_{un})) = 0$ [Serre (1967a), I.2, Lemma 3].

Remark 3.9. The perceptive reader will already have observed that the proof of the proposition becomes much simpler if one assumes that A has good reduction.

Remark 3.10. (a) Let R be an excellent Henselian discrete valuation ring with finite residue field, and let K be the field of fractions of R. For any abelian variety A over K, let A(K)^ be the completion of A(K) for the topology defined by K. Let $\hat{K}$ be the completion of K. Then

(i) the map $A(K)^{\wedge} \longrightarrow A(\hat{K})$ is an isomorphism;

(ii) the map $H^1(K,A) \longrightarrow H^1(\hat{K},A)$ is an isomorphism.

Therefore the augmented cup-product pairings

$$H^r(K,A^t) \times H^{1-r}(K,A) \longrightarrow H^2(K,\mathbb{G}_m)$$

induce isomorphisms $A^t(K)^{\wedge} \longrightarrow H^1(K,A)^*$ and $H^1(K,A^t) \longrightarrow A(K)^*$.

To prove (i) we have to show that every element of $A(\hat{K})$ can be approximated arbitrarily closely by an element of $A(K)$, but Greenberg's approximation theorem [Greenberg (1966)] says that every element of $\mathcal{A}(\hat{R})$ can be approximated arbitrarily closely by an element of $\mathcal{A}(R)$, and $\mathcal{A}(\hat{R}) = \mathcal{A}(\hat{K})$.

The injectivity of $H^1(K,A) \longrightarrow H^1(\hat{K},A)$ also follows from Greenberg's theorem, because an element of $H^1(K,A)$ is represented by a torsor P over K, which extends to a flat projective scheme $\mathcal{P}$ over R; if $P(\hat{K})$ is nonempty, then $\mathcal{P}(R/m^i)$ is nonempty for all i, which (by Greenberg's theorem) implies that $\mathcal{P}(R)$ is nonempty. For the surjectivity, one endows $H^1(\hat{K},A)$ with its natural topology, and observes that $H^1(K,A)$ is dense in it (because, for any finite Galois extension L of K, $Z^1(L/K,A)$ has a a natural structure as an algebraic group [Milne (1980), p115], and so Greenberg's theorem can be applied again). Proposition 3.8 then shows that the topology on $H^1(\hat{K},A)$ is discrete.

Exercise 3.11. Investigate to what extent the results of this section continue to hold when K is replaced by a complete local field with quasi-finite residue field.

Notes: The duality between $H^1(K,A^t)$ and $H^1(K,A)$ in (3.4) was the first major theorem of the subject (see [Tate (1957/58)]; it was proved before (2.3)), and so can be regarded as the forerunner of the rest of the results in this chapter. The proof of Theorem 3.2 is modelled on a proof of Tate's of (3.4) (cf. [Milne (1970/72), p276]). The description of the pairing in (3.4) given in (3.5) that of Tate's original paper. Proposition 3.8 can be found in [Tate (1962)] in the case of good reduction; the stronger form given here is wellknown.

§4 Global fields

Throughout this section, K will be a global field, and S will be a nonempty set of primes of K, containing the archimedean primes in the case the K is a number field. If $F \supset K$, then the set of primes of F lying over primes in S will also be denoted by S (or, occasionally, by S_F). We write K_S for the maximal subfield of K_s that is ramified over K only at primes in S, and G_S for $\mathrm{Gal}(K_S/K)$. Also

$$R_{K,S} = \bigcap_{v \notin S} \mathcal{O}_v = \{a \in K \mid \mathrm{ord}_v(a) \geq 0 \text{ for all } v \notin S\}$$

denotes the ring of S-integers in K. For each prime v we choose an embedding (over K) of K_s into $K_{v,s}$, and consequently an extension w of v to K_s and an identification of $G_v \overset{\mathrm{df}}{=} \mathrm{Gal}(K_{v,s}/K_v)$ with the decomposition group of w in G_K.

Let P denote the set of prime numbers ℓ such that ℓ^∞ divides the degree of K_S over K. If K is a function field, then P contains all prime numbers because K_S contains Kk_s where k_s is the separable closure of the field of constants of K. If K is a number field, then P contains at least all the primes ℓ such that $\ell R_{K,S} = R_{K,S}$ (that is, such that S contains all primes dividing ℓ) because for such primes, K_S contains the ℓ^mth roots of 1 for all m. (It seems not to be known how large P is in the number field case; for example, if $K = \mathbb{Q}$ and S

$= \{\ell,\infty\}$, is P the set of all prime numbers?)

For a finite extension F of K contained in K_S, we use the following notations:

J_F = the group of idèles of F;

$J_{F,S} = \{(a_w) \in J_F \mid a_w = 1 \text{ for } w \notin S\} \approx \prod'_{w\in S} F_w^\times$ (restricted topological product relative to the subgroups $\hat{\mathcal{O}}_w^\times$);

$R_{F,S} = \bigcap_{w\notin S} \mathcal{O}_w$ = ring of S_F-integers (= integral closure of $R_{K,S}$ in F);

$E_{F,S} = R_{F,S}^\times$ = group of S_F-units;

$C_{F,S} = J_{F,S}/E_{F,S}$ = group of S_F-idèle classes;

$U_{F,S} = \{(a_w) \in J_F \mid a_w \in \hat{\mathcal{O}}_w^\times \text{ for } w \notin S,\ a_w = 1 \text{ otherwise}\}$
$\approx \prod_{w\notin S} \hat{\mathcal{O}}_w^\times$.

Define

$J_S = \varinjlim J_{F,S}$, $R_S = \varinjlim R_{F,S}$, $E_S = \varinjlim E_{F,S}$,

$C_S = \varinjlim C_{F,S}$, $U_S = \varinjlim U_{F,S}$,

where the limit in each case is over all finite extensions F of K contained in K_S.

When S contains all primes of K, we usually drop it from the notation. In this case $K_S = K_s$, $G_S = G_K$, and P contains all prime numbers. Moreover $J_{F,S} = J_F$, $R_{F,S} = F$, $E_{F,S} = F^\times$, and $C_{F,S} = C_F$ is the idèle class group of F. Since everything becomes much simpler in this case, the reader is invited to assume S contains all primes on a first reading.

A duality theorem for the P-class formation (G_S,C_S)

Let $C_S(F) = C_F/U_{F,S}$; we shall show that (G_S,C_S) is a P-class formation with $C_S^{\mathrm{Gal}(K_S/F)} = C_S(F)$. Note that when S contains all primes, (G_S,C_S) is the class formation (G,C) considered in (1.6c), and $C_S(F) = C_F$.

Lemma 4.1. *There is an exact sequence*

$$0 \longrightarrow C_{F,S} \longrightarrow C_F/U_{F,S} \longrightarrow Id_{F,S} \longrightarrow 0,$$

where $Id_{F,S}$ *is the ideal class group of* $R_{F,S}$. *In particular, if* S *omits only finitely many primes, then* $Id_{F,S} = 1$ *and* $C_{F,S} \xrightarrow{\approx} C_F/U_{F,S}$.

Proof: Note that $F^\times \cap U_{F,S} = \{1\}$ and $J_{F,S} \cap (F^\times . U_{F,S}) = E_{F,S}$ (intersections inside J_F). Therefore $U_{F,S}$ can be regarded as a subgroup of C_F and the injection $J_{F,S} \hookrightarrow J_F$ induces an injection $C_{F,S} \hookrightarrow C_F/U_{F,S}$. The cokernel of this last map is $J_F/J_{F,S}\cdot U_{F,S}\cdot F^\times \approx (\bigoplus_{v \notin S} \mathbb{Z})/\mathrm{Im}(F^\times)$, which can be identified with the ideal class group of $R_{F,S}$. If S omits only finitely many primes, then $R_{K,S}$ is a Dedekind domain with only finitely many prime ideals, and any such ring is principal.

Proposition 4.2. *The pair* (G_S, C_S) *is a P-class formation and* $C_S^{G_S} = C_K/U_{K,S}$.

Proof: As we observed in (1.6c), (G,C) is a class formation. Therefore (G_S, C^{H_S}), where $H_S = \mathrm{Gal}(K_s/K_S)$, is a P-class formation (see the discussion preceding 1.13). The next two lemmas show that there is a canonical isomorphism $H^r(G_S, C^{H_S}) \xrightarrow{\approx} H^r(G_S, C_S)$ for all $r \geq 1$, and since the same is true for any open subgroup of G_S, it follows that (G_S, C_S) is also a P-class formation.

Lemma 4.3. *There is a canonical exact sequence*

$$0 \longrightarrow U_S \longrightarrow C^{H_S} \longrightarrow C_S \longrightarrow 0.$$

Proof: When S is finite, on passing to the direct limit over the isomorphisms $C_{F,S} \xrightarrow{\approx} C_F/U_{F,S}$ we obtain an isomorphism $C_S \xrightarrow{\approx} C^{H_S}/U_S$, which gives the exact sequence. In the general case, we have to show that $\varinjlim \mathrm{Id}_{F,S} = 0$. Let L be the maximal unramified extension of F (in K_S) in which all primes of S split, and let F' be the maximal abelian subextension of L/F. Thus F' is the maximal abelian unramified extension of F in which all primes of S split (that is, such that all primes in S are mapped to 1 by the reciprocity map). Class field theory [Tate (1967a), 11.3] gives us a commutative diagram

$$\begin{array}{ccc} \mathrm{Id}_{F,S} & \xrightarrow{\approx} & \mathrm{Gal}(L/F)^{ab} = \mathrm{Gal}(F'/F) \\ \downarrow & & \downarrow V \\ \mathrm{Id}_{F',S} & \xrightarrow{\approx} & \mathrm{Gal}(L/F')^{ab} \end{array}$$

with V the transfer (that is, Verlagerung) map. The principal ideal theorem [Artin and Tate (1961), XIII.4] shows that V is zero. Since similar remarks hold for all finite extensions F of K contained in K_S, we see that $\varinjlim \mathrm{Id}_{F,S} = 0$ (direct limit over such F), and this completes the proof.

Lemma 4.4. *With the above notations,* $H^r(G_S, U_S) = 0$ *for* $r \geq 1$. *Therefore the cohomology sequence of the sequence in* (4.3) *gives isomorphisms* $C_S(K) \xrightarrow{\approx} C_S^{G_S}$, *and* $H^r(G_S, C^{H_S}) \xrightarrow{\approx} H^r(G_S, C_S)$, $r \geq 1$.

Proof: By definition,

$$H^r(G_S, U_S) = \varinjlim_F H^r(\mathrm{Gal}(F/K), U_{F,S}) = \varinjlim_F H^r(\mathrm{Gal}(F/K), \prod_{w \notin S_F} \hat{\mathcal{O}}_w^\times).$$

The cohomology of finite groups commutes with products, and so

$$\begin{aligned} H^r(\mathrm{Gal}(F/K), \prod_{w \notin S_F} \hat{\mathcal{O}}_w^\times) &= \prod_{v \notin S_K} H^r(\mathrm{Gal}(F/K), \prod_{w|v} \hat{\mathcal{O}}_w^\times), \\ &= \prod_{v \notin S_K} H^r(\mathrm{Gal}(F_w/K_v), \hat{\mathcal{O}}_w^\times), \end{aligned}$$

where in the last product w denotes the chosen prime w lying over v. Now $H^r(\mathrm{Gal}(F_w/K_v), \hat{\mathcal{O}}_w^\times) = 0$ for $r \geq 1$ because v is unramified in F (cf.

[Serre (1967a), Pptn 1]), and this completes the proof because $(C^{H_S})^{G_S} = C_K$ [Tate (1967a), 8.1].

We write $D_S(F)$ and D_F for the identity components of $C_S(F)$ and C_F. When K is a function field, the idèle groups are totally disconnected, and so their identity components reduce to the identity element.

Lemma 4.5. *Assume that K is a number field. Then* $D_S(K) = D_K U_{K,S}/U_{K,S}$. *It is divisible, and there is an exact sequence*

$$0 \longrightarrow D_S(K) \longrightarrow C_S(K) \xrightarrow{\text{rec}} G_S^{ab} \longrightarrow 0.$$

Proof: When S contains all primes of K, this is a standard part of class field theory; in fact D_K is the group of divisible elements in C_K [Artin and Tate (1961), VII, IX]. The identity component of $C_S(K)$ is the closure of the image of the identity component of C_K. As $U_{K,S}$ is compact, $C_K \longrightarrow C_S(K)$ is a proper map, and so the image of the identity component is already closed. This proves the first statement, and $D_S(K)$ is divisible because it is a quotient of a divisible group. The image of $U_{K,S}$ in G^{ab} is the subgroup fixing $K_S \cap K_{ab}$, which is also the kernel of $G^{ab} \longrightarrow G_S^{ab}$, and the existence of the exact sequence follows from applying the snake lemma to the diagram

$$\begin{array}{ccccccc} & & U_{K,S} & \longrightarrow & \mathrm{Gal}(K_{ab}/K_S \cap K_{ab}) & \longrightarrow & 0 \\ & & \downarrow & & \downarrow & & \\ 0 \longrightarrow D_K & \longrightarrow & C_K & \longrightarrow & \mathrm{Gal}(K_{ab}/K) & \longrightarrow & 0. \end{array}$$

Theorem 4.6. *Let M be a finitely generated* G_S*-module, and let* $\ell \in P$.

(a) *The map*

$$\alpha^r(G_S,M)(\ell)\colon \operatorname{Ext}^r_{G_S}(M,C_S)(\ell) \longrightarrow H^{2-r}(G_S,M)^*(\ell)$$

is an isomorphism for all $r \geq 1$.

(b) *Let* K *be a number field, and choose a finite totally imaginary Galois extension* L *of* K *contained in* K_S *and such that* $\mathrm{Gal}(K_S/L)$ *fixes* M; *if* P *contains all primes numbers or if* M *is a finite module such that* $[M]R_S = R_S$, *then there is an exact sequence*

$$\operatorname{Hom}(M,D_S(L)) \xrightarrow{N_{L/K}} \operatorname{Hom}_{G_S}(M,C_S) \xrightarrow{\alpha^0} H^2(G_S,M)^* \longrightarrow 0.$$

(c) *Let* K *be a function field; for any finitely generated* G_S*-module, there is an isomorphism*

$$\operatorname{Hom}_{G_S}(M,C_S)^\wedge \longrightarrow H^2(G_S,M)^*$$

where ^ *denotes the completion relative to the topology of open subgroups of finite index.*

Proof: Assume first that K is a number field. Lemma 4.5 shows that, for all $\ell \in P$ and all m, $\alpha^1(G_S,\mathbb{Z}/\ell^m\mathbb{Z})$ is bijective and $\alpha^0(G_S,\mathbb{Z}/\ell^m\mathbb{Z})$ is surjective. Thus it follows from (1.13) that part (a) of the theorem is true for number fields and that $\alpha^0(G_S,M)(\ell)$ is surjective for finite M.

For (b), note first that when $M = \mathbb{Z}$ and $L = K$, the sequence becomes that in the lemma. It follows easily that the sequence is exact whenever G_S acts trivially on M and $L = K$. Let M and L be as in (b), and consider the diagram (1.9.1) in the proof of (1.8):

$$\begin{array}{ccccccccc}
\operatorname{Hom}_{G_S}(M_1,C_S) & \longrightarrow & \operatorname{Hom}_U(M,C_S) & \longrightarrow & \operatorname{Hom}_{G_S}(M,C_S) & \longrightarrow & \cdot & \longrightarrow & \cdots \\
\downarrow & & \downarrow & & \downarrow & & \downarrow\approx & & \\
H^2(G_S,M_1)^* & \longrightarrow & H^2(U,M)^* & \longrightarrow & H^2(G_S,M)^* & \longrightarrow & \cdot & \longrightarrow & \cdots .
\end{array}$$

Here $U = \mathrm{Gal}(K_S/L)$. All vertical maps in the diagram are surjective, and so we get an exact sequence of kernels:

$$\operatorname{Ker}(\alpha^0(G_S,M_1)) \longrightarrow \operatorname{Ker}(\alpha^0(U,M)) \longrightarrow \operatorname{Ker}(\alpha^0(G_S,M)) \longrightarrow 0.$$

We have already observed that the kernel of $\alpha^0(U,M)$ is $\mathrm{Hom}(M,D_S(L))$, and therefore the kernel of $\alpha^0(G_S,M)$ is the image $N_{L/K}(\mathrm{Hom}(M,D_S(L))$ of this in $\mathrm{Hom}_G(M,C_S)$.

When K is a function field, $\mathrm{rec}_G\colon C_K \to G^{ab}$ is injective with dense image. More precisely, there is an exact sequence

$$0 \to C_K \to G^{ab} \to \hat{\mathbb{Z}}/\mathbb{Z} \to 0,$$

and the first arrow induces a topological isomorphism of $\{a \in C_K \mid |a| = 1\}$ onto the open subgroup $\mathrm{Gal}(K_S/Kk_s)$ of G^{ab} [Artin and Tate (1961), 8.3]. From this we again get an exact sequence

$$0 \to C_S^{G_S} \to G_S^{ab} \to \hat{\mathbb{Z}}/\mathbb{Z} \to 0.$$

As $\hat{\mathbb{Z}}/\mathbb{Z}$ is uniquely divisible, part (a) of the theorem follows in this case directly from (1.8). Part (c) can be proved by a similar argument to that which completes the proof of (2.1).

We next reinterprete (4.6) as a statement about the cohomology of an algebraic torus T over K. Let $\mathbb{A}_{F,S} = \prod'_{w\in S} F_w$ be the ring of S-adèles of F, and let $\mathbb{A}_S = \varinjlim \mathbb{A}_{F,S}$, where the limit is again over finite extensions of K contained in K_S. As for any algebraic group over K, it is possible to define the set $T(\mathbb{A}_{F,S})$ of points of T with values in $\mathbb{A}_{F,S}$, and we let $T(\mathbb{A}_S) = \varinjlim T(\mathbb{A}_{F,S})$. If T is split by K_S, then $T(\mathbb{A}_S) = X_*(T)\otimes_{\mathbb{Z}} J_S$. This suggests the definition $T(R_S) = X_*(T)\otimes_{\mathbb{Z}} E_S$. A cocharacter $\chi \in X^*(T)$ defines compatible maps

$$T(R_S) \to E_S, \qquad T(\mathbb{A}_S) \to J_S,$$

and hence a map $T(\mathbb{A}_S)/T(R_S) \to J_S/E_S^{\times} = C_S$. We have therefore a pairing

$$X^*(T) \times T(\mathbb{A}_S)/T(R_S) \to C_S,$$

which induces cup-product pairings

$$H^r(G_S, X^*(T)) \times H^{2-r}(G_S, T(\mathbb{A}_S)/T(R_S)) \longrightarrow H^2(G_S, C_S) \longrightarrow \mathbb{Q}/\mathbb{Z}.$$

Corollary 4.7. *Let T be a torus over K split by K_S, and let $\ell \in P$. Then the cup-product pairings defined above induce dualities between:*

the compact group $H^0(G_S, X^*(T))^\wedge$ (ℓ-adic completion) and the discrete group $H^2(G_S, T(\mathbb{A}_S)/T(R_S))(\ell)$;

the finite groups $H^1(G_S, X^*(T))(\ell)$ and $H^1(G_S, T(\mathbb{A}_S)/T(R_S))(\ell)$;

and, when P contains all prime numbers,

the discrete group $H^2(G_S, X^*(T))$ and the compact group $H^0(G_S, T(\mathbb{A}_S)/T(R_S))^\wedge$ (completion for the topology of open subgroups of finite index).

Proof: As we saw in (1.11), $\mathrm{Ext}^r_{G_S}(X^*(T), C_S) = H^r(G_S, X_*(T) \otimes C_S)$. On tensoring the exact sequence

$$0 \longrightarrow E_S \longrightarrow J_S \longrightarrow C_S \longrightarrow 0$$

with $X_*(T)$, we find that $X_*(T) \otimes C_S = T(\mathbb{A}_S)/T(R_S)$. Therefore $\mathrm{Ext}^r_{G_S}(X^*(T), C_S) = H^r(G_S, T(\mathbb{A}_S)/T(R_S))$, and part (a) of the theorem gives us an isomorphism $H^r(G_S, T(\mathbb{A}_S)/T(R_S))(\ell) \xrightarrow{\approx} H^{2-r}(G_S, X^*(T))^*(\ell)$ for $r \geq 1$. As $H^s(G_S, X^*(T))$ is obviously finite for $s = 1$ and is finitely generated for $s = 0$, this proves the first two assertions. In the function field case, we also have an isomorphism $H^0(G_S, T(\mathbb{A}_S)/T(R_S))^\wedge \longrightarrow H^2(G_S, X^*(T))^*$. In the number field case, completing the exact sequence

$$\mathrm{Hom}(X_*(T), D_S(L)) \longrightarrow H^0(G_S, T(\mathbb{A}_S)/T(R_S)) \longrightarrow H^2(G_S, X^*(T))^* \longrightarrow 0$$

given by (4.6b) yields the required isomorphism because the first group is divisible.

Statement of the main theorem

The rest of this section is devoted to stating and proving Tate's theorem [Tate (1962), Thm 3.1], which combines the dualities so far obtained for local and global fields. From now on, M is a finitely generated G_S-module the order of whose torsion subgroup is a unit in R_S.

For v a prime of K, let $G_v = \mathrm{Gal}(K_{v,s}/K_v)$. In the nonarchimedean case, we write $k(v)$ for the residue field at v, and $g_v = \mathrm{Gal}(k(v)_s/k(v)) = G_v/I_v$. The choice of the embedding $K_s \hookrightarrow K_{v,s}$ determines maps $G_v \longrightarrow G_K \twoheadrightarrow G_S$, and using these maps we obtain localization maps $H^r(G_S,M) \longrightarrow H^r(G_v,M)$ for each G_S-module M. We write $H^r(K_v,M) = H^r(G_v,M)$ *except in the case that* v *is archimedean, in which case we set* $H^r(K_v,M) = H^r_T(G_v,M)$. Thus $H^0(\mathbb{R},M) = M^{\mathrm{Gal}(\mathbb{C}/\mathbb{R})}/N_{\mathbb{C}/\mathbb{R}}M$ and $H^0(\mathbb{C},M) = 0$. When v is nonarchimedean and M is unramified at v, we write $H^r_{un}(K_v,M)$ for the image of $H^r(g_v,M)$ in $H^r(G_v,M)$. Thus $H^0_{un}(K_v,M) = H^0(K_v,M)$, $H^1_{un}(K_v,M) \approx H^1(g_v,M)$, and, unless M has elements of infinite order, $H^2_{un}(K_v,M) = 0$. A finitely generated G_S-module M is unramified for all but finitely many v in S, and we define $P^r_S(K,M)$ to be the restricted topological product of the $H^r(K_v,M)$ relative to the subgroups $H^r_{un}(K_v,M)$. Thus

$P^0_S(K,M) = \prod_{v\in S} H^0(K_v,M)$ with the product topology (it is compact if M is finite);

$P^1_S(K,M) = \prod'_{v\in S} H^1(K_v,M)$ with the restricted product topology (it is always locally compact because each $H^1(K_v,M)$ is finite by (2.1)) .

If M is finite, then

$P^r_S(K,M) = \bigoplus_{v\in S} H^r(K_v,M)$ (discrete topology) for $r \neq 0,1$.

Lemma 4.8. *For any finitely generated* G_S*-module* M*, the image of*

$H^r(G_S,M) \longrightarrow \prod_{v\in S} H^r(K_v,M)$ *is contained in* $P_S^r(K,M)$.

Proof: Let $\gamma \in H^r(G_S,M)$. Then γ arises from an element γ' of $H^r(\mathrm{Gal}(L/K),M)$ for some finite Galois extension L of K contained in K_S, and for all v that are unramified in L, the image of γ in $H^r(K_v,M)$ lies in $H^r_{un}(K_v,M)$.

The lemma provides us with maps β^r: $H^r(G_S,M) \longrightarrow P_S^r(K,M)$ for all r. When necessary, we write $\beta_S^r(K,M)$ for β^r.

Lemma 4.9. *Assume that M is finite. Then the inverse image of any compact subset of* $P_S^1(K,M)$ *under the map* $\beta_S^1(K,M)$ *is finite (in other words, the map is proper when* $H^1(G_S,M)$ *is given the discrete topology).*

Proof: After replacing K with a finite extension contained in K_S, we can assume that G_S acts trivially on M. Let T be a subset of S omitting only finitely many elements, and let

$$P(T) = \prod_{v\in S-T} H^1(K_v,M) \times \prod_{v\in T} H^1_{un}(K_v,M).$$

Then P(T) is compact by Tikhonov's theorem, and every compact neighbourhood of 1 in $P_S^1(K,M)$ is contained in such a set. It suffices therefore to show that the inverse image of P(T) is finite. An element of this set is a homomorphism f: $G_S \longrightarrow M$ such that $K_S^{\mathrm{Ker}(f)}$ is unramified at all primes v in T. Therefore $K_S^{\mathrm{Ker}(f)}$ is an extension of K of degree dividing the fixed integer [M] and unramified outside the finite set S − T. It is a wellknown consequence of Hermite's theorem (see for example [Serre (1964), pII-48]) that there are only finitely many such extension fields, and therefore there are only finitely many maps f.

Define

$$Ш^r_S(K,M) = \operatorname{Ker}(\beta^r\colon H^r(G_S,M) \longrightarrow P^r_S(K,M)).$$

For a finite G_S-module M, we write $M^D = \operatorname{Hom}(M,K_S^\times) = \operatorname{Hom}(M,E_S)$. It is again a finite G_S-module, and if the order of M is a unit in $R_{K,S}$, then $M^D = \operatorname{Hom}(M,K_s^\times)$ and M^{DD} is canonically isomorphic to M.

The results (2.3), (2.6), and (2.13) combine to show that for all $r \in \mathbb{Z}$, $P^r_S(K,M)$ is the algebraic and topological dual of $P^{2-r}_S(K,M^D)$. Therefore there are continuous maps

$$\gamma^r = \gamma^r_S(K,M^D)\colon P^r_S(K,M^D) \longrightarrow H^{2-r}(G_S,M)^*$$

with γ^r the dual of β^{2-r}.

Theorem 4.10. *Let M be a finite G_S-module whose order is a unit in $R_{K,S}$.*

(a) *The groups $Ш^1_S(K,M)$ and $Ш^2_S(K,M^D)$ are finite and there is a canonical nondegenerate pairing*

$$Ш^1_S(K,M) \times Ш^2_S(K,M^D) \longrightarrow \mathbb{Q}/\mathbb{Z}.$$

(b) *The map $\beta^0_S(K,M)$ is injective and $\gamma^2_S(K,M^D)$ is surjective; for r = 0,1,2, $\operatorname{Im}(\beta^r_S(K,M)) = \operatorname{Ker}(\gamma^r_S(K,M^D))$.*

(c) *For $r \geq 3$, β^r is a bijection $\beta^r\colon H^r(G_S,M) \xrightarrow{\approx} \prod_{v \text{ real}} H^r(K_v,M)$.*

Consequently, there is an exact sequence of locally compact groups and continuous homomorphisms

$$\begin{array}{ccccc} 0 \longrightarrow H^0(G_S,M) & \xrightarrow{\beta^0} & P^0_S(K,M) & \xrightarrow{\gamma^0} & H^2(G_S,M^D)^* \\ & & & & \downarrow \\ H^1(G_S,M^D)^* & \xleftarrow{\gamma^1} & P^1_S(K,M) & \xleftarrow{\beta^1} & H^1(G_S,M) \\ \downarrow & & & & \\ H^2(G_S,M) & \xrightarrow{\beta^2} & P^2_S(K,M) & \xrightarrow{\gamma^2} & H^0(G_S,M^D)^* \longrightarrow 0. \end{array}$$

The groups in this sequence have the following topological proper-

ties:

finite	*compact*	*compact*
compact	*locally compact*	*discrete*
discrete	*discrete*	*finite.*

The finiteness of $Ш^1_S(K,M)$ is contained in (4.9); that of $Ш^2_S(K,M^D)$ will follow from the existence of the nondegenerate pairing in (a). The vertical arrows in the above diagram will be defined below; alternatively they can be deduced from the nondegenerate pairings in (a) because the cokernels of γ^0 and γ^1 are $Ш^2_S(K,M^D)^*$ and $Ш^1_S(K,M^D)^*$ respectively.

Example 4.11. (i) For any integer $m > 1$ and any set of primes S of density greater than 1/2, $Ш^1_S(K,\mathbb{Z}/m\mathbb{Z}) = 0$; consequently, $Ш^2_S(K,\mu_m) = 0$ under the same condition provided m is a unit in $R_{K,S}$.

(ii) If S omits only finitely many primes of K and m is a unit in $R_{K,S}$, then $Ш^1_S(K,\mu_m) = 0$ or $\mathbb{Z}/2\mathbb{Z}$; consequently, $Ш^2_S(K,\mathbb{Z}/m\mathbb{Z}) = 0$ or $\mathbb{Z}/2\mathbb{Z}$ under the same conditions.

To see (i), note that $H^1(G_S,\mathbb{Z}/m\mathbb{Z}) = \mathrm{Hom}(G_S,\mathbb{Z}/m\mathbb{Z})$ and $H^1(K_v,\mathbb{Z}/m\mathbb{Z}) = \mathrm{Hom}(G_v,\mathbb{Z}/m\mathbb{Z})$. Therefore an element of $Ш^1_S(K,\mathbb{Z}/m\mathbb{Z})$ corresponds to a cyclic extension of K in which all primes of S split. The Chebotarev density theorem shows that such an extension must be trivial when S has density greater that 1/2.

To see (ii), note that $Ш^1_S(K,\mu_m)$ is the kernel of $K^\times/K^{\times m} \longrightarrow \bigoplus_{v\in S} K_v^\times/K_v^{\times m}$, that is, it is the set of elements of $K^\times$ that are local m^{th} powers modulo those that are global m^{th} powers. This set is described in [Artin and Tate (1961), X.1], where the "special case" in which $Ш^1_S(K,\mu_m) \neq 0$ is also determined.

Proof of the main theorem

The proof of the theorem will consist of identifying the exact sequence in the statement of the theorem with the $\mathrm{Ext}_{G_S}(M^D,-)$-sequence of

$$0 \longrightarrow E_S \longrightarrow J_S \longrightarrow C_S \longrightarrow 0,$$

except that, in the number field case, $\mathrm{Hom}_{G_S}(M^D,J_S)$ and $\mathrm{Hom}_{G_S}(M^D,C_S)$ must be replaced by their quotients by $N_{L/K}\mathrm{Hom}(M^D, \prod_{v\ \mathrm{arch}} L_w^{\times})$ and $N_{L/K}\mathrm{Hom}(M^D,D_{L,S})$ for any field L as in (4.6b).

In fact we shall consider more generally a finitely generated G_S-module M. In this case we write M^d for the dual of M loosely regarded as a group scheme over $\mathrm{Spec}(R_{K,S})$. More precisely, when M is being regarded as a G_S-module, we let $M^d = \mathrm{Hom}(M,E_S)$. For $v \notin S$, M is a g_v-module, and we write $M^d = \mathrm{Hom}(M,\hat{\mathcal{O}}^{\times}_{v,un})$; for $v \in S$, M is a G_v-module, and we write $M^d = \mathrm{Hom}(M,K^{\times}_{v,s})$. It will always be clear from the context, which of these three we mean.

Lemma 4.12. *Let M be a finitely generated G_S-module such that the order of M_{tors} is a unit in $R_{K,S}$.*

(a) *The group* $\mathrm{Ext}^r_{G_S}(M,E_S) = H^r(G_S,M^d)$, *all* $r \geq 0$.

(b) *For* $v \notin S$, $H^r(g_v,M^d) = \mathrm{Ext}^r_{g_v}(M,\hat{\mathcal{O}}^{\times}_{v,un})$; *for* $r \geq 2$, *both groups are* 0.

Proof: (a) As E_S is divisible by all integers that are units in $R_{K,S}$, this is a special case of (0.8).

(b) As $\hat{\mathcal{O}}^{\times}_{v,un}$ is divisible by all integers dividing the order of M_{tors}, this is again a special case of (0.8). The g_v-module $\hat{\mathcal{O}}^{\times}_{v,un}$ is cohomologically trivial [Serre (1967a), 1.2], and so an easy generalization to profinite groups of [Serre (1962), IX.6, Thm 11] shows that there exists a short exact sequence

$$0 \longrightarrow \hat{\mathcal{O}}^{\times}_{v,un} \longrightarrow I^0 \longrightarrow I^1 \longrightarrow 0$$

of g_v-modules with I^0 and I^1 injective. It is obvious from this that $\mathrm{Ext}^r_{g_v}(M, \hat{\mathcal{O}}^{\times}_{v,un}) = 0$ for $r \geq 2$.

Lemma 4.13. *In addition to the hypotheses of* (4.12), *assume either that* M *is finite or that* S *omits only finitely many primes. Then*

$$\mathrm{Hom}_{G_S}(M, J_S) = \prod_{v \in S} H^0(G_v, M^d) \quad (= P^0_S(K, M^d) \textit{ if } K \textit{ is a function field})$$

and

$$\mathrm{Ext}^r_{G_S}(M, J_S) = P^r_S(K, M^d), \quad r \geq 1.$$

Proof: We consider finite subsets T of S satisfying the same hypotheses as S relative to M, namely, T contains all archimedean primes plus those nonarchimedean primes at which M is ramified, and the order of M_{tors} is a unit in $R_{K,T}$. Let $J_{F,S\supset T} = \prod_{w \in T} F^{\times}_w \times \prod_{w \in S-T} \hat{\mathcal{O}}^{\times}_v$.

Then $J_S = \varinjlim_{F,T} J_{F,S\supset T}$ (limit over F and T with $F \subset K_T$ and splitting M), and so (0.10) shows that

$$\mathrm{Ext}^r_{G_S}(M, J_S) = \varinjlim_{F,T} \mathrm{Ext}^r_{\mathrm{Gal}(F/K)}(M, J_{F,S\supset T}).$$

Since Exts commute with products in the second place (to see this, compute them by taking a projective resolution of the term in the first place), on applying (0.11) we find that

$$\mathrm{Ext}^r_{G_{F/K}}(M, J_{F,S\supset T}) = (\prod_{v \in T} \mathrm{Ext}^r_{G_{F_w/K_v}}(M, F^{\times}_w) \times (\prod_{v \in S-T} \mathrm{Ext}^r_{G_{F_w/K_v}}(M, \hat{\mathcal{O}}^{\times}_{F_w})).$$

As $\hat{\mathcal{O}}^{\times}_{v,un}$ is cohomologically trivial, (0.9) shows that for $v \in S - T$,

$$\mathrm{Ext}^r_{G_{F_w/K_v}}(M, \hat{\mathcal{O}}^{\times}_{F_w}) = \mathrm{Ext}^r_{g_v}(M, \hat{\mathcal{O}}^{\times}_{v,un}),$$

and we have already seen that

$$\mathrm{Ext}^r_{g_v}(M, \hat{\mathcal{O}}^{\times}_{v,un}) = H^r(g_v, M^d).$$

On combining these statements, we find that

$$\mathrm{Ext}^r_{G_S}(M,J_S) = \varinjlim_{F,T} \Big(\prod_{v\in T} \mathrm{Ext}^r_{G_{F_w/K_v}}(M,F_w^\times) \times \Big(\prod_{v\in S-T} H^r(g_v,M^d)\Big).$$

For $r \le 1$, (0.9) shows that we can replace $\mathrm{Ext}^r_{G_{F_w/K_v}}(M,F_w^\times)$ with $\mathrm{Ext}^r_{G_v}(M,K_{v,s}^\times)$, which equals $H^r(G_v,M^d)$ by (0.8). Hence

$$\mathrm{Ext}^r_{G_S}(M,J_S) = \varinjlim_{T} \Big(\prod_{v\in T} H^r(G_v,M^d) \times \prod_{v\in S-T} H^r(g_v,M^d)\Big),$$

which equals $\prod_{v\in S} H^0(G_v,M^d)$ in the case that $r = 0$, and equals $\prod'_{v\in S} H^1(G_v,M) \overset{\mathrm{df}}{=} P^1_S(K,M)$ for $m = 1$.

For $r \ge 2$, $H^r(g_v,M^d) = 0$ by (4.12), and so

$$\begin{aligned}\mathrm{Ext}^r_{G_S}(M,J_S) &= \varinjlim_{F} \Big(\bigoplus_{v\in S} \mathrm{Ext}^r_{G_{F_w/K_v}}(M,F_w^\times)\Big) \quad (\text{limit over all } F \subset K_S,\ F \supset K)\\ &= \bigoplus_{v\in S} \Big(\varinjlim_{F} \mathrm{Ext}^r_{G_{F_w/K_v}}(M,F_w^\times)\Big).\end{aligned}$$

In the case that S contains almost all primes, $\varinjlim_F F_w = K_{v,s}$ and so $\varinjlim_F \mathrm{Ext}^r_{G_{F_w/K_v}}(M,F_w^\times) = \mathrm{Ext}^r_{G_v}(M,K_{v,s}^\times)$. In the case that M is finite, we know that if ℓ divides the order of M, then S contains all primes lying over ℓ. Therefore $\varinjlim H^2(\mathrm{Gal}(K_{v,s}/F_w),K_{v,s}^\times)(\ell) = \varinjlim \mathrm{Br}(F_w)(\ell) = 0$, and the spectral sequence (0.9)

$$\mathrm{Ext}^r_{\mathrm{Gal}(F_w/K_v)}(M,H^s(\mathrm{Gal}(K_{v,s}/F_w),K_{v,s}^\times) \Rightarrow \mathrm{Ext}^r_{G_v}(M,K_{v,s}^\times)$$

shows that again $\varinjlim_F \mathrm{Ext}^r_{G_{F_w/K_v}}(M,F_w^{\times}) = \mathrm{Ext}^r_{G_v}(M,K_{v,s}^\times)$. From (0.8) we know that $\mathrm{Ext}^r_{G_v}(M,K_{v,s}^\times) = H^r(G_v,M^d)$ $(= H^r(K_v,M^d))$, and so this completes the proof of the lemma.

Remark 4.14. Without the additional hypotheses, (4.13) is false. For example, let $K = \mathbb{Q}$, $S = \{\infty\}$, and $M = \mathbb{Z}$. Then $G_S = \{1\}$, and so

$\mathrm{Ext}^r_{G_S}(\mathbb{Z}, J_S) = 0$ for $r > 0$, but $P^2_S(K, M^d) = H^2(\mathrm{Gal}(\mathbb{C}/\mathbb{R}), \mathbb{C}^\times) = \frac{1}{2}\mathbb{Z}/\mathbb{Z}$.

Now assume that M is finite. On using (4.12), (4.13), and (4.6) to replace terms in the sequence

$$\cdots \longrightarrow \mathrm{Ext}^r_{G_S}(M^D, E_S) \longrightarrow \mathrm{Ext}^r_{G_S}(M^D, J_S) \longrightarrow \mathrm{Ext}^r_{G_S}(M^D, C_S) \longrightarrow \cdots,$$

we obtain an exact sequence

$$0 \longrightarrow H^0(G_S, M) \longrightarrow \prod_{v \in S} H^0(G_S, M) \longrightarrow \mathrm{Hom}_{G_S}(M^D, C_S)$$

$$\longrightarrow H^1(G_S, M) \longrightarrow P^1_S(K, M) \longrightarrow H^1(G_S, M^D)^*$$

$$\longrightarrow H^2(G_S, M) \longrightarrow P^2_S(K, M) \longrightarrow H^0(G_S, M^D)^* \longrightarrow H^3(G_S, M) \longrightarrow \bigoplus_{v \text{ real}} H^3(G_v, M) \to 0$$

and isomorphisms

$$H^r(G_S, M) \xrightarrow{\approx} \bigoplus_{v \text{ real}} H^r(K_v, M), \quad r \geq 4.$$

This is the required exact sequence except for the first three terms in the number field case and the surjectivity of $P^2_S(K, M^D) \longrightarrow H^0(G_S, M)^*$. But this last map is dual to $H^0(G_S, M) \longrightarrow P^0_S(K, M)$, which is injective. (Note that if $M \neq 0$ in the number field case, then S must contain at least one nonarchimedean prime.) For the first three terms of the sequence in the number field case, consider the exact commutative diagram:

$$\begin{array}{ccccccccc}
 & & \mathrm{Hom}(M^D, \prod_{v \text{ arch}} L_v^\times) & \longrightarrow & \mathrm{Hom}(M^D, C_S(L)) & & & & \\
 & & \downarrow N_{L/K} & & \downarrow N_{L/K} & & & & \\
0 \longrightarrow H^0(G_S, M) & \longrightarrow & \prod_{v \in S} H^0(G_v, M) & \longrightarrow & \mathrm{Hom}_{G_S}(M^D, C_S) & \longrightarrow & \mathrm{Ker}(\beta^1) & \longrightarrow & 0. \\
 & & \downarrow & & \downarrow & & & & \\
 & & P^0_S(K, M) & \longrightarrow & H^2(G_S, M)^* & & & & \\
 & & \downarrow & & \downarrow & & & & \\
 & & 0 & & 0. & & & &
\end{array}$$

The map the map $\mathrm{Hom}(M^D, \prod_{v \text{ arch}} L_v^\times) \longrightarrow \mathrm{Hom}(M^D, D_S(L))$ is always an isomorphism on torsion, and therefore it is an isomorphism in our case. The snake lemma now gives us an exact sequence

$$0 \longrightarrow H^0(G_S, M) \longrightarrow P_S^0(K, M) \longrightarrow H^2(G_S, M^D)^* \longrightarrow \mathrm{Ker}(\beta^1) \longrightarrow \dots ,$$

which completes the proof of the theorem. (An alternative approach is to note that the first half of the sequence can be obtained as the algebraic and topological dual of the second half.)

Consequences

Corollary 4.15. *If S is finite and M is a finite G_S-module whose order is a unit in $R_{K,S}$, then the groups $H^r(G_S, M)$ are finite for all r.*

Proof: In this case the groups $P_S^r(K, M)$ are finite, and so the finiteness of $H^0(G_S, M)$ is obvious and that of $H^1(G_S, M)$ and $H^2(G_S, M)$ follows from the finiteness of $Ш_S^1(K, M)$ and $Ш_S^2(K, M)$.

Corollary 4.16. *Let M be a finite G_S-module whose order is a unit in $R_{K,S}$. Then, for any finite subset T of S omitting at least one finite prime of S, the map*

$$H^2(G_S, M) \longrightarrow \bigoplus_{v \in T} H^2(G_v, M)$$

is surjective. In particular, in the number field case the map $H^2(K, M) \longrightarrow \bigoplus_{v \text{ real}} H^2(K_v, M)$ is surjective.

Proof: Let v_0 be a finite prime of K not in T. In order to prove the corollary, it suffices to show that for any element $a = (a_v)$ of $P_S^2(K, M)$, it is possible to modify a_{v_0} so as to get an element in the image of β^2. Theorem 4.10 shows that, in the duality between

$P^0_S(K,M^D)$ and $P^2_S(K,M)$, the image of β^2 is the orthogonal complement of the image of β^0. Let χ be the character of $P^0_S(K,M^D)$ defined by a, and let χ' be its restriction to $H^0(G_S,M^D)$. The map

$$H^0(G_S,M) \longrightarrow H^0(K_{v_0},M)$$

is injective, and every character of $H^0(G_S,M^D)$ extends to one of $H^0(K_{v_0},M^D)$. Choose such an extension of χ' and let a'_{v_0} be the element of $H^2(K_{v_0},M)$ corresponding to it by duality. When the component a_{v_0} of a is replaced by $a_{v_0} - a'_{v_0}$, then a becomes orthogonal to Im β^0 and is therefore in the image of β^2.

Corollary 4.17. *For any number field* K,

$H^0(G_K,\mathbb{Z}) = \mathbb{Z}$,

$H^2(G_K,\mathbb{Z}) = \mathrm{Hom}(C_K/D_K,\mathbb{Q}/\mathbb{Z})$,

$H^{2r}(G_K,\mathbb{Z}) = (\mathbb{Z}/2\mathbb{Z})^t$ *for* $2r \geq 4$, *where* t *is the number of real primes of* K, *and*

$H^r(G_K,\mathbb{Z}) = 0$ *for* r *odd*.

Proof: The assertions for $r \leq 2$ are obvious. According to (1.12), $G \overset{df}{=} G_K$ contains an open subgroup U of index 2 having strict cohomological dimension 2. Therefore $H^r(G,\mathbb{Z}[G/U]) = H^r(U,\mathbb{Z}) = 0$ for $r \geq 3$. Let σ generate G/U. The exact sequence

$$0 \longrightarrow \mathbb{Z} \xrightarrow{1+\sigma} \mathbb{Z}[G/U] \xrightarrow{1-\sigma} \mathbb{Z}[G/U] \xrightarrow{\sigma\mapsto 1} \mathbb{Z} \longrightarrow 0$$

gives rise to isomorphisms $H^r(G,\mathbb{Z}) \to H^{r+2}(G,\mathbb{Z})$ for $r \geq 3$. For $r \geq 4$,

$$H^r(G,\mathbb{Z}) = H^{r-1}(G,\mathbb{Q}/\mathbb{Z}) = \varinjlim H^{r-1}(G,\tfrac{1}{n}\mathbb{Z}/\mathbb{Z}) \overset{(4.10)}{=} \varinjlim \prod_{v \text{ real}} H^{r-1}(K_v,\tfrac{1}{n}\mathbb{Z}/\mathbb{Z})$$

$= \prod_{v \text{ real}} H^r(K_v,\mathbb{Z})$. (We applied (4.10) with S the set of all primes of K.) If r is odd, $H^r(\mathbb{R},\mathbb{Z}) = 0$, and if r is even, $H^r(\mathbb{R},\mathbb{Z}) \approx \mathbb{Z}/2\mathbb{Z}$, and so this completes the proof.

Corollary 4.18. *For any prime* ℓ *that is a unit in* R_S,

$$H^r(G_S,E_S)(\ell) \longrightarrow \bigoplus_{v \text{ real}} H^r(G_v,K^{\times}_{v,s})(\ell)$$

is an isomorphism, all $r \geq 3$. *In particular, when* $r \geq 3$, $H^r(G_S,E_S)(\ell) = 0$ *if* ℓ *or* r *is odd.*

Proof: From the sequence

$$0 \longrightarrow E_S \longrightarrow K_S^{\times} \longrightarrow \bigoplus_{v \notin S} \mathbb{Z} \longrightarrow 0$$

(here S denotes the set of primes of K_S lying over a prime of S), we get an exact sequence

$$H^2(G_S,E_S) \longrightarrow \mathrm{Br}(K) \longrightarrow \bigoplus_{v \notin S} \mathrm{Br}(K_v)$$

(cf. A.7). Therefore, the map $H^2(G_S,E_S) \longrightarrow \bigoplus_{v \text{ real}} \mathrm{Br}(K_v)$ is surjective. The Kummer sequence

$$0 \longrightarrow \mu_{\ell^n} \longrightarrow E_S \xrightarrow{\ell^n} E_S \longrightarrow 0$$

gives us the first row of the next diagram

$$\begin{array}{ccccccc}
H^2(G_S,E_S) & \longrightarrow & H^3(G_K,\mu_{\ell^n}) & \longrightarrow & H^3(G_K,E_S)_{\ell^n} & \longrightarrow & 0 \\
\downarrow \text{surj} & & \downarrow \approx & & \downarrow & & \\
\bigoplus_{v \text{ real}} \mathrm{Br}(K_v) & \longrightarrow & \bigoplus_{v \text{ real}} H^3(G_v,\mu_{\ell^n}) & \longrightarrow & \bigoplus_{v \text{ real}} H^3(G_v,K^{\times}_{v,s})_{\ell^n} & \longrightarrow & 0,
\end{array}$$

and the five-lemma proves our assertion for $r = 3$. One now proceeds by induction, using the continuation of the diagram.

Remark 4.19. Let F be a finite extension of K contained in K_S, and let $H_S = \mathrm{Gal}(K_S/F)$. Then the following diagram is commutative:

$$\begin{array}{ccccccccc}
0 \longrightarrow H^0(G_S,M) & \longrightarrow & P_S^0(K,M) & \longrightarrow & H^2(G_S,M^D)^* & \longrightarrow & H^1(G_S,M) & \longrightarrow & \dots \\
\downarrow \mathrm{Res} & & \downarrow \mathrm{Res} & & \downarrow \mathrm{Cor}^* & & \downarrow \mathrm{Res} & & \\
0 \longrightarrow H^0(H_S,M) & \longrightarrow & P_S^0(F,M) & \longrightarrow & H^2(H_S,M^D)^* & \longrightarrow & H^1(H_S,M) & \longrightarrow & \dots .
\end{array}$$

This follows from the commutativity of the following diagrams:

$$\begin{array}{ccccccc} \cdots \longrightarrow \mathrm{Ext}^r_{G_S}(M,E_S) & \longrightarrow & \mathrm{Ext}^r_{G_S}(M,J_S) & \longrightarrow & \mathrm{Ext}^r_{G_S}(M,C_S) & \longrightarrow & \cdots \\ \downarrow & & \downarrow & & \downarrow & & \\ \cdots \longrightarrow \mathrm{Ext}^r_{H_S}(M,E_S) & \longrightarrow & \mathrm{Ext}^r_{H_S}(M,J_S) & \longrightarrow & \mathrm{Ext}^r_{H_S}(M,C_S) & \longrightarrow & \cdots , \end{array}$$

and

$$\begin{array}{ccc} \mathrm{Ext}^r_{G_S}(M,C_S) & \longrightarrow & H^{2-r}(G_S,M^D)^* \\ \downarrow & & \downarrow \mathrm{Cor}^* \\ \mathrm{Ext}^r_{H_S}(M,C_S) & \longrightarrow & H^{2-r}(H_S,M^D)^*. \end{array}$$

An explicit description of the pairing between $Ш^1$ and $Ш^2$

Finally we shall give an explicit description of the pairing

$$Ш^1_S(K,M) \times Ш^2_S(K,M^D) \longrightarrow \mathbb{Q}/\mathbb{Z}.$$

Represent $a \in Ш^1_S(K,M)$ and $a' \in Ш^2_S(K,M^D)$ by cocycles $\alpha \in Z^1(G_S,M)$ and $\alpha' \in Z^2(G_S,M^D)$. Write α_v and α'_v for the restrictions of α and α' to G_v. Then for each $v \in S$, we have a 0-cochain β_v and 1-cochain β'_v such that $d\beta_v = \alpha_v$ and $d\beta'_v = \alpha'_v$. The cup-product $\alpha\cup\alpha' \in Z^3(G_S,E_S)$, and as $H^3(G_S,E_S)$ has no nonzero elements of order dividing the $[M]$, there is a 2-cochain ϵ (for G_S) with coefficients in E_S such that $\alpha\cup\alpha' = d\epsilon$. Then $d(\beta_v\cup\alpha'_v) = d\epsilon_v = d(\alpha_v\cup\beta'_v)$ and $d(\beta_v\cup\beta'_v) = \alpha_v\cup\beta'_v - \beta_v\cup\alpha'_v$, and so for each v, $(\alpha_v\cup\beta'_v) - \epsilon_v$ and $(\alpha'_v\cup\beta_v) - \epsilon_v$ are cocycles representing the same class, say c_v, in $H^2(G_v,K^\times_{v,s})$. Set $\langle a,a'\rangle = \sum \mathrm{inv}_v(c_v)$. It is easy to see that this element is independent of the choices made, and one can show that it is equal to the image of (a,a') under the pairing constructed in the proof of the theorem.

Generalization to finitely generated modules

We note that in the course of the proof of (4.10) we have shown the following result.

Theorem 4.20. *Assume that S omits only finitely many primes of K, and let M be a finitely generated module over G_S such that the order of M_{tors} is a unit in $R_{K,S}$.*

(a) *The group $Ш_S^2(K,M^d)$ is finite and is dual to $Ш_S^1(K,M)$.*

(b) *There is an exact sequence of continuous homomorphisms*

$$\begin{array}{ccccc} H^1(G_S,M)^* & \longleftarrow & \prod H^1(K_v,M^d) & \longleftarrow & H^1(G_S,M^d) \\ \downarrow & & & & \\ H^2(G_S,M^d) & \longrightarrow & \oplus\, H^2(G_v,M^d) & \longrightarrow & H^0(G_S,M^d)^* \longrightarrow 0, \end{array}$$

and for $r \geq 3$ *there are isomorphisms*

$$H^r(G_S,M^d) \xrightarrow{\approx} \prod_{v\ \mathrm{real}} H^r(K_v,M^d).$$

(c) *In the function field case, the sequence in* (b) *can be extended by*

$$0 \longrightarrow H^0(G_S,M^d)^\wedge \longrightarrow \prod H^0(G_v,M^d)^\wedge \longrightarrow H^2(G_S,M)^* \longrightarrow \cdots$$

where ^ denotes completion with respect to the topology of open subgroups of finite index.

Proof: To obtain (b) and (c), write down the Ext(M,-)-sequence of

$$0 \longrightarrow E_S \longrightarrow J_S \longrightarrow C_S \longrightarrow 0$$

and use (4.6), (4.12), and (4.13) to replace various of the terms. Part (a) is a restatement of the fact that the sequence in (b) is exact at $H^2(G_S,M^d)$.

Corollary 4.21. *Let T be a torus over K. If S omits only finitely many primes, then there are isomorphisms* $H^r(G_S,T) \xrightarrow{\approx} \bigoplus_{v\ \mathrm{real}} H^r(K_v,T)$ *for all* $r \geq 3$. *In particular* $H^r(G_S,\mathbb{G}_m) = 0$ *for all odd* $r \geq 3$.

Proof: Take $M = X^*(T)$ in the theorem.

I do not know to what extent Theorem 4.20 holds with M and M^d interchanged, but R. Kottwitz has shown that for any torus T over a number field K, and r = 1, 2, there is a canonical nondegenerate pairing of finite groups

$$Ш^r(K,T) \times Ш^{3-r}(K,X^*(T)) \longrightarrow \mathbb{Q}/\mathbb{Z}.$$

For $r \neq 1,2$, $Ш^r(K,T)$ and $Ш^r(K,X^*(T))$ are zero. Here $Ш^r(K,T)$ is defined to be the kernel of $H^r(K,T) \longrightarrow \prod_{\text{all } v} H^r(K_v,T)$. See also (II.4) below.

Notes: Theorem 4.10 is due to Tate (see [Tate (1962)] for an announcement with a brief indication of proof). Parts of the theorem were found independently by Poitou ([Poitou (1966), (1967)]). The above proof of (4.10) generalizes that in [Tate (1966)], which treats only the case that S contains all primes of K. There is also a proof in [Haberland (1978)] similarly generalizing [Poitou (1967)]. Corollaries 4.16 and 4.17 are also due to Tate (cf. [Borel and Harder (1978), 1.6] and [Serre (1977), 6.4]).

Proofs of parts of the results in this section can also be found in [Takahashi (1969)], [Uchida (1969)], [Bashmakov (1972)], and [Langlands (1983), VII.2].

§5 Global Euler-Poincaré characteristics

Let K be a global field, and let S be a finite nonempty set of primes including all archimedean primes. As in §4, we write K_S for the maximal subfield of K_s that is ramified over K only at primes in S, G_S for $\mathrm{Gal}(K_S/K)$, and $R_{K,S}$ for the ring of S-integers $\bigcap_{v \notin S} \mathcal{O}_v$. Let

M be a finite G_S-module whose order is a unit in R_S. We know from (4.15) that the groups $H^r(G_S,M)$ are finite for all r, and we would like to define $\chi(G_S,M)$ to be the alternating product of their orders. However, when K is a real number field, the cohomology groups will in general be nonzero for an infinite number of values of r (see 4.10c), and so this is not possible. Instead, we abuse notation, and set

$$\chi(G_S,M) = \frac{[H^0(G_S,M)][H^2(G_S,M)]}{[H^1(G_S,M)]}.$$

Theorem 5.1. *With the above definition,*

$$\chi(G_S,M) = \prod_{v \text{ arch}} \frac{[H^0(G_v,M)]}{|[M]|_v}.$$

Remark 5.2. (a) In the function field case, the theorem says simply that $\chi(G_S,M) = 1$. In the number field case, (2.13c) shows that

$$[H^0(G_v,M)]/|[M]|_v = [H^1(G_v,M)]/[H^0(G_v,M^D)],$$

and (2.13a) shows that $[H^1(G_v,M)] = [H^1(G_v,M^D)]$, which equals $[H^0_T(G_v,M^D)]$ because the Herbrand quotient of a finite module is 1. Therefore the formula can also be written as

$$\chi(G_S,M) = \prod_{v \text{ arch}} \frac{[H^0_T(G_v,M^D)]}{[H^0(G_v,M^D)]}.$$

(b) Because S is finite, all groups in the complex in Theorem 4.10 are finite, and so the exactness of the complex implies that

$$\chi(G_S,M)\ \chi(G_S,M^D) = \prod_{v\in S} \chi(K_v,M) \tag{5.2.1}$$

where $\chi(K_v,M) = [H^0(K_v,M)][H^1(K_v,M)]^{-1}[H^2(K_v,M)]$ (notations as in §4). According to (2.8), $\chi(K_v,M) = |[M]|_v$ if v is nonarchimedean, and obviously $\chi(K_v,M) = [H^0(K_v,M)] = [H^0_T(G_v,M)]$ if v is archimedean. By assumption $|[M]|_v = 1$ if $v \notin S$, and so the product formula shows that

$$\prod_{v\in S}\chi(K_v,M) = \prod_{v\text{ arch}} [H^0(K_v,M)]/|[M]|_v.$$

Now (2.13c) allows us to rewrite this as

$$\prod_{v\text{ arch}} \frac{[H^0_T(G_v,M)][H^0_T(G_v,M^D)]}{[H^0(G_v,M)][H^0(G_v,M^D)]}.$$

Therefore (5.2.1) is also implied by (5.1), and conversely, in the case that $M \approx M^D$, (5.2.1) implies the theorem.

(c) The theorem can sometimes be useful in computing the order of $H^1(G_S,M)$. It says that

$$[H^1(G_S,M)] = [H^0(G_S,M)][H^2(G_S,M)] \prod_{v\text{ arch}} |[M]|_v/[H^0(G_v,M)],$$

and we know by (4.10) that $H^2(G_S,M)$ fits into an exact sequence

$$0 \longrightarrow Ш^2_S(K,M) \longrightarrow H^2(G_S,M) \longrightarrow \bigoplus_{v\in S} H^2(K_v,M) \longrightarrow H^0(G_S,M^D)^* \longrightarrow 0.$$

By duality, $[Ш^2_S(K,M)] = [Ш^1_S(K,M^D)]$ and $[H^2(K_v,M)] = [H^0(K_v,M^D)]$, and so the theorem is equivalent to the statement

$$[H^1(G_S,M)] = [Ш^1_S(K,M^D)] \frac{[H^0(G_S,M)]}{[H^0(G_S,M^D)]} \prod_{v\in S}[H^0(K_v,M^D] \prod_{v\text{ arch}} \frac{|[M]|_v}{[H^0(G_v,M)]}.$$

Proof (of 5.1): The method of proof is similar to that of (2.8). Let $\varphi(M)$ be the quotient of $\chi(G_S,M)$ by the right hand side of the equation. We have to show that $\varphi(M) = 1$. The argument in (5.2b) shows that (4.10) implies that $\varphi(M)\varphi(M^D) = 1$, and so in order to prove the theorem for a module M, it suffices to show that $\varphi(M) = \varphi(M^D)$

Lemma 5.3. *The map φ from the category of finite G_S-modules to $\mathbb{Q}_{>0}$ is additive.*

Proof: Let

$$0 \longrightarrow M' \longrightarrow M \longrightarrow M'' \longrightarrow 0$$

be a short exact sequence, and consider the truncated cohomology sequence

$$0 \longrightarrow H^0(G_S,M') \longrightarrow \dots \longrightarrow H^4(G_S,M'') \longrightarrow H^5(G_S,M')' \longrightarrow 0,$$

where $H^5(G_S,M')'$ is the kernel of the boundary map $H^5(G_S,M') \longrightarrow H^5(G_S,M)$. According to (4.10), for $r \geq 3$, we can replace $H^r(G_S,-)$ with $P_S^r(K,-) = \bigoplus_{v \text{ arch}} H^r(G_v,-)$. Now $[P_S^3(K,M)] = [P_S^4(K,M)]$ because the Herbrand quotient of a finite module is one, and so the sequence leads to the equality

$$\chi(M')\chi(M'') = \chi(M).[P_S^5(K,M')'],$$

where $P_S^5(K,M')'$ denotes the kernel of the map $P_S^5(K,M') \longrightarrow P_S^5(K,M)$. Because of the periodicity of the cohomology of a finite cyclic group, $[P_S^5(K,M')'] = [C]$, where

$$C = \operatorname{Ker}\Big(\bigoplus_{v \text{ real}} H^1(G_v,M') \longrightarrow \bigoplus_{v \text{ real}} H^1(G_v,M)\Big).$$

From the exact sequence

$$0 \longrightarrow \bigoplus_{v \text{ arch}} H^0(G_v,M') \longrightarrow \bigoplus_{v \text{ arch}} H^0(G_v,M) \longrightarrow \bigoplus_{v \text{ arch}} H^0(G_v,M'') \longrightarrow C \longrightarrow 0,$$

we see that

$$[C] = \prod_{v \text{ arch}} \frac{[H^0(G_v,M')]\,[H^0(G_v,M'')]}{[H^0(G_v,M)]}.$$

As $[M'][M''] = [M]$, it is now clear that $\varphi(M')\varphi(M'') = \varphi(M)$.

The lemma shows that it suffices to prove the theorem for a module M killed by some prime p, and the assumptions on M require that p be a unit in R_S. Choose a finite Galois extension L of K, $L \subset K_S$, that splits M and contains a primitive p^{th} root of 1 (primitive 4th root in the case that p = 2). Let $\bar{G}$ be Gal(L/K). We need only consider modules M split by L. Note that φ defines a homomorphism from the Grothendieck group $R_{\mathbb{F}_p}(\bar{G})$ to $\mathbb{Q}_{>0}$. An argument as in the proof of Theorem 2.8 (using 2.10) allows us to replace K by a larger field, and consequently assume that $\bar{G}$ is a cyclic group of order prime to p. Note that L is totally imaginary, and so $H^r(\mathrm{Gal}(K_S/L),M) = 0$ for $r \geq 3$ (by 4.10). It follows that there is a well-defined homomorphism $\chi'\colon R_{\mathbb{F}_p}(\bar{G}) \longrightarrow R_{\mathbb{F}_p}(\bar{G})$ sending the class [M] of M in $R_{\mathbb{F}_p}(\bar{G})$ to

$$[H^0(\mathrm{Gal}(K_S/L,M)] - [H^1(\mathrm{Gal}(K_S/L),M)] + [H^2(\mathrm{Gal}(K_S/L),M)].$$

As $\mathrm{Hom}(-,\mathbb{F}_p)$ is exact, it also defines a functor $*\colon R_{\mathbb{F}_p}(\bar{G}) \longrightarrow R_{\mathbb{F}_p}(\bar{G})$.

Lemma 5.4. *For a finite $\mathbb{F}_p[\bar{G}]$-module M, there are the following formulas:*

(a) $\chi'(M^D) = [M]^*.\chi'(\mu_p)$.

(b) $[M].[\mathbb{F}_p[\bar{G}]] = \dim_{\mathbb{F}_p}(M).[\mathbb{F}_p[\bar{G}]]$.

Proof: (a) On tensoring a resolution of μ_p by $\mathrm{Hom}(M,\mathbb{F}_p)$, we see that the cup-product pairing arising from

$$(\zeta,f) \mapsto (x \mapsto \zeta^{f(x)})\colon \mu_p \times \mathrm{Hom}(M,\mathbb{F}_p) \longrightarrow M^D$$

defines an isomorphisms

$$H^r(\mathrm{Gal}(K_S/L),\mu_p) \otimes \mathrm{Hom}(M,\mathbb{F}_p) \longrightarrow H^r(\mathrm{Gal}(K_S/L),M^D)$$

for all r (recall that $\mathrm{Gal}(K_S/L)$ acts trivially on M and μ_p). This gives the formula.

(b) Let M_0 denote M regarded as a $\bar{G}$-module with the trivial action. As we observed in §2, $\sigma\otimes m \mapsto \sigma\otimes\sigma m$ extends to an isomorphism

$\mathbb{F}_p[\bar{G}]\otimes M_0 \longrightarrow \mathbb{F}_p[\bar{G}]\otimes M$, and this gives (b).

On applying both parts of the lemma, we see that

$$\chi'(M^D).[\mathbb{F}_p[\bar{G}]]^* = [M]^*.[\mathbb{F}_p[\bar{G}]]^*.\chi'(\mu_p) = \dim(M).[\mathbb{F}_p[\bar{G}]]^*.\chi'(\mu_p).$$

Similarly $\chi'(M).[\mathbb{F}_p[\bar{G}]]^* = \dim(M).[\mathbb{F}_p[\bar{G}]]^*.\chi'(\mu_p)$. Let θ be the homomorphism $R_{\mathbb{F}_p}(\bar{G}) \longrightarrow \mathbb{Q}_{>0}$ sending the class of a module N to the order of $N^{\bar{G}}$. Then $\theta\circ\chi' = \chi$, and so on applying θ to the above equalities, we find that $\chi(M) = \chi(M^D)$.

Let v be a real prime of K. If $L_w \neq K_v$, then p must be odd, and so $[M^{G_v}] = [M_{G_v}]$, which equals $[(M^D)^{G_v}]$. This shows that the factors of $\varphi(M)$ and $\varphi(M^D)$ corresponding to v are equal. It is now clear that $\varphi(M) = \varphi(M^D)$, and we have already noted that this is implies that $\varphi(M) = 1$.

Remark 5.5. In the function field case there is a completely different approach to the theorem. Let $\bar{K} = Kk_s$ (composite inside K_S), and let $H = \mathrm{Gal}(K_S/\bar{K})$. Let g(K) be the genus of K, and let s be the order of the set of primes of $\bar{K}$ lying over primes in S. Then H is an extension of a group H' having $2g(K) + s$ generators and a single well-known relation (the tame fundamental group of the curve over k_s obtained by omitting the points of S) by a pro-p group, $p = \mathrm{char}(K)$. Using this, or a little étale cohomology, it is possible to show that $H^r(H,M)$ is finite for all finite H-modules M of order prime to p (cf. [Milne (1980), V.2]). Also, it follows from (4.10) that $H^r(H,M) = 0$ for $r > 2$. The Hochschild-Serre spectral sequence for $H \subset G$ reduces to short exact sequences

$$0 \longrightarrow H^{r-1}(H,M)_g \longrightarrow H^r(G_S,M) \longrightarrow H^r(H,M)^g \longrightarrow 0,$$

in which $g = G/H = \mathrm{Gal}(k_s/k) = \langle\sigma\rangle$ and the two end groups are defined by the exactness of

$$0 \longrightarrow N^g \longrightarrow N \xrightarrow{\sigma-1} N \longrightarrow N_g \longrightarrow 0.$$

It follows from the first set of exact sequences that

$$\chi(G_S,M) = \frac{[H^0(H,M)^g][H^1(H,M)_g][H^2(H,M)^g]}{[H^0(H,M)_g][H^1(H,M)^g][H^2(H,M)_g]},$$

and from the second that this product is equal to 1.

An extension to infinite S

As we observed above, in the case that S is finite, all groups in the complex in (4.10) are finite, and therefore the alternating product of their orders is one. It is shown in [Oesterlé (1982/83)] that, when S is infinite, it is possible to define natural Haar measures on the groups in the complex, and prove that (in an appropriate sense) the alternating product of the measures is again one. For example, the measure to take on $P^1_S(K,M)$ is the Haar measure for which the compact subgroup $\prod H^1(g_v,M^{I_v})$ (product over all nonarchimedean v) in has measure 1 (note that $H^1(g_v,M^{I_v}) = H^1_{un}(K_v,M)$ if M is unramified at v). The main result of [Oesterlé (1982/83)] can be stated as follows.

Theorem 5.6. *Let K be a global field, let S be a (possibly infinite) set of primes of K, and let M be a finite G_S-module. Assume that S contains all archimedean primes and all primes for which [M] is not a unit. Relative to the Haar measure on $P^1_S(K,M)$ defined above, a fundamental domain for $P^1_S(K,M)$ modulo the action of the discrete subgroup $H^1(G_S,M)/Ш^1_S(K,M)$ has finite measure*

$$\frac{[\text{Ш}^1_S(K,M)]\,[H^0(G_S,M^D)]}{[\text{Ш}^1_S(K,M^D)][H^0(G_S,M)]} \prod_{v\ \text{arch}} [H^0(G_v,M)].$$

Proof: Suppose first that S is finite. Then the groups are all finite, and the measure of the fundamental domain in question is

$$\frac{\prod [H^1(K_v,M)]}{\prod [H^1(g_v,M^I)]}\,\frac{[\text{Ш}^1_S(K,M)]}{[H^1(G_S,M)]}.$$

From (5.2c) we know that this is equal to

$$\frac{[\text{Ш}^1_S(K,M)]\,[H^0(G_S,M^D)]}{[\text{Ш}^1_S(K,M^D)][H^0(G_S,M)]} \prod_{v\in S} \frac{[H^1(K_v,M)]}{[H^1(g_v,M^I)]\,[H^0(K_v,M^D)]} \prod_{v\ \text{arch}} \frac{[H^0(G_v,M)]}{|[M]|_v}.$$

As $[H^1(g_v,M^I)] = [H^0(g_v,M^I)] = [H^0(G_v,M)]$ for v nonarchimedean (we set it to zero for v archimedean) and $[H^0(K_v,M^D)] = [H^2(K_v,M)]$, we see that the middle term is

$$\prod_{v\in S} \chi(K_v,M)^{-1} \times \prod_{v\ \text{arch}} [H^0_T(G_v,M)].$$

In (5.2b) we showed that $\prod_{v\in S} \chi(K_v,M)^{-1} = \prod_{v\ \text{arch}} |[M]|_v/[H^0_T(G_v,M)]$. This verifies the theorem in this case. For an infinite set S, one chooses a suitably large finite subset S' of S and shows that the theorem for S is equivalent to the theorem for S' (see [Oesterlé (1982/83), §7]).

Notes: Theorem 5.1 is due to Tate (see [Tate (1965/66), 2.2] for the statement together with hints for a proof). Detailed proofs are given in [Kazarnovskii (1972)] and [Haberland (1978), §3]. The above proof differs from previous proofs in that it avoids any calculation of the cohomology of μ_n.

In his original approach to Theorem 4.10, Tate proved it first in the case that S is finite by making use of a counting argument involving (presumably) the formula (5.2.1) for $\chi(G_S,M)\chi(G_S,M^D)$ in order to show that $Ш^1_S(K,M)$ and $Ш^2_S(K,M^D)$ have the same order. He deduced it for an infinite S by passing to the limit. (See [Tate (1962), p192].)

Theorem 5.6 is taken from [Oesterlé (1982/83)].

§6 Abelian varieties over global fields

Throughout this section K will be a global field, and A will be an abelian variety over K. The letter S will always denote a non-empty set of primes of K containing all archimedean primes and all primes at which A has bad reduction. We continue to write K_S for the maximal subfield of K_s containing K that is ramified only at primes in S, G_S for $\mathrm{Gal}(K_S/K)$, and $R_{K,S}$ for the subring $\bigcap_{v\notin S} \mathcal{O}_v$ of K. The letter m is reserved for an integer that is a unit in $R_{K,S}$; thus $|m|_v = 1$ for all $v \notin S$. For example, it is always permitted to take S to be the set of all primes of K, and in that case m can be any integer prime to char(K). As usual, we fix an embedding of K_s into $K_{v,s}$ for each prime v of K.

For an abelian group M, M^ denotes the m-adic completion $\varprojlim M/m^nM$. If X is an algebraic group over K, then we often write $H^r(G_S,X)$ for $H^r(G_S,X(K_S))$ (equal to $H^r(K,X) \overset{\mathrm{df}}{=} H^r(G_K,X(K_s))$ in the case that S contains all primes of K). When X is an algebraic group over K_v, we set $H^r(K_v,X) = H^r(G_v,A(K_{v,s}))$ except when v is archimedean, in which case we set it equal to $H^r_T(G_v,X(K_{v,s}))$. By $H^r(-,X(m))$ we mean $\varinjlim_n H^r(-,X_{m^n})$ and by $H^r(-,T_mX)$ we mean $\varprojlim_n H^r(-,X_{m^n})$.

The weak Mordell-Weil theorem

The Mordell-Weil theorem says that $A(K)$ is finitely generated. The first step in its proof is the weak Mordell-Weil theorem: for some integer $n > 1$, $A(K)/nA(K)$ is finite. We prove a stronger result in (6.2) below.

Lemma 6.1. *Let A and B be abelian varieties over K having good reduction outside S, and let* $f\colon A \longrightarrow B$ *be an isogeny whose degree is a unit in* $R_{K,S}$. *Write* A_f *for* $\mathrm{Ker}(f)$. *Then all points in* $A_f(K_s)$ *have their coordinates in* K_S, *and there is an exact sequence*

$$0 \longrightarrow A_f(K_S) \longrightarrow A(K_S) \xrightarrow{f} B(K_S) \longrightarrow 0.$$

In particular, there is an exact sequence

$$0 \longrightarrow A_m(K_S) \longrightarrow A(K_S) \xrightarrow{m} A(K_S) \longrightarrow 0.$$

Proof: Let $P \in B(K)$; its inverse image $f^{-1}(P)$ in A is a finite subscheme of A. We shall show that this finite subscheme splits over K_S, which implies that P lies in the image of $A(K_S) \longrightarrow B(K_S)$. When P is taken to be zero, $f^{-1}(P)$ is A_f, and so this shows that A_f is split over K_S, i.e., that $A_f(K_S) = A_f(K_s)$.

By assumption, A and B extend to abelian schemes $\mathcal{A}$ and $\mathcal{B}$ over $\mathrm{Spec}(R_{K,S})$. The map f extends to a finite flat map $f\colon \mathcal{A} \longrightarrow \mathcal{B}$ which, because its degree is prime to the residue characteristics of $R_{K,S}$, is also étale. Our point P extends to a section $\mathcal{P}$ of $\mathcal{B}$ over $\mathrm{Spec}\, R_S$, and $f^{-1}(\mathcal{P})$ is a finite étale subscheme of $\mathcal{A}$ over $\mathrm{Spec}(R_S)$. Any such scheme splits over R_S, which implies that $f^{-1}(P)$ splits, and proves the lemma. (For more details on such things, see [Milne (1986b), §20].)

The lemma yields exact sequences

$$\cdots \longrightarrow H^r(G_S,A_f) \longrightarrow H^r(G_S,A) \xrightarrow{f} H^r(G_S,B) \to \cdots$$

$$\cdots \longrightarrow H^r(G_S,A_m) \longrightarrow H^r(G_S,A) \xrightarrow{m} H^r(G_S,A) \to \cdots .$$

Proposition 6.2. (Weak Mordell-Weil theorem) *For any integer* n *prime to* char(K), A(K)/nA(K) *is a finite group.*

Proof: Given n, we can choose a finite set S of primes of K satisfying the conditions in the first paragraph and such that n is a unit in $R_{K,S}$. Then (6.1) provides us with an exact sequence

$$0 \to A_n(K_S) \longrightarrow A(K_S) \xrightarrow{n} A(K_S) \to 0.$$

The cohomology sequence of this gives an injection $A(K_S)^{(n)} \hookrightarrow H^1(G_S,A_n)$, and we have seen in (4.15) that this last group is finite.

To deduce the full Mordell-Weil theorem from (6.2), one uses heights (see [Lang (1983), V]).

The Selmer and Tate-Shafarevich groups

The Tate-Shafarevich group A classifies the forms of A for which the Hasse principle fails. The Selmer group gives a computable upper bound for the rank of A(K). The difference between the upper bound and the actual rank is measured by Tate-Shafarevich group.

Lemma 6.3. *Let a be an element of* $H^1(K,A)$. *Then for all but finitely many primes* v *of* K, *the image of* a *in* $H^1(K_v,A)$ *is zero.*

Proof: As $H^1(K,A)$ is torsion, na = 0 for some n, and as $H^1(K,A_n) \to H^1(K,A)_n$ is surjective, there is a $b \in H^1(K,A_n)$ mapping to a. For almost all v, $A_n(K_s)$ is an unramified G_v-module and b maps into $H^1_{un}(K_v,A_n)$ (see 4.8). Therefore a maps into $H^1(g_v,A(K_{v,un}))$ for almost all v, but (3.8) shows that this last group is zero unless v is one of the finitely many primes at which A has bad reduction.

The *Tate-Shafarevich group* $Ш_S(K,A)$ is defined to be the kernel of

$$H^1(G_S,A) \longrightarrow \bigoplus_{v\in S} H^1(K_v,A).$$

The *Selmer groups* $S_S(K,A)_m$ and $S_S(K,A,m)$ are defined by the exact sequences

$$0 \longrightarrow S_S(K,A)_m \longrightarrow H^1(G_S,A_m) \longrightarrow \bigoplus_{v\in S} H^1(K_v,A)$$

$$0 \longrightarrow S_S(K,A,m) \longrightarrow H^1(G_S,A(m)) \longrightarrow \bigoplus_{v\in S} H^1(K_v,A).$$

The second sequence can be obtained by replacing m with m^n in the first sequence and passing to the direct limit. Therefore

$$S_S(K,A,m) = \varinjlim_n S_S(K,A)_{m^n}.$$

When S contains all primes of K, we drop it from the notation. Thus,

$$Ш(K,A) = \mathrm{Ker}(H^1(K,A) \longrightarrow \prod_{\text{all } v} H^1(K_v,A))$$

$$S(K,A,m) = \mathrm{Ker}(H^1(K,A(m)) \longrightarrow \prod_{\text{all } v} H^1(K_v,A)).$$

Proposition 6.4. *There is an exact sequence*

$$0 \longrightarrow A(K)^{(m)} \longrightarrow S_S(K,A)_m \longrightarrow Ш_S(K,A)_m \longrightarrow 0.$$

Proof: Apply the snake lemma to the diagram

$$\begin{array}{ccccccc} & & 0 & & 0 & & \\ & & \downarrow & & \downarrow & & \\ & & S_S(K,A)_m & \longrightarrow & Ш_S(K,A)_m & & \\ & & \downarrow & & \downarrow & & \\ 0 \longrightarrow A(K)^{(m)} & \longrightarrow & H^1(G_S,A_m) & \longrightarrow & H^1(G_S,A)_m & \longrightarrow 0 & \\ \downarrow & & \downarrow & & \downarrow & & \\ 0 & \longrightarrow & \oplus H^1(K_v,A) & \xrightarrow{\approx} & \oplus H^1(K_v,A) & \longrightarrow 0. & \end{array}$$

Proposition 6.5. *There are exact sequences*

$$0 \longrightarrow H^1(G_S,A(m)) \longrightarrow H^1(K,A(m)) \longrightarrow \bigoplus_{v \notin S} H^1(K_v,A),$$

$$0 \longrightarrow H^1(G_S,A)(m) \longrightarrow H^1(K,A)(m) \longrightarrow \bigoplus_{v \notin S} H^1(K_v,A).$$

Proof: For $v \notin S$, there is a commutative diagram

$$\begin{array}{ccc} H^1(G_S,A(m)) & \longrightarrow & H^1(g_v,A(K_{v,un})) \\ \downarrow & & \downarrow \\ H^1(K,A(m)) & \longrightarrow & H^1(G_v,A(K_{v,s})) \end{array} .$$

According to (3.8), $H^1(g_v,A(K_{v,un})) = 0$, and so the diagram shows that the image of $H^1(G_S,A(m))$ in $H^1(K,A(m))$ is contained in the kernel of $H^1(K,A(m)) \longrightarrow \bigoplus_{v \notin S} H^1(K_v,A)$.

Conversely, let a lie in this kernel. We may assume that a is the image of an element b of $H^1(K,A_m)$ (after possibly replacing m by a power). To prove that the first sequence is exact, it suffices to show that b (hence a) is split by a finite extension of K unramified outside S. After replacing K by such an extension, we can assume that $A_m(K) = A_m(K_s)$ (because of 6.1). Then b corresponds to a homomorphism $f\colon Gal(K_s/K) \longrightarrow A_m(K)$, and it remains to show that the subfield K_f of K_s fixed by the kernel of f is unramified outside S. This can be checked locally. If $v \notin S$, then, by assumption, the image of b in $H^1(K_v,A_m)$ maps to zero in $H^1(K_v,A)$. It therefore arises from an element c_v of $A(K_v)$. The closure of K_f in $K_{v,s}$ is $K_v(m^{-1}c_v)$, which is unramified by (6.1).

The exactness of the second exact sequence can be derived from the first. In the diagram

$$\begin{array}{ccccccc}
 & & \varinjlim A(K)^{(m)} & = & \varinjlim A(K)^{(m)} & & \\
 & & \downarrow & & \downarrow & & \\
0 & \longrightarrow & H^1(G_S, A(m)) & \longrightarrow & H^1(K, A(m)) & \longrightarrow & \bigoplus_{v \notin S} H^1(K_v, A) \\
 & & \downarrow & & \downarrow & & \downarrow \\
0 & \longrightarrow & H^1(G_S, A)(m) & \longrightarrow & H^1(K, A)(m) & \longrightarrow & \bigoplus_{v \notin S} H^1(K_v, A) \\
 & & \downarrow & & \downarrow & & \\
 & & 0 & & 0 & &
\end{array}$$

the exactness of the bottom row follows from the exactness of the rest of the rest of the diagram (use the snake lemma, for example).

Corollary 6.6. *For all* S *and* m *(as in the first paragraph)*,

$$Ш_S(K,A)(m) = Ш(K,A)(m)$$

$$S_S(K,A,m) = S(K,A,m).$$

Proof: The kernel-cokernel sequence (see 0.24) of the pair of maps

$$H^1(K,A)(m) \xrightarrow{\beta} \bigoplus_{\text{all } v} H^1(K_v,A)(m) \xrightarrow{pr} \bigoplus_{v \notin S} H^1(K_v,A)(m)$$

is

$$0 \longrightarrow Ш(K,A)(m) \longrightarrow H^1(G_S,A)(m) \longrightarrow \bigoplus_{v \in S} H^1(K_v,A)(m) \longrightarrow \dots,$$

because (6.5) allows us to replace $\operatorname{Ker}(pr \circ \beta)$ with $H^1(G_S,A)(m)$. This sequence identifies $Ш(K,A)(m)$ with $Ш_S(K,A)(m)$. The second equality is proved by replacing $H^1(K,A)(m)$ in the proof with $H^1(K,A(m))$.

Remark 6.7. Recall (4.15) that $H^1(G_S,A_m(K_S))$ is finite when S is finite. Therefore its subgroup $S_S(K,A)_m$ is finite when S is finite, and (6.4) then shows that $Ш_S(K,A)_m$ is finite. It follows now from (6.6) that $Ш(K,A)_m$ is finite, and (6.4) in turn shows that $S(K,A)_m$ is finite. Consequently, $S(K,A)(m)$ and $Ш(K,A)(m)$ are extensions of finite groups by divisible groups isomorphic to direct sums of copies of $\mathbb{Q}_\ell/\mathbb{Z}_\ell$, ℓ dividing m. It is widely conjectured that $Ш(K,A)$ is in fact finite.

Definition of the pairings

The main results in this section will concern the continuous homomorphisms

$$\beta^0\colon A(K)^\wedge \longrightarrow \prod_{v\in S} H^0(K_v,A)^\wedge \qquad \text{(compact groups)}$$

$$\beta^r\colon H^r(G_S,A)(m) \longrightarrow \bigoplus_{v\in S} H^r(K_v,A)(m),\ r\neq 0,\ \text{(discrete groups).}$$

Write $Ш^r_S(K,A,m) = \mathrm{Ker}(\beta^r)$. Thus $Ш^1_S(K,A,m) = Ш_S(K,A)(m)$, which we have shown to be independent of S. We also write

$$Ш^r_S(K,A(m)) = \varinjlim_n Ш^r_S(K,A_{m^n}) = \mathrm{Ker}(H^r(G_S,A(m)) \longrightarrow \prod_{v\in S} H^r(K_v,A(m)))$$

$$Ш^r_S(K,T_mA) = \varprojlim_n Ш^r_S(K,A_{m^n}) = \varprojlim_n \mathrm{Ker}(H^r(G_S,A_{m^n}) \longrightarrow \prod_{v\in S} H^r(K_v,A_{m^n})).$$

Lemma 6.8. *For any* $r \geq 2$, *there is a canonical isomorphism*

$$Ш^r_S(K,A,m) \xrightarrow{\approx} Ш^r_S(K,A(m)).$$

Proof: For each $r \geq 2$ there is an exact commutative diagram:

$$\begin{array}{ccccccccc}
0 \longrightarrow & H^{r-1}(G_S,A)\otimes\mathbb{Q}_m/\mathbb{Z}_m & \longrightarrow & H^r(G_S,A(m)) & \longrightarrow & H^r(G_S,A)(m) & \longrightarrow 0 \\
& \downarrow \beta^{r-1}(A)\otimes 1 & & \downarrow \beta^r(A(m)) & & \downarrow \beta^r(A)(m) & \\
0 \longrightarrow & \bigoplus_{v\in S} H^{r-1}(K_v,A)\otimes\mathbb{Q}_m/\mathbb{Z}_m & \longrightarrow & \bigoplus_{v\in S} H^r(K_v,A(m)) & \longrightarrow & \bigoplus_{v\in S} H^r(K_v,A)(m) & \longrightarrow 0.
\end{array}$$

As $r-1 \geq 1$, the groups $H^{r-1}(G_S,A)$ and $H^{r-1}(K_v,A)$ are both torsion, and so their tensor products with $\mathbb{Q}_m/\mathbb{Z}_m$ are both zero. The diagram therefore becomes

$$\begin{array}{ccc} H^r(G_S, A(m)) & \xrightarrow{\approx} & H^r(G_S, A)(m) \\ \downarrow \beta^r(A(m)) & & \downarrow \beta^r(A)(m) \\ \bigoplus_{v\in S} H^r(K_v, A(m)) & \xrightarrow{\approx} & \bigoplus_{v\in S} H^r(K_v, A)(m), \end{array}$$

from which the result is obvious.

Proposition 6.9. *For* $r = 0, 1, 2$, *there are canonical pairings*

$$\langle\ ,\ \rangle\colon Ш^r_S(K, A, m) \times Ш^{2-r}_S(K, A^t, m) \longrightarrow \mathbb{Q}/\mathbb{Z}.$$

Proof: There is a unique pairing making the diagram

$$\begin{array}{ccccc} Ш^0_S(K, A, m) & \times & Ш^2_S(K, A^t, m) & \longrightarrow & \mathbb{Q}/\mathbb{Z} \\ \downarrow & & \uparrow \approx & & \| \\ Ш^1_S(K, T_m A) & \times & Ш^2_S(K, A^t(m)) & \longrightarrow & \mathbb{Q}/\mathbb{Z}. \end{array}$$

commute. Here the bottom pairing is induced by the e_m-pairing and the pairings in §4, the first vertical arrow is induced by the map $H^r(G_S, A) \longrightarrow \varprojlim H^{r+1}(G_S, A_m)$, and the second vertical map is the isomorphism in (6.8). This defines the pairing in the case $r = 0$, and the case $r = 2$ can be treated similarly.

The definition of the pairing in the case $r = 1$ is more difficult. We will in fact define a pairing

$$\langle\ ,\ \rangle\colon Ш_S(K, A)_m \times Ш_S(K, A^t)_m \longrightarrow \mathbb{Q}/\mathbb{Z}.$$

Since the Tate-Shafarevich groups are independent of S, we take S to be the set of all primes of K. If $*$ is a global cohomology class, cocycle, or cochain, we write $*_v$ for the corresponding local object.

Let $a \in Ш(K, A)_m$ and $a' \in Ш(K, A^t)_m$. Choose elements b and b' of $H^1(G_K, A_m)$ and $H^1(G_K, A^t_m)$ mapping to a and a' respectively. For each v, a maps to zero in $H^1(K_v, A)$, and so it is obvious from the diagram

$$\begin{array}{ccccc} A(K_v) & \longrightarrow & H^1(K_v,A_m) & \longrightarrow & H^1(K_v,A) \\ \| & & \uparrow & & \\ A(K_v) & \longrightarrow & H^1(K_v,A_{m^2}) & & \end{array}$$

that we can lift b_v to an element $b_{v,1} \in H^1(G_v,A_{m^2})$ that is in the image of $A(K_v)$.

Suppose first that a is divisible by m in $H^1(G_K,A)$, say $a = ma_1$, and choose an element $b_1 \in H^1(G_K,A_{m^2})$ mapping to a_1. Then $b_{v,1} - b_{1,v}$ maps to zero under $H^1(K_v,A_{m^2}) \longrightarrow H^1(K_v,A_m)$, and so it is the image of an element c_v in $H^1(K_v,A_m)$. We define

$$\langle a,a'\rangle = \sum inv_v(c_v \cup b'_v) \in \mathbb{Q}/\mathbb{Z}$$

where the cup-product is induced by the e_m-pairing $A_m \times A_m^t \longrightarrow \mathbb{G}_m$, and inv_v is the canonical map $H^2(K_v,\mathbb{G}_m) \xrightarrow{\approx} \mathbb{Q}/\mathbb{Z}$.

In the general case, let β be a cocycle representing b, and lift it to a cochain $\beta_1 \in C^1(G_K,A_{m^2})$. Choose a cocycle $\beta_{v,1} \in Z^1(G_v,A_{m^2})$ representing $b_{v,1}$, and a cocycle $\beta' \in Z^1(G_K,A_m)$ representing b'. The coboundary $d\beta_1$ of β_1 takes values in A_m, and $d\beta_1 \cup \beta'$ represents an element of $H^3(G_K,K_s^\times)$. But this last group is zero (by 4.18 or 4.21), and so $d\beta_1 \cup \beta' = d\epsilon$ for some 2-cochain ϵ. Now $(\beta_{v,1} - \beta_{1,v}) \cup \beta'_v - \epsilon_v$ is a 2-cocycle, and we can define

$$\langle a,a'\rangle = \sum inv_v((\beta_{v,1} - \beta_{1,v}) \cup \beta'_v - \epsilon_v) \in \mathbb{Q}/\mathbb{Z}.$$

It is not difficult to check that the pairing is independent of the choices made.

Remark 6.10. (a) If B is a second abelian variety over K having good reduction outside S and $f\colon A \longrightarrow B$ is an isogeny, then

$$\langle f(a),b\rangle = \langle a,f^t(b)\rangle, \quad a \in Ш^r_S(K,A,m), \quad b \in Ш^{2-r}_S(K,B^t,m).$$

This follows from the fact that the local pairings are functorial.

(b) Let D be a divisor on A rational over K, and let $\varphi_D: A \longrightarrow A^t$ be the corresponding homomorphism sending $a \in A(K_s)$ to the class of $D_a - D$, where D_a is the translate D + a of D. Then $\langle c, \varphi_D(c)\rangle = 0$ for all $c \in Ш^1_S(K,A,m)$. See [Tate (1962), Thm 3.3]. This can be proved by identifying the pairing defined in (6.9) with that defined in (6.11) below, which we check has this property. See also (II.5).

Remark 6.11. There is a more geometric description of a pairing on the Tate-Shafarevich groups, which in the case of elliptic curves reduces to the original definition of [Cassels (1962), §3]. An element a of Ш(K,A) can be represented by a locally trivial principal homogeneous space X over K. Let $K_s(X)$ be the function field of $X \otimes_K K_s$. Then the exact sequence

$$0 \longrightarrow K_s^\times \longrightarrow K_s(X)^\times \longrightarrow Q \longrightarrow 0$$

leads to a commutative diagram

$$\begin{array}{ccccccc} & & Br(K) & \longrightarrow & H^1(G_K, K_s(X)^\times) & \longrightarrow & H^2(G_K, Q) \longrightarrow 0 \\ & & \downarrow & & \downarrow & & \downarrow \\ 0 & \longrightarrow & \bigoplus_{\text{all } v} Br(K_v) & \longrightarrow & \bigoplus_{\text{all } v} H^1(G_v, K_{v,s}(X)^\times) & \longrightarrow & \bigoplus_{\text{all } v} H^2(G_v, Q). \end{array}$$

The zero at top right comes from the fact that $H^3(G_K, K_s^\times) = 0$ (see 4.21). The zero at lower left is a consequence of the local triviality of X. Indeed, consider an arbitary smooth variety Y over a field k. The map $Br(Y) \longrightarrow Br(k(Y))$ is injective [Milne (1980), II.2.6]. The structure map $Y \longrightarrow Spec(k)$ induces a map $Br(k) \longrightarrow Br(Y)$, and any element of Y(k) defines a section to this map, which is then injective. In our situation, we have a diagram

$$\begin{array}{ccccc} 0 \longrightarrow Br(K_v) & \longrightarrow & Br(X_{K_v}) & \longrightarrow & Br(X_{K_{v,s}}) \\ \| & & \downarrow \text{inj} & & \downarrow \text{inj} \\ Br(K_v) & \longrightarrow & Br(K_v(X)) & \longrightarrow & Br(K_{v,s}(X)) \end{array}$$

from which the claimed injectivity is obvious.

The exact sequence

$$0 \longrightarrow Q \longrightarrow \mathrm{Div}^0(X \otimes K_s) \longrightarrow \mathrm{Pic}^0(X \otimes K_s) \longrightarrow 0$$

yields a cohomology sequence

$$H^1(G_K, \mathrm{Div}^0(X\otimes K_s)) \longrightarrow H^1(G_K, \mathrm{Pic}^0(X\otimes K_s)) \longrightarrow H^2(G_K, Q) \longrightarrow \dots .$$

A trivialization $A\otimes K_s \xrightarrow{\approx} X\otimes K_s$ determines an isomorphism $\mathrm{Pic}^0(X\otimes K_s) \xrightarrow{\approx} \mathrm{Pic}^0(A\otimes K_s)$. Because the trivialization is uniquely determined up to translation by an element of $A(K_s)$ and translations by elements in $A(K_s)$ act trivially on $\mathrm{Pic}^0(A\otimes K_s)$ [Milne (1986b), 9.2], the isomorphism is independent of the choice of the trivialization. A similar argument shows that it is a G_K-isomorphism. Therefore the sequence gives a map $H^1(G_K, A^t) \longrightarrow H^2(G_K, Q)$. Let $a' \in Ш(K, A^t)$, and let b' be its image in $H^2(G_K, Q)$. Then b' lifts to an element of $H^1(G_K, K_s(X)^\times)$, and the image of this in $\oplus H^1(K_v, K_{v,s}(X)^\times))$ lifts to an element $(c_v) \in \oplus Br(K_v)$. Define $\langle a, a'\rangle = \sum inv_v(c_v) \in \mathbb{Q}/\mathbb{Z}$. Note that the cokernel of $Br(K) \longrightarrow \oplus Br(K_v)$ is $\mathbb{Q}/\mathbb{Z}$, and so $\langle a, a'\rangle$ can also be described as the image of b' under the map defined be the snake lemma. As the principal homogeneous space X is uniquely determined up to isomorphism by a, this shows that $\langle a, a'\rangle$ is well-defined.

It is easy to prove that this pairing is alternating. Let $P \in X(K_s)$; then $\sigma P = P + \alpha(\sigma)$ where $(\alpha(\sigma))$ is a cocycle representing a. The map φ_D sends $Q \in A(K_s)$ to the class of $D_Q - D$ in $\mathrm{Pic}^0(A)$, and so a' is represented by the cocycle $(\alpha'(\sigma)) \in Z^1(G_K, A^t)$, where $\alpha'(\sigma)$

is represented by the divisor $E_\sigma = D_{\alpha(\sigma)} - D$. Now use the trivialization $Q \mapsto P + Q$: $A\otimes K_s \xrightarrow{\approx} X\otimes K_s$ to identify $\mathrm{Pic}^0(A)$ with $\mathrm{Pic}^0(X)$. Then one sees immediately that (α'_σ), regarded as a crossed homomorphism into $\mathrm{Pic}^0(X_{K_s})$, lifts to a crossed homomorphism into $\mathrm{Div}^0(X_{K_s})$. Therefore the image of a' in $H^2(G_K,Q)$ is zero, and so $\langle a,a'\rangle = 0$.

We leave it to the reader to check that this pairing agrees with that defined in (6.9).

Remark 6.12. When A is the Jacobian of a curve X over K, there is yet another description of a pairing on the Tate-Shafarevich groups. Write S for the canonical map $\mathrm{Div}^0(X\otimes K_s) \longrightarrow A(K_s)$.

Let $a \in Ш(K,A)$ be represented by $\alpha \in Z^1(G_K,A(K_s))$, and let $\alpha_v = d\beta_v$ with $\beta_v \in Z^0(G_v,A(K_{v,s}))$. Write

$$\alpha = S(\mathfrak{a}), \qquad \mathfrak{a} \in C^1(G_K,\mathrm{Div}^0(X\otimes K_s))$$

$$\beta_v = S(\mathfrak{b}_v), \qquad \mathfrak{b}_v \in C^0(G_v,\mathrm{Div}^0(X\otimes K_{v,s})).$$

Then $\mathfrak{a}_v = d\mathfrak{b}_v + (f_v)$ in $C^1(G_v,\mathrm{Div}^0(X\otimes K_{v,s}))$, where and $f_v \in C^1(G_v,K_{v,s}(X)^\times)$. Moreover $d\mathfrak{a} = (f)$, $f \in Z^2(G_K,K_s(X)^\times)$. Let a' be a second element of $Ш(K,A)$ and define $\mathfrak{a}'$, $\mathfrak{b}'_v$, f'_v, and f' as for a. Set

$$\langle a,a'\rangle = \sum \mathrm{inv}_v(c_v), \qquad c_v = \text{class of } g_v \cup \mathfrak{a} - \mathfrak{b}_v \cup f$$

where $\cup$ denotes the cup-product pairing induced by $(h,\mathfrak{c}) \mapsto h(\mathfrak{c})$. One shows without serious difficulty that f, $\mathfrak{a}$, g_v, and $\mathfrak{b}_v$ can be chosen so that $f(\mathfrak{b}_v)$ and $g_v(\mathfrak{a})$ are defined. Moreover, $\langle a,a\rangle = 0$.

The main theorem

We shall need to consider the duals of the maps β^r. Recall (3.4, 3.6, and 3.7) that $H^r(K_v,A)$ is dual to $H^{1-r}(K_v,A^t)$, except possibly for the p-components in characteristic p. Therefore there exist maps,

$$\gamma^1\colon \bigoplus_{v\in S} H^1(K_v,A)(m) \longrightarrow A^t(K)^*(m) \qquad \text{(discrete groups)}$$

$$\gamma^0\colon \prod_{v\in S} H^0(K_v,A)^\wedge \longrightarrow H^1(G_S,A^t)(m)^*, \qquad \text{(compact groups)}$$

such that $\gamma^r(A) = \beta^{1-r}(A^t)^*$.

Theorem 6.13. (a) *The left and right kernels of the canonical pairing*

$$Ш^1(K,A)(m) \times Ш^1(K,A^t)(m) \longrightarrow \mathbb{Q}/\mathbb{Z}$$

are the divisible subgroups of $Ш^1(K,A)(m))$ *and* $Ш^1(K,A^t)(m)$.

(b) *The following statements are equivalent:*

(i) $Ш^1(K,A)(m)$ *is finite;*

(ii) $\mathrm{Im}(\beta^0) = \mathrm{Ker}(\gamma^0)$ *and the pairing between* $Ш^0_S(K,A,m)$ *and* $Ш^2_S(K,A^t,m)$ *is nondegenerate.*

(c) *The map* β^2 *is surjective with kernel the divisible subgroup of* $H^2(G_S,A)(m)$, *and for* $r > 2$, β^r *is an isomorphism*

$$H^r(G_S,A)(m) \xrightarrow{\approx} \bigoplus_{v \text{ real}} H^r(K_v,A)(m).$$

Remark 6.14. (a) Much of the above theorem is summarized by the following statement: if $Ш(K,A)(m)$ and $Ш(K,A^t)(m)$ are finite, then there is an exact sequence with continuous maps

$$\bigoplus_{v \text{ real}} H^2(K_v,A^t)(m)^* \hookrightarrow H^2(G_S,A^t)(m)^* \longrightarrow H^0(G_S,A)^\wedge$$

$$\downarrow \beta^0$$

$$H^1(G_S,A)(m) \longleftarrow H^1(G_S,A^t)(m)^* \xleftarrow{\gamma^0} \prod_{v\in S} H^0(K_v,A)^\wedge$$

$$\downarrow \beta^1$$

$$\bigoplus_{v\in S} H^1(K_v,A)(m) \xrightarrow{\gamma^1} H^0(G_S,A^t)^{\wedge *} \longrightarrow H^2(G_S,A)(m) \xrightarrow{\beta^2} \bigoplus_{v \text{ real}} H^2(K_v,A)(m).$$

The unnamed arrows exist because of the nondegeneracy of the pairings defined in (6.9).

(b) We shall see in (6.23) and (6.24) below that if S contains almost all primes of K, then β^0 and β^2 are both injective. In this case, the above sequence can be shortened to a four-term sequence:

$$0 \longrightarrow Ш(K,A)(m) \longrightarrow H^1(G_S,A)(m) \longrightarrow \oplus\, H^1(K_v,A)(m) \longrightarrow H^0(G_S,A^t)^{\wedge *} \longrightarrow 0.$$

In particular, when S contains all primes of K and the Tate-Shafarevich groups are finite, then the dual of the exact sequence

$$0 \longrightarrow Ш(K,A) \longrightarrow H^1(K,A) \longrightarrow \bigoplus_{\text{all } v} H^1(K_v,A) \longrightarrow Б \longrightarrow 0$$

is an exact sequence

$$0 \longleftarrow Ш(K,A^t) \longleftarrow H^1(K,A)^* \longleftarrow \prod_{\text{all } v} H^0(K_v,A^t) \longleftarrow A^t(K)^\wedge \longleftarrow 0,$$

except possibly for the p-components in characteristic $p \neq 0$. Here Б is defined to be the cokernel of the preceding map. In the second sequence, $H^0(K_v,A^t) = A^t(K_v)$ unless v is archimedean, in which case it equals the quotient of $A(K_v)$ by its identity component (see 3.7). The term $A^t(K)^\wedge$ is the profinite completion of $A^t(K)$, which is equal to its closure in $\prod H^0(K_v,A)$ (see 6.23b).

(c) If Ш(K,A) is finite, then so also is $Ш(K,A^t)$. To see this note that there is an integer m and maps $f\colon A \longrightarrow A^t$ and $g\colon A^t \longrightarrow A$ such that $f\circ g = m = g\circ f$. Therefore there are maps $Ш(f)\colon Ш(K,A) \longrightarrow Ш(K,A^t)$ and $Ш(g)\colon Ш(K,A^t) \longrightarrow Ш(K,A)$ whose composites are both multiplication by m. It follows that the kernel of Ш(g) is contained in $Ш(K,A^t)_m$. When m is prime to the characteristic, we observed in (6.7) that $Ш(K,A^t)_m$ is finite, and an elementary proof of the same statement for m a power of char(K) can be found in [Milne (1970b)] (see also Chapter III). Hence the kernel of Ш(g) is finite, and this shows that $Ш(A^t)$ is finite.

We begin the proof of (6.13) with part (c). As we saw in the proof of (6.8), when $r \geq 2$, there is a commutative diagram

$$\begin{array}{ccc} H^r(G_S, A(m)) & \xrightarrow{\approx} & H^r(G_S, A)(m) \\ \downarrow \beta^r(A(m)) & & \downarrow \beta^r(A)(m) \\ \bigoplus_{v \in S} H^r(K_v, A(m)) & \xrightarrow{\approx} & \bigoplus_{v \in S} H^r(K_v, A)(m) \end{array} .$$

As $H^r(K_v, A)$ is zero when $r \geq 2$ and v is nonarchimedean (see 3.2), the sum at lower right needs to be taken only over the real primes. When $r > 2$, $\beta^r(A(m))$ is an isomorphism (see 4.10c), and so $\beta^r(A)$ is an isomorphism. When $r = 2$, (4.10) shows that the cokernel of $\beta^2(A(m))$ is

$$\varinjlim_m A^t_{m^n}(K)^* = (\varprojlim_m A^t_{m^n}(K))^* = (T_m A^t(K))^*,$$

which is zero because $A^t(K)$ is finitely generated (by the Mordell-Weil theorem). Consider the diagram

$$\begin{array}{ccccccc} H^2(G_S, A_m) & \longrightarrow & H^2(G_S, A)(m) & \xrightarrow{m} & H^2(G_S, A)(m) & \longrightarrow & H^3(G_S, A_m) \\ \downarrow \beta^2(A_m) & & \downarrow \beta^2(A) & & \downarrow \beta^2(A) & & \downarrow \beta^3(A_m) \\ \bigoplus_{v \text{ real}} H^2(K_v, A_m) & \longrightarrow & \bigoplus_{v \text{ real}} H^2(K_v, A)(m) & \longrightarrow & \bigoplus_{v \text{ real}} H^2(K_v, A)(m) & \longrightarrow & \bigoplus_{v \text{ real}} H^3(K_v, A_m). \end{array}$$

The first vertical arrow is surjective by (4.16). We have just shown that $\beta^2(A)$ is surjective, and we know that $\beta^3(A_m)$ is an isomorphism by (4.10c). Therefore we have a surjective map of complexes, and so the sequence of kernels is exact, from which it follows that $\mathrm{Ker}(\beta^2(A))$ is divisible by m. On repeating this argument with m replaced by m^n we find that $\mathrm{Ker}(\beta^2(A))$ is divisible by all powers of m. Since it obviously contains $H^2(G_S, A)(m)_{div}$, this shows that it equals $H^2(G_S, A)(m)_{div}$. This completes the proof of part (c).

We next prove part (b). Let $v \in S$, and consider the diagram

$$\begin{array}{ccccccccc} 0 \to & A(K)^{(m)} & \to & H^1(G_S,A_m) & \to & H^1(G_S,A)_m & \to 0 \\ & \downarrow & & \downarrow & & \downarrow & \\ 0 \to & H^0(K_v,A)^{(m)} & \to & H^1(K_v,A_m) & \to & H^1(K_v,A)_m & \to 0. \end{array}$$

On replacing m with m^n and passing to the inverse limit, and then replacing the bottom row by the restricted product over all v in S, we obtain an exact commutative diagram

$$\begin{array}{ccccccccc} 0 \to & A(K)\hat{} & \to & H^1(G_S,T_mA) & \to & T_mH^1(G_S,A) & \to 0 \\ & \downarrow \beta^0 & & \downarrow & & \downarrow \beta^1 & \\ 0 \to & \prod_{v\in S} H^0(K_v,A)\hat{} & \to & \prod'_{v\in S} H^1(K_v,T_mA) & \to & \prod'_{v\in S} T_mH^1(K_v,A) & \to 0. \end{array}$$

The snake lemma now gives an exact sequence

$$0 \to Ш^0_S(K,A) \to Ш^1_S(K,T_mA) \longrightarrow T_mШ(K,A) \longrightarrow \frac{\prod H^0(K_v,A)\hat{}}{\operatorname{Im}(\beta^0)} \xrightarrow{\epsilon} (H^1(G_S,A^t(m))^*.$$

Here we have used (4.10) to identify the cokernel of the middle vertical map with a subgroup of $\varprojlim H^1(G_S,A^t_{m^n})^* = (\varinjlim H^1(G_S,A^t_{m^n}))^* = H^1(G_S,A^t(m))^*$.

Consider the maps

$$\prod_{v\in S} H^0(K_v,A)\hat{} \xrightarrow{\gamma^0} (H^1(G_S,A^t)(m))^* \xrightarrow{\epsilon'} H^1(G_S,A^t(m))^*,$$

the second of which is the dual of $H^1(G_S,A^t(m)) \twoheadrightarrow H^1(G_S,A^t)(m)$ and is therefore injective; consequently, $\operatorname{Ker}(\epsilon'\circ\gamma^0) = \operatorname{Ker}(\gamma^0)$. The composite $\epsilon'\circ\gamma^0$ is the composite of the projection

$$\prod_{v\in S} H^0(K_v,A)\hat{} \longrightarrow \frac{\prod H^0(K_v,A)\hat{}}{\operatorname{Im}(\beta^0)}$$ with ϵ. Since $\operatorname{Im}(\beta^0)$ is goes to zero under $\epsilon'\circ\gamma^0$, we see that it must also be mapped to zero by γ^0, that is, $\gamma^0\circ\beta^0 = 0$ (without any assumptions). We also see that $\operatorname{Ker}(\gamma^0) = \operatorname{Im}(\beta^0)$ if and only if ϵ is injective, which is equivalent to $Ш^1_S(K,T_mA) \longrightarrow T_mШ(K,A)$ being surjective.

Consider on the other hand the first part

$$0 \longrightarrow Ш^0_S(K,A) \longrightarrow Ш^1_S(K,T_mA) \longrightarrow T_mШ(K,A)$$

of the above exact sequence and the isomorphism

$$Ш^2_S(K,A)(m) \xleftarrow{\approx} Ш^2_S(K,A(m))$$

in (6.8). Clearly the duality between $Ш^1_S(K,T_mA)$ and $Ш^2_S(K,A(m))$ arising from (4.10) induces a duality between $Ш^0(K,A)$ and $Ш^2(K,A)(m)$ if and only if the map $Ш^1_S(K,T_mA) \longrightarrow T_mШ(K,A)$ is zero.

On combining the conclusions of the last two paragraphs, we find that the following two statements are equivalent:

(*) $Ш^0_S(K,A)$ and $Ш^2_S(K,A^t)$ are dual and $\mathrm{Im}(\beta^0) = \mathrm{Ker}(\gamma^0)$

(**) $Ш^1_S(K,T_mA) \longrightarrow T_mШ(K,A)$ is both surjective and zero.

Clearly (**) is equivalent to $T_mШ(K,A)$ being zero, but $T_mШ(K,A) = 0$ if and only if the m-divisible subgroup of $Ш(K,A)(m)$ is zero, in which case the group is finite. This proves the equivalence of statements (i) and (ii) in (b).

In preparing for the proof of (a), we shall need a series of lemmas. Since the statement of (a) does not involve S, we can choose it to be any set we wish provided it satisfies the conditions in the first paragraph. We always take it to be finite.

Lemma 6.15. *Let* $a \in \bigoplus_{v\in S} H^1(K_v,A_m)$, *and consider the pairing*

$\sum \langle\ ,\ \rangle_v\colon \prod H^1(K_v,A_m) \times \oplus H^1(K_v,A^t_m) \longrightarrow \mathbb{Q}/\mathbb{Z}$, $\langle a_v,a'_v\rangle_v = \mathrm{inv}_v(a_v \cup a'_v)$.

Then $\langle a,a'\rangle = 0$ *for all* a' *in the image of* $S_S(K,A^t)_m \longrightarrow \bigoplus_{v\in S} H^1(K_v,A^t_m)$ *if and only if* a *can be written* $a = a_1 + a_2$ *with* a_1 *and* a_2 *in the images of* $\prod H^0(K_v,A) \longrightarrow \prod H^1(K_v,A_m)$ *and* $H^1(G_S,A_m) \longrightarrow \prod H^1(K_v,A_m)$ *respectively.*

Proof: The dual of the diagram

$$\begin{array}{ccccccc} \oplus H^1(K_v, A^t_m) & & & & & & \\ \downarrow & \nwarrow \beta^1 & & & & & \\ \oplus H^1(K_v, A^t) & \longleftarrow & H^1(G_S, A^t_m) & \longleftarrow & S_S(K, A^t)_m & \longleftarrow & 0 \end{array}$$

is

$$\begin{array}{ccccccc} \prod H^1(K_v, A_m) & & & & & & \\ \uparrow & \searrow \gamma^1 & & & & & \\ \prod H^0(K_v, A) & \longrightarrow & H^1(G_S, A^t_m)^* & \longrightarrow & (S_S(K, A^t)_m)^* & \longrightarrow & 0. \end{array}$$

Let $a \in \prod H^1(K_v, A_m)$. If a maps to zero in $(S_S(K,A^t)_m)^*$, then $\gamma^1(a)$ is the image of an element b in $\prod H^0(K_v, A)$. Let a_1 denote the image of b in $\prod H^1(K_v, A_m)$; then $a - a_1$ is in the kernel of γ^1. But according to (4.10), the kernel of γ^1 is the image of $H^1(G_S, A_m)$, and so $a - a_1 = a_2$ for some $a_2 \in H^1(G_S, A_m)$.

Lemma 6.16. *Let* Ш'(A) *be the subgroup of* Ш(K,A) *of elements that become divisible by* m *in* $H^1(G_S, A)$. *Then there is an exact sequence*

$$0 \longrightarrow Ш'(K,A) \longrightarrow Ш(K,A) \longrightarrow Ш^2_S(K,A_m).$$

Proof: Consider

$$\begin{array}{ccccc} Ш(K,A) & \xrightarrow{m} & Ш(K,A) & \longrightarrow & Ш^2_S(K,A_m) \\ \downarrow & & \downarrow & & \downarrow \\ H^1(G_S,A) & \xrightarrow{m} & H^1(G_S,A) & \longrightarrow & H^2(G_S,A_m) \\ \downarrow & & \downarrow & & \downarrow \\ \bigoplus_{v\in S} H^1(K_v,A) & \xrightarrow{m} & \bigoplus_{v\in S} H^1(K_v,A) & \longrightarrow & \bigoplus_{v\in S} H^2(K_v,A_m). \end{array}$$

An element a in Ш(K,A) maps to zero in $Ш^2_S(K,A)$ if and only if it maps to zero in $H^2(G_S, A_m)$, and this occurs if and only if its image in $H^1(G_S, A)$ is divisible by m.

Lemma 6.17. *Let* $a \in Ш'(K,A)$. *Then* $a \in mШ(K,A)$ *if and only if* $\langle a,a' \rangle$

$= 0$ *for all* $a' \in Ш(K,A^t)_m$.

Proof: If $a = ma_0$ with $a_0 \in Ш(K,A)$, then

$$\langle a,a'\rangle = \langle ma_0,a'\rangle = \langle a_0,ma'\rangle = 0$$

for all $a' \in Ш(K,A^t)_m$. Conversely, assume that a satisfies the second condition, and let $a_1 \in H^1(G_S,A)$ be such that $ma_1 = a$; we have to show that a_1 can be modified to lie in $Ш(K,A)$. Choose a finite set S satisfying the conditions at the start of this section and containing all v for which $a_{1,v} \neq 0$. If a_1 is replaced by its sum with an element of $H^r(G_S,A)$, then it is still zero outside S (see the proof of 6.5). Define b_1, $b_{v,1}$, and c_v as in (6.9); thus $b_1 \in H^1(G_S,A_{m^2})$ and maps to a_1, $b_{v,1} \in H^1(K_v,A_{m^2})$ and maps to b_v, and $c_v \in H^1(K_v,A_m)$ and maps to $b_{v,1} - b_{1,v}$. We shall show that there is an element $b_0 \in H^1(G_S,A_m)$ such that

$$b_{0,v} \equiv c_v \equiv -b_{1,v} \mod A(K_v)^{(m)}$$

for all v. This will complete the proof, because then $a_1 + a_0$, with a_0 the image of b_0 in $H^1(G_S,A)$, lies in $Ш(K,A)$ and is such that $m(a_1 - a_0) = ma_1 = a$.

According to (6.14), an element b_0 will exist if and only if $\sum \langle c_v,b'_v\rangle = 0$ for all b' in $S_S(K,A)_m$. But, by definition of the pairing on the Tate-Shafarevich groups (6.9), $\sum \langle c_v,b'_v\rangle = \langle a,a'\rangle$ where a' is the image of b' in $Ш(K,A^t)_m$, and our assumption on a is that this last term is zero.

We now complete the proof of part (a) of the theorem. Note that because the groups are torsion, the pairing must kill the divisible subgroups. Consider the diagram

$$\begin{array}{ccccc}
0 \longrightarrow & Ш'(K,A)/mШ(K,A)) & \longrightarrow Ш(K,A)/mШ(K,A) & \longrightarrow Ш^2_S(K,A_m) \\
& \downarrow & \downarrow & \downarrow \approx \\
0 \longrightarrow & (Ш(K,A^t)_m/\mathrm{Im}\ Ш^1(K,A^t_m))^* \longrightarrow & (Ш(K,A^t)_m)^* & \longrightarrow Ш^1_S(K,A^t_m)^*.
\end{array}$$

The top row comes from (6.16) and the bottom row is the dual of an obvious sequence

$$Ш^1_S(K,A^t_m) \longrightarrow Ш(K,A^t)_m \longrightarrow \mathrm{Coker} \longrightarrow 0.$$

The first vertical map is the injection given by Lemma 6.17, and the third vertical arrow is the isomorphism of (4.10). A diagram chase now shows that the middle vertical arrow is also injective. On passing to the limit over powers of m, we obtain an injection $Ш(K,A)^\wedge \hookrightarrow Ш(K,A^t)(m)^*$. But $Ш(K,A)^\wedge = Ш(K,A)/Ш(K,A)_{m\text{-div}}$, and so the left kernel in the pairing

$$Ш(K,A)(m) \times Ш(K,A^t)(m) \longrightarrow \mathbb{Q}/\mathbb{Z}$$

is $Ш(K,A)_{m\text{-div}}$. Therefore

$$[Ш(K,A)/Ш(K,A)_{m\text{-div}}] \leq [Ш(K,A^t)/Ш(K,A^t)_{m\text{-div}}].$$

Since this holds for all A, we also have

$$\begin{aligned}
[Ш(K,A^t)/Ш(K,A^t)_{m\text{-div}}] &\leq [Ш(K,A^{tt})/Ш(K,A^{tt})_{m\text{-div}}] \\
&= [Ш(K,A)]/Ш(K,A)_{m\text{-div}}].
\end{aligned}$$

It follows that all these orders are equal, and therefore that the right kernel is $Ш(K,A^t)_{m\text{-div}}$.

Remark 6.18. If A has dimension one and m is prime, then $Ш^2_S(K,A_m) = 0$ (see 9.6), and so $Ш'(K,A)_m = Ш(K,A)_m$ (see 6.17). Therefore in this case it is significantly easier both to define the pairing on the Tate-Shafarevich groups and to prove its nondegeneracy.

Complements in the case that S contains almost all primes

We shall now show how a theorem of [Serre (1964/71)] can be used to improve some of these results when S omits only finitely many primes.

Proposition 6.19. *Let* m *be an integer prime to* char(K), *and let* $\bar{G}$ *be the image of* $Gal(K_S/K)$ *in* T_mA. *Then the group* $H^1(\bar{G}, T_mA)$ *is finite.*

Proof: When m is prime, this is proved in [Serre (1964/71), II.2], and the result for a composite m follows immediately.

We give a second proof of (6.19) based on a theorem of Bogomolov and a lemma of Sah. Note that (6.1) shows that the action of G_K on T_mA factors through G_S. Also that, because T_mA is a $\mathbb{Z}_m$-module, $\mathbb{Z}_m$ is a subring of $End(T_mA)$ and $\mathbb{Z}_m^\times$ is a subgroup of $Aut(T_mA)$.

Lemma 6.20. *For any prime* $\ell \neq char(K)$, *the image of* G_S *in* $Aut(T_\ell A)$ *contains an open subgroup of* $\mathbb{Z}_\ell^\times$.

Proof: Theorem 3 of [Bogomolov (1981)] shows that (at least when K is a number field), for any prime ℓ, the Lie algebra of the image of G_K in $Aut(T_\ell A)$ contains the scalars. This implies that the image of G_K is open in $\mathbb{Z}_\ell^\times$.

Lemma 6.21. *Let* G *be a profinite group and* M *a* G*-module. For any element* σ *of the centre of* G, $H^r(G,M)$ *is annihilated by* $x \mapsto \sigma x - x$.

Proof: We first allow σ to be any element of G, not necessarily a central element. The pair of maps

$$g \mapsto \sigma g \sigma^{-1}: G \longrightarrow G, \qquad m \mapsto \sigma^{-1} m: M \longrightarrow M$$

are compatible, and so define automorphisms $\alpha^r: H^r(G,M) \longrightarrow H^r(G,M)$.

According to (0.15) α^r is the identity map. In the case that σ central, one sees by looking on cochains that α^r is the map induced by the G-homomorphism $\sigma^{-1}: M \longrightarrow M$. Consequently, σ acts as the identity map on $H^r(G,M)$, as claimed by the lemma.

We now (re-)prove 6.19. It follows from (6.20) that there is an integer i and an element $\sigma \in \bar{G}$ such that $\sigma x = (\ell^i - 1)x$, all $x \in T_\ell A$. Now (6.21) shows that $\ell^i H^1(\bar{G},T_\ell A) = 0$. Corollary 4.15 implies that $H^1(\bar{G},A_{\ell^n})$ is finite for all n, and so the inverse limit of the exact sequences

$$H^1(\bar{G},A_{\ell^n}) \xrightarrow{\ell^i} H^1(\bar{G},A_{\ell^{n+i}}) \longrightarrow H^1(\bar{G},A_{\ell^i})$$

is an exact sequence

$$H^1(\bar{G},T_\ell A) \xrightarrow{0} H^1(\bar{G},T_\ell A) \longrightarrow H^1(\bar{G},A_{\ell^i}),$$

and so $H^1(\bar{G},T_\ell A)$ is a subgroup of the finite group $H^1(\bar{G},A_{\ell^i})$.

Proposition 6.22. *If S omits only finitely many primes of K, then the map*

$$H^1(G_S,T_mA) \longrightarrow \prod_{v \in S} H^1(K_v,T_mA)$$

is injective.

Proof: Write $Ш^1_S(K,T_mA)$ for the kernel of the map in the statement of the proposition. Then there is an exact commutative diagram

$$\begin{array}{ccccccc} 0 \longrightarrow & H^1(\bar{G},T_mA) & \longrightarrow & H^1(G_S,T_mA) & \longrightarrow & H^1(G',T_mA) \\ & \downarrow & & \downarrow & & \downarrow \\ 0 \longrightarrow & \bigoplus_{v \in S} H^1(\bar{G}_v,T_mA) & \longrightarrow & \bigoplus_{v \in S} H^1(G_v,T_mA) & \longrightarrow & \bigoplus_{v \in S} H^1(G'_v,T_mA) \end{array}$$

in which $\bar{G}$ and $\bar{G}_v$ are the images of G and G_v in $\mathrm{Aut}(T_mA)$ and G' and G'_v are the kernels of $G \twoheadrightarrow \bar{G}$ and $G_v \twoheadrightarrow \bar{G}_v$. The two right hand groups consist of continuous homomorphisms, and so the Chebotarev density theorem shows that the right hand vertical map is injective. It follows that the subgroup $Ш^1_S(K,T_mA)$ of $H^1(G_S,T_mA)$ is contained in $H^1(\bar{G},T_mA)$, and is therefore torsion. Since $T_mH^1(G_S,A)$ is torsion-free, the sequence

$$0 \longrightarrow A(K)^\wedge \longrightarrow H^1(G_S,T_mA) \longrightarrow T_mH^1(G_S,A)$$

now shows that any element c of $Ш^1_S(K,T_mA)$ is in $A(K)^\wedge$. But for any nonarchimedean prime v, the map $A(K)^\wedge \longrightarrow A(K_v)^\wedge$ is injective on torsion points, and so $c = 0$.

Corollary 6.23. *Assume S omits only finitely many primes.*

(a) *There is an injection*

$$T_mШ^1(K,A) \longrightarrow \prod_{v\in S} H^0(K_v,A)^\wedge/A(K)^\wedge$$

(b) *There is a sequence of injective maps*

$$A(K)^\wedge \longrightarrow \varprojlim S_S(K,A)_m \longrightarrow \prod_{v\in S} H^0(K_v,A)^\wedge.$$

In particular, $Ш^0_S(K,A) = 0$. *The kernel of* $A(K) \longrightarrow A(K)^\wedge$ *is the subgroup of elements with finite order prime to* m.

Proof: Consider the diagram

$$\begin{array}{ccccccccc}
0 \to & A(K)^\wedge & \to & H^1(G_S,T_mA) & \xrightarrow{a} & T_mH^1(G_S,A) & \to 0 \\
 & \downarrow & & \downarrow c & & \downarrow b & \\
0 \to & \prod_{v\in S} H^0(K_v,A)^\wedge & \to & \prod_{v\in S} H^1(K_v,T_mA) & \xrightarrow{d} & \prod_{v\in S} T_mH^1(K_v,A) & \to 0.
\end{array}$$

The vertical arrow marked c is injective, and that marked b has kernel $T_mШ(K,A)$. Therefore part (a) follows from the snake lemma. The first map in part (b) is the inclusion $\mathrm{Ker}(a) \hookrightarrow \mathrm{Ker}(b\circ a)$. The second is the injection $\mathrm{Ker}(d\circ c) \hookrightarrow \mathrm{Ker}(d)$.

Corollary 6.24. *Let S be as in the proposition. The map* $H^2(G_S,A) \longrightarrow \bigoplus_{v \text{ real}} H^2(K_v,A)$ *is an isomorphism; in particular,* $Ш^2_S(K,A) = 0$.

Proof: We have a commutative diagram

$$\begin{array}{ccccccccc} 0 & \longrightarrow & H^1(G_S,A)\otimes\mathbb{Q}_m/\mathbb{Z}_m & \longrightarrow & H^2(G_S,A(m)) & \longrightarrow & H^2(G_S,A)(m) & \longrightarrow & 0 \\ & & \downarrow & & \downarrow & & \downarrow & & \\ 0 & \longrightarrow & \oplus H^1(K_v,A)\otimes\mathbb{Q}_m/\mathbb{Z}_m & \longrightarrow & \oplus H^2(K_v,A(m)) & \longrightarrow & \oplus H^2(K_v,A)(m) & \longrightarrow & 0. \end{array}$$

Because $H^1(K_v,A)$ is torsion, its tensor product with $\mathbb{Q}_m/\mathbb{Z}_m$ is zero. Therefore a nonzero element of $Ш^2_S(K,A)(m)$ would give rise to a nonzero element of $Ш^2_S(K,A(m))$, but this group is dual to $Ш^1_S(K,T_mA^t)$, which the proposition shows to be zero. Therefore the map is injective, and it was shown to be surjective in (6.13c).

Remark 6.25. Note that (6.23b) solves the congruence subgroup problem for subgroups of $A(K)$ of index prime to the characteristic of K: any such subgroup contains a subgroup defined by congruence conditions. (In fact, that was Serre's purpose in proving (6.19).)

On combining the above results with Theorem 6.13, we obtain the following theorem.

Theorem 6.26. *Assume that* S *omits only finitely many primes of* K.

(a) *The left and right kernels of the canonical pairing*

$$Ш(K,A)(m) \times Ш(K,A^t)(m) \longrightarrow \mathbb{Q}/\mathbb{Z}$$

are the divisible subgroups of $Ш(K,A)(m)$ *and* $Ш(K,A^t)(m)$.

(b) *The Tate-Shafarevich group* $Ш(K,A)(m)$ *is finite if and only if* $\mathrm{Im}(\beta^0) = \mathrm{Ker}(\gamma^0)$; *in this case there is an exact sequence*

$$0 \longrightarrow Ш(K,A)(m) \longrightarrow H^1(G_S,A)(m) \longrightarrow \bigoplus_{v\in S} H^1(K_v,A)(m) \longrightarrow A^t(K)^{\wedge *} \longrightarrow 0.$$

(c) *The groups* $Ш^r_S(K,A,m)$ *are zero for* $r \neq 1$, *and for* $r \geq 2$, β^r *is an isomorphism*

$$H^r(G_S,A)(m) \xrightarrow{\approx} \bigoplus_{v \text{ real}} H^r(K_v,A)(m).$$

Remark 6.27. The Tate-Shafarevich group is not known to be finite for a single abelian variety over a number field. However there are numerous examples where it has been shown that some component Ш(K,A)(m) is finite. The first examples of abelian varieties over global fields known to have finite Tate-Shafarevich groups are to be found in [Milne (1967)] and [Milne (1968)]. There it is shown that, for constant abelian varieties over a function field K, the Tate-Shafarevich group is finite and has the order predicted by the conjecture of Birch and Swinnerton-Dyer (see the next section for a statement of the conjecture; an abelian variety over a function field K is *constant* if it is obtained by base change from an abelian variety over the field of constants of K). See also [Milne (1975)], where (among other things) it is shown that the same conjecture is true for the elliptic curve

$$Y^2 = X(X - 1)(X - T)$$

over k(T), k finite.

Notes: Theorem 6.13 was proved by Cassels in the case of elliptic curves [Cassels (1962), (1964)] and by Tate in the general case (announcement [Tate (1962)]). So far as I know, no complete proof of it has been published before. The survey article [Bashmakov (1972)] contains proofs of parts of it, and [Wake (1986)] shows how to deduce (6.22), (6.23), and (6.24) from (6.19); both works have been helpful in the writing of this section in the absence of Tate's original proofs.

§7 An application to the conjecture of Birch and Swinnerton-Dyer

The results of the preceding two sections will be applied to show that the conjecture of Birch and Swinnerton-Dyer, as generalized to abelian varieties by Tate, is compatible with isogenies (except possibly for isogenies whose degree is divisible by the characteristic of K). We begin by reviewing the statement of the conjecture in [Tate (1965/66), §1]. Throughout, A and B will be abelian varieties of dimension d over a global field K, and $G = \mathrm{Gal}(K_s/K)$.

L-series. Let v be a nonarchimedean prime of K, and let k(v) be the corresponding residue field. If A has good reduction at v, then it gives rise to an abelian variety A(v) over k(v). The characteristic polynomial of the Frobenius endomorphism of A(v) is a polynomial $P_v(T)$ of degree 2d with coefficients in $\mathbb{Z}$ such that, when we factor it as $P_v(T) = \prod (1 - a_i T)$, then $\prod (1 - a_i^m)$ is the number of points on A(v) with coordinates in the finite field of degree m over k(v) (see [Milne (1986b), §19]). It can be described also in terms of $V_\ell A \overset{\mathrm{df}}{=} \mathbb{Q}_\ell \otimes T_\ell A$. Let $D_v \supset I_v$ be the decomposition and inertia groups at v, and let Fr_v be the Frobenius element of D_v/I_v. Then (6.1) shows that I_v acts trivially on $T_\ell A$, and it is known (ibid.) that

$$P_v(A,T) = \det(1 - (\mathrm{Fr}_v)T \mid V_\ell A), \quad \ell \neq \mathrm{char}\ k.$$

For any finite set S of primes of K including the archimedean primes and those where A has bad reduction, we define the L-series $L_S(s,A)$ by the formula

$$L_S(s,A) = \prod_{v \notin S} P_v(A, Nv^{-s})^{-1}$$

where $Nv = [k(v)]$. Because the inverse roots a_i of $P_v(T)$ have absolute value $q^{1/2}$, the product is dominated by $\zeta_K(s-1/2)^{2d}$, and it therefore converges for $\mathrm{Re}(s) > 3/2$. It is widely conjectured that

$L_S(s,A)$ can be analytically continued to a meromorphic function on the whole complex plane. This is known in the function field case, but in the number field case it has been verified only for modular elliptic curves, abelian varieties with potential complex multiplication, and some other abelian varieties.

Let ω be a nonzero global differential d-form on A. As $\Gamma(A,\Omega_A^d)$ has dimension 1, ω is uniquely determined up to multiplication by an element of $K^\times$. For each nonarchimedean prime v of K, let μ_v be the Haar measure on K_v for which $\mathcal{O}_v$ has measure 1, and for each archimedean prime, take μ_v to be the usual Lebesgue measure on K_v. With these choices, we have $\mu_v(cU) = |c|_v\mu_v(U)$ for any $c \in K^\times$ and compact $U \subset K_v$. Just as a differential on a manifold and a measure on $\mathbb{R}$ define a measure on the manifold, ω and μ_v define a measure on $A(K_v)$, and we set $\mu_v(A,\omega) = \int_{A(K_v)} |\omega|_v\mu_v^d$ (see [Weil, (1961)]). Let μ be the measure $\prod \mu_v$ on the adèle ring $\mathbb{A}_K$ of K, and set $|\mu| = \int_{\mathbb{A}_K/K} \mu$. For any finite set S of primes of K including all archimedean primes and those nonarchimedean primes for which A has bad reduction or such that ω does not reduce to a nonzero differential d-form on A(v), we define

$$L_S^*(s,A) = L_S(s,A)\,\frac{|\mu|^d}{\prod_{v\in S}\mu_v(A,\omega)}.$$

This function is independent of the choice of ω; if $\omega' = c\omega$ is a second differential d-form on A having good reduction outside S, then c must be a unit at all primes outside S, and so the product formula shows that $\prod_{v\in S}\mu_v(A,\omega') = \prod_{v\in S}\mu_v(A,\omega)$. The function $L_S^*(s,A)$ depends on the choice of S, but its asymptotic behaviour as s approaches 1 does not, because if v is a prime at which A and ω have good reduction at v, then it is known that $\mu_v(A,\omega) = [A(k(v))]/(Nv)^d$ (ibid., 2.2.5), and it is easy to see that this equals $P_v(A, Nv^{-1})$.

Heights. The logarithmic height of a point $x = (x_0:\ldots:x_m)$ in $\mathbb{P}^m(K)$ is defined by

$$h(x) = \log \prod_{\text{all } v} \max_{0 \le i \le m} \{|x_i|_v\}.$$

The product formula shows that this is independent of the representation of x. Let D be a very ample divisor on A. After replacing D with $D + (-1)^*D$, we may assume that D is linearly equivalent to $(-1)^*D$. Let $f\colon A \hookrightarrow \mathbb{P}^n$ be the embedding defined by D, and for $a \in A(K)$, let $\varphi_D(a)$ be the point in $A^t(K)$ represented by the divisor $(D + a) - D$. Then there is a unique bi-additve pairing

$$\langle\ ,\ \rangle : A^t(K) \times A(K) \longrightarrow \mathbb{R}$$

such that $\langle\varphi_D(a),a\rangle + 2h(f(a))$ is bounded on A(K). The discriminant of the pairing is known to be nonzero. The pairing is functorial in the sense that if $f\colon A \longrightarrow B$ is an isogeny, then the diagram

$$\begin{array}{ccccc} A^t(K) & \times & A(K) & \longrightarrow & \mathbb{R} \\ \uparrow f^t & & \downarrow f & & \| \\ B^t(K) & \times & B(K) & \longrightarrow & \mathbb{R} \end{array}$$

commutes. (See [Lang (1983), Chapter V].)

Statement. In order to state the conjecture of Birch and Swinnerton-Dyer we need to assume that the following two conjectures hold for A:

(a) the function $L_S(s,A)$ has an analytic continuation to a neighbourhood of 1;

(b) the Tate-Shafarevich group Ш(K,A) of A is finite.

The conjecture then asserts:

$$\text{(B-S/D)} \qquad \lim_{s\to 1} \frac{L_S^*(s,A)}{(s-1)^r} = \frac{[Ш(K,A)]\,|\det\langle a_i', a_j\rangle|}{(A^t(K):\sum \mathbb{Z}a_i')\;(A(K):\sum \mathbb{Z}a_i)}$$

where r is the common rank of A(K) and $A^t(K)$, and $(a_i')_{1\le i\le r}$ and $(a_i)_{1\le i\le r}$ are families of elements of $A^t(K)$ and A(K) that are linearly independent over $\mathbb{Z}$.

Lemma 7.1. *Let A and B be isogenous abelian varieties over a global field K, and let S be a finite set of primes including all archimedean primes and all primes at which A or B has bad reduction.*

(a) *The functions $L_S(s,A)$ and $L_S(s,B)$ are equal. In particular, if one function can be continued to a neighbourhood of s = 1, then so also can the other.*

(b) *Assume that the isogeny has degree prime to the char(K). If one of Ш(A) of Ш(B) is finite, then so also is the other.*

Proof: (a) An isogeny A ⟶ B defines an isomorphism $V_\ell A \xrightarrow{\approx} V_\ell B$, and so the polynomials $P_v(T)$ are the same for A and for B.

(b) Let f: A ⟶ B be the isogeny, and let A_f be the kernel of f. Enlarge S so that deg(f) is a unit in $R_{K,S}$. Then (6.1) gives us an exact sequence

$$\cdots \longrightarrow H^1(G_S,A_f) \longrightarrow H^1(G_S,A) \xrightarrow{f} H^1(G_S,B) \longrightarrow \cdots .$$

According to (4.15), $H^1(G_S,A_f)$ is finite, and so the kernels of f: $H^1(G_S,A) \longrightarrow H^1(G_S,B)$ and *a fortiori* Ш(f): Ш(K,A) ⟶ Ш(K,B) are finite. Therefore if Ш(K,B) is finite, so also is Ш(K,A), and the reverse implication follows by the same argument from the fact there exists an isogeny g: B ⟶ A such that g∘f = deg(f).

Before stating the main theorem of this section, it is convenient to make another definition. If $f\colon X \longrightarrow Y$ is a homomorphism of abelian groups with finite kernel and cokernel, we define

$$z(f) = \frac{[\mathrm{Ker}(f)]}{[\mathrm{Coker}(f)]}.$$

Lemma 7.2. (a) *If X and Y are finite, then* $z(f) = [X]/[Y]$.

(b) *Consider maps of abelian groups* $X \xrightarrow{f} Y \xrightarrow{g} Z$; *if any two of* $z(f)$, $z(g)$, *and* $z(gf)$ *are defined, then so also is the third, and* $z(gf) = z(g)z(f)$.

(c) *If* $X^{\cdot} = (0 \longrightarrow X^0 \longrightarrow \dots \longrightarrow X^n \longrightarrow 0)$ *is a complex of finite groups, then*

$$\prod [X^r]^{(-1)^r} = \prod [H^r(X^{\cdot})]^{(-1)^r}.$$

(d) *If* $f^{\cdot}\colon X^{\cdot} \longrightarrow Y^{\cdot}$ *is a map of exact sequences of finite length, and* $z(f^r)$ *is defined for all r, then* $\prod z(f^r)^{(-1)^r} = 1$.

Proof: Part (b) is obvious from the kernel-cokernel sequence of the two maps. Part (d) is obvious from the snake lemma when $X^{\cdot}$ and $Y^{\cdot}$ are short exact sequences, and the general case reduces to that case. The remaining statements are even easier.

Theorem 7.3. *Assume that the abelian varieties A and B are isogenous by an isogeny of degree prime to the* char(K). *If the conjecture of Birch and Swinnerton-Dyer is true for one of A or B, then it is true for both.*

Proof: We assume that the conjecture is true for B and prove that it is then true for A. Let $f\colon A \longrightarrow B$ be an isogeny of degree prime to the characteristic of K, and let $f^t\colon B^t \longrightarrow A^t$ be the dual isogeny. Choose an element $\omega_B \in \Gamma(B, \Omega^d_{B/K})$, and let ω_A be its inverse image $f^*\omega_B$ on A. Fix a finite set S of primes of K including all archimedean primes, all primes at which A or B has bad redution, all primes whose residue characteristic divides the degree of f, and all primes at which ω_B or ω_A does not reduce to a nonzero global differential form. Finally choose linearly independent families of elements $(a_i)_{1\le i\le r}$ of A(K) and $(b'_i)_{1\le i\le r}$ of $B^t(K)$, where r is the common rank of the groups of K-rational points on the four abelian varieties, and let $b_i = f(a_i)$ and $a'_i = f^t(b'_i)$. Then $(a'_i)_{1\le i\le r}$ and $(b_i)_{1\le i\le r}$ are linearly independent families of elements of $A^t(K)$ and B(K). The proof will proceed by comparing the corresponding terms in the conjectured formulas for A and for B.

The functoriality of the height pairings shows that $\langle f^t(b'_j), a_i\rangle = \langle b'_j, f(a_i)\rangle$, and this can be rewritten as $\langle a'_j, a_i\rangle = \langle b'_j, b_i\rangle$. Therefore

$$\boxed{\det \langle a'_j, a_i\rangle = \det \langle b'_j, b_i\rangle}\ .$$

From the diagram

$$\begin{array}{ccccccccc} 0 \longrightarrow & \sum \mathbb{Z}a_i & \longrightarrow & A(K) & \longrightarrow & A(K)/\sum \mathbb{Z}a_i & \longrightarrow 0, \\ & \downarrow \approx & & \downarrow f(K) & & \downarrow & \\ 0 \longrightarrow & \sum \mathbb{Z}b_i & \longrightarrow & B(K) & \longrightarrow & B(K)/\sum \mathbb{Z}b_i & \longrightarrow 0 \end{array}$$

and its analogue for f^t, we see that

$$\boxed{z(f(K)) = \frac{(A(K):\sum \mathbb{Z}a_i)}{(B(K):\sum \mathbb{Z}b_i)}, \quad z(f^t(K)) = \frac{(B^t(K):\sum \mathbb{Z}b'_i)}{(A^t(K):\sum \mathbb{Z}a'_i)}}\ .$$

We have seen in (7.1) and (6.14c) that the finiteness of Ш(B) implies that of Ш(A), Ш(A^t), and Ш(B^t), and so (6.13a) shows that the two pairings in the following diagram are nondegenerate,

$$\begin{array}{ccccc} Ш(A) & \times & Ш(A^t) & \longrightarrow & \mathbb{Q}/\mathbb{Z}. \\ \downarrow Ш(f) & & \uparrow Ш(f^t) & & \| \\ Ш(B) & \times & Ш(B^t) & \longrightarrow & \mathbb{Q}/\mathbb{Z} \end{array}$$

Therefore [Coker Ш(f)] = [Ker Ш(f^t)], and so we have the equalities

$$\boxed{\frac{[Ш(A)]}{[Ш(B)]} = z(Ш(f)) = \frac{[\text{Ker } Ш(f)]}{[\text{Ker } Ш(f^t)]}}.$$

Finally consider the map $f(K_v)$: $A(K_v) \longrightarrow B(K_v)$. By definition $\omega_A = f^*\omega_B$, and so $\mu_v(U,\omega_A) = \mu_v(fU,\omega_B)$ for any subset U of $A(K_v)$ that is mapped injectively into $B(K_v)$. Therefore

$$\mu_v(A(K_v),\omega_A) = [\text{Ker } f(K_v)]\ \mu_v(f(A(K_v)),\omega_B).$$

Since $\mu_v(f(A(K_v)),\omega_B) = [\text{Coker } f(K_v)]^{-1}\mu_v(B(K_v),\omega_B)$, we see that $z(f(K_v)) = \mu_v(A,\omega_A)/\mu_v(B,\omega_B)$, and so

$$\boxed{\frac{L^*(s,A)}{L^*(s,B)} = \frac{\prod \mu_v(B,\omega_B)}{\prod \mu_v(A,\omega_A)} = \prod_{v\in S} z(f(K_v))^{-1}}$$

On combining all the boxed formulas, we find that to prove the theorem it suffices to show that

$$\prod_{v\in S} z(f(K_v)) = \frac{[\text{Ker } Ш(f^t)]}{[\text{Ker } Ш(f)\ ]}\,\frac{z(f(K))}{z(f^t(K))} \qquad \textbf{(7.3.1)}.$$

Consider the commutative diagram

$$\begin{array}{ccccccccc}
 & & 0 \longrightarrow H^0(G_S,M) & \longrightarrow & \bigoplus_{v\in S} H^0(K_v,M) & & & & \\
 & & & & \downarrow & & & & \\
 & & & & H^2(G_S,M^D)^* & & & & \\
 & & & & \downarrow & & & & \\
0 \longrightarrow & & \text{Coker}(f(K)) & \longrightarrow & H^1(G_S,M) & \longrightarrow & H^1(G_S,A)_f & \longrightarrow 0 \\
 & & \downarrow\varphi' & & \downarrow\varphi & & \downarrow\varphi'' & \\
0 \longrightarrow & & \bigoplus_{v\in S}\text{Coker}(f(K_v)) & \longrightarrow & \bigoplus_{v\in S} H^1(K_v,M) & \longrightarrow & \bigoplus_{v\in S} H^1(K_v,A)_f & \longrightarrow 0 \\
 & & \downarrow\psi' & & \downarrow\psi & & \downarrow\psi'' & \\
0 \longrightarrow & & H^1(G_S,B^t)^*_{f^t} & \longrightarrow & H^1(G_S,M^D)^* & \longrightarrow & (\text{Coker } f^t(K))^* & \longrightarrow 0
\end{array}$$

in which the rows are extracted from the cohomology sequences of

$$0 \longrightarrow M \longrightarrow A(K_S) \xrightarrow{f} B(K_S) \longrightarrow 0,$$

$$0 \longrightarrow M \longrightarrow A(K_{v,s}) \xrightarrow{f} B(K_{v,s}) \longrightarrow 0,$$

and

$$0 \longrightarrow M^D \longrightarrow B^t(K_S) \xrightarrow{f^t} A^t(K_S) \longrightarrow 0$$

respectively, and the middle column is part of the exact sequence in Theorem 4.10. The rows are exact. The duality between $B(K_v)$ and $H^1(K_v,B^t)$ induces a duality between $B(K_v)/fA(K_v)$ and $H^1(K_v,B^t)_{f^t}$, and the map ψ' is the dual of the composite

$$H^1(G_S,B^t)_{f^t} \longrightarrow \bigoplus_{v\in S} H^1(K_v,B^t)_{f^t} \xrightarrow{\approx} \bigoplus_{v\in S} (B(K_v)/fA(K_v))^*.$$

The map ψ'' is the dual of the composite

$$A^t(K)/f^tB^t(K) \longrightarrow \bigoplus_{v\in S} A^t(K_v)/f^tB^t(K_v) \xrightarrow{\approx} \bigoplus_{v\in S} (H^1(K_v,A)_f)^*.$$

The two outside columns need not be exact, but it is clear from the diagram that they are complexes.

The serpent lemma and a small diagram chase give us an exact sequence

$$0 \longrightarrow \mathrm{Ker}(\varphi') \longrightarrow \mathrm{Ker}(\varphi) \longrightarrow \mathrm{Ker}(\varphi'') \longrightarrow \mathrm{Ker}(\psi')/\mathrm{Im}(\varphi') \longrightarrow 0.$$

As $\mathrm{Ker}(\varphi'') = \mathrm{Ker}\,Ш(f)$, we obtain the formula

$$\frac{[\mathrm{Ker}\ \varphi']}{[\mathrm{Ker}\ \varphi\ ]}\,\frac{[\mathrm{Ker}\ Ш(f)]}{[\mathrm{Ker}\ \psi'/\mathrm{Im}\ \varphi']} = 1.$$

From the first column, we get (using (7.2c) and that $\mathrm{Coker}(\psi') = (\mathrm{Ker}\ Ш(f^t))^*$)

$$\frac{[\mathrm{Coker}\ f(K)]}{\prod_{v\in S}[\mathrm{Coker}\ f(K_v)]}\,[H^1(G_S,B^t)_{f^t}] = \frac{[\mathrm{Ker}\ \varphi']}{[\mathrm{Ker}\ \psi'/\mathrm{Im}\ \varphi']}\ [\mathrm{Ker}\ Ш(f^t)].$$

From the third row, we get

$$1 = \frac{[H^1(G_S, B^t)_{f^t}]}{[H^1(G_S, M^D]}\,[\mathrm{Coker}\ f^t(K)].$$

From the middle column we get (using that $H^0(G_S, M) = \mathrm{Ker}\ f(K), \dots$)

$$1 = \frac{[\mathrm{Ker}\ f(K)]}{\prod [\mathrm{Ker}\ f(K_v)]} \prod_{v\ \mathrm{arch}} \frac{[H^0(K_v, M)]}{[H^0_T(K_v, M)]}\ \frac{[H^2(G_S, M^D)]}{[\mathrm{Ker}\ \varphi]}.$$

Finally, we have the obvious equality

$$[\mathrm{Ker}\ f^t(K)] = [H^0(G_S, M^D)].$$

On multiplying these five equalities together, we find that

$$\prod_{v\in S} z(f(K_v)) = \frac{[\mathrm{Ker}\ Ш(f^t)]}{[\mathrm{Ker}\ Ш(f)]}\ \frac{z(f(K))}{z(f^t(K)}\ \chi(G_S, M^D) \prod_{v\ \mathrm{arch}} \frac{[H^0(K_v, M)]}{[H^0_T(K_v, M)]}.$$

Theorem 5.1 (in the form (5.2a)) shows that the product of the last two terms on the right of the equation is 1, and so this completes the proof of the theorem.

Remark 7.4. Since in the number field case the conjecture of Birch and Swinnerton-Dyer is not known for a single abelian variety, it is worth pointing out that the above arguments apply to the m-primary components of the groups involved: if $Ш(K, A)(m)$ is finite and has the order predicted by the conjecture, then the same is true of any abelian variety isogenous to A.

Remark 7.5. We mention two results of a similar (but simpler) nature to (7.3).

Let A be an abelian variety over a finite separable extension F of the global field K. Then A gives rise to an abelian variety A_* over K by restriction of scalars. The conjecture of Birch and Swinnerton-Dyer holds for A over F if and only if it holds for A_* over K (see [Milne (1972), Thm 1]).

Let A be an abelian variety over a number field K, and assume that it acquires complex multiplication over F, and that F is the smallest extension of K for which this is true. Under certain hypotheses on A, it is known that the conjecture of Birch and Swinnerton-Dyer holds for A over K if and only if it holds for A_F over F (ibid. Corollary to Thm 3).

Notes: For elliptic curves, Theorem 7.3 was proved by Cassels [Cassels (1965)]. The general case was proved by Tate (announcement [Tate (1965/66), Theorem 2.1]). The above proof was explained to me by Tate in 1967

§8 Abelian class field theory, in the sense of Langlands

Abelian class field theory for a global field K defines a reciprocity map $\text{rec}_K\colon C_K \longrightarrow \text{Gal}(K_s/K)^{ab}$ that classifies the finite abelian extensions of K. Dually, one can regard it as associating a character $\chi\circ\text{rec}_K$ of C_K with each (abelian) character χ of $\text{Gal}(K_s/K)$ of finite order; the correspondence is such that the L-series of χ and $\chi\circ\text{rec}_K$ are equal. It is this second interpretation that generalizes to the nonabelian situation. For any reductive group G over a local or global field K, Langlands has conjectured that it is possible to associate an automorphic representation of G with each "admissable" homomorphism of the Weil group W_K of K (Weil-Deligne group in the case of a local field) into a certain complex group ${}^LG$; the L-series of the automorphic representation is to equal that of the Weil-group representation. In the case that $G = \mathbb{G}_m$, the correspondence is simply that noted above. For a general reductive group, the conjecture is difficult even to state since it requires a knowledge of representation theory over adèle groups (see [Borel (1979)]). For a torus however the statement of the conjecture is simple, and we shall prove it in this case. First we prove a duality theorem (8.6), and then we explain the relation of the theorem to Langlands's conjectural class field theory.

In contrast to the rest of these notes, in this section we shall consider cohomology groups $H^r(G,M)$ in which G is not a profinite group. The symbol $H^r(G,M)$ will denote the group constructed without regard for topologies, and $H^r_{cts}(G,M)$ will denote the group defined using continuous cochains. As usual, when G is finite, $H^r_T(G,M)$, $r \in \mathbb{Z}$, denotes the Tate group. For a topological group M

$M^* = \mathrm{Hom}_{cts}(M,\mathbb{Q}/\mathbb{Z})$ = group of characters of M of finite order;

$M^u = \mathrm{Hom}_{cts}(M,\mathbb{R}/\mathbb{Z})$ = group of characters of M (the Pontryagin dual of M);

$M' = \mathrm{Hom}_{cts}(M,\mathbb{C}/\mathbb{Z}) = \mathrm{Hom}_{cts}(M,\mathbb{C}^\times)$ = group of generalized characters of M;

$M^\dagger = \mathrm{Hom}(M,\mathbb{C}^\times)$ = group of generalized (not necessarily continuous) characters of M. When M is discrete, $M' = M^\dagger$.

As usual, when K is a global field, we write C_K for the idèle class group of K. In order to be able to give uniform statements, we sometimes write C_K for $K^\times$ when K is a local field.

Weil groups

First we need to define the *Weil group* of a local or global field K. This is a triple $(W_K,\varphi,(r_F))$ comprising a topological group W_K, a continuous homomorphism $\varphi\colon W_K \longrightarrow \mathrm{Gal}(K_s/K)$ with dense image, and a family of isomorphisms $r_F\colon C_F \xrightarrow{\approx} W_F^{ab}$, one for each finite extension $F \subset K_s$ of K, where $W_F = \varphi^{-1}(G_F)$. (Here, as always, W_K^{ab} is the quotient of W_K by the *closure* W_K^c of its commutator subgroup.)

For any finite extension F of K, define $W_{F/K} = W_K/W_F^c$; then, if F is Galois over K, there is an exact sequence

$$0 \longrightarrow C_F \longrightarrow W_{F/K} \longrightarrow G_{F/K} \longrightarrow 0$$

whose class in $H^2(G_{F/K}, C_F)$ is the canonical class (that is, the element denoted by $u_{G_{F/K}}$ in the second paragraph of §1). The topology on $W_{F/K}$ is such that C_F receives its usual topology and is an open subgroup of $W_{F/K}$. The full Weil group W_K is equal to the inverse limit $\varprojlim W_{F/K}$ (as a topological group).

Examples 8.1. (a) Let K be a nonarchimedean local field. The Weil group W_K is the dense subgroup of G_K consisting of elements that act as an integral multiple of the Frobenius automorphism on the residue field. It therefore contains the inertia subgroup I_K of G_K, and the quotient W_K/I_K is $\mathbb{Z}$. The topology on W_K is that for which I_K receives the profinite topology and is an open subgroup of W_K. The map φ is the inclusion map, and r_F is the unique isomorphism $F^\times \longrightarrow W_F^{ab}$ such that r_F followed by φ is the reciprocity map.

(b) Let K be an archimedean local field. If $K = \mathbb{C}$, then W_K is $\mathbb{C}^\times$, φ is the trivial map $\mathbb{C}^\times \longrightarrow \mathrm{Gal}(\mathbb{C}/\mathbb{C})$, and r_K is the identity map. If K is real, then $W_K = K_s^\times \cup jK_s^\times$ (disjoint union) with the rules $j^2 = -1$ and $jzj^{-1} = \bar{z}$ (complex conjugate). The map φ sends $K_s^\times$ to 1 and j to the nontrivial element of G_K. The map r_{K_s} is the identity map, and r_K is characterized by

$$r_K(-1) = jW_K^c$$

$$r_K(x) = x^{1/2}W_K^c \text{ for } x \in K,\ x > 0.$$

(c) Let K be a function field in one variable over a finite field. The Weil group W_K is the dense subgroup of $\mathrm{Gal}(K_s/K)$ of elements that act as an integral multiple of the Frobenius automorphism on the algebraic closure of the field of constants. It therefore contains the geometric Galois group $G_{Kk_s} = \mathrm{Gal}(K_s/Kk_s) \subset G_K$, and the quotient of W by G_{Kk_s} is $\mathbb{Z}$. The topology on W_K is that for which G_{Kk_s} receives the profinite topology and is an open subgroup of W_K. The map φ is the inclusion map, and r_F is the unique isomorphism $C_F \longrightarrow W_F^{ab}$ such that r_F followed by φ is the reciprocity map.

(d) Let K be an algebraic number field. Only in this case, which of course is the most important, is there no explicit description of the Weil group. It is constructed as the inverse limit of the extensions corresponding to the canonical classes $u_{G_{F/K}}$ (see [Artin and Tate (1961), XV], where the Weil group is constructed for any class formation, or [Tate (1979)]).

Let K be a global field. For each prime v of K, it is possible to construct a commutative diagram

$$\begin{array}{cccc} \varphi_v\colon & W_{K_v} & \longrightarrow & G_{K_v} \\ & \downarrow & & \downarrow \\ \varphi\colon & W_K & \longrightarrow & G_K \end{array},$$

(see [Tate (1979), 1.6.1]). We shall assume in the following that one such diagram has been selected for each v.

Some cohomology

We regard the cohomology and homology groups as being constructed using the standard complexes. For example, $H^r(G,M) = H^r(C^{\cdot}(G,M))$ where $C^{\cdot}(G,M)$ consists of maps $[g_1,\ldots,g_r] \mapsto \alpha(g_1,\ldots,g_r)\colon G^r \longrightarrow M$. When G is finite, the groups $H_T^{-1}(G,M)$ and $H_T^{0}(G,M)$ are determined by the exact sequence

$$0 \longrightarrow H_T^{-1}(G,M) \longrightarrow M_G \xrightarrow{N_G} M^G \longrightarrow H_T^0(G,M) \longrightarrow 0.$$

Lemma 8.2. *Let G be a finite group, and let Q be an abelian group regarded as a G-module with the trivial action. If Q is divisible, then for all G-modules M, the cup-product pairing*

$$H_T^{r-1}(G,\mathrm{Hom}(M,Q)) \times H_T^{-r}(G,M) \longrightarrow H_T^{-1}(G,Q) \subset Q$$

induces an isomorphism

$$H_T^{r-1}(G,\mathrm{Hom}(M,Q)) \longrightarrow \mathrm{Hom}(H_T^{-r}(G,M),Q),$$

all r.

Proof: This is proved in [Cartan and Eilenberg (1956), XII.6.4].

Let (G,C) be a class formation, and let $\bar{G}$ be the quotient of G by an open normal subgroup H. The pairing

$$(f,c) \mapsto f(c\otimes -): (C^H \otimes M)^\dagger \times C^H \longrightarrow M^\dagger$$

and the canonical class $u \in H^2(\bar{G},C^H)$ define maps

$$a \mapsto a\cup u: H_T^r(\bar{G},(C^H\otimes M)^\dagger) \longrightarrow H_T^{r+2}(\bar{G},M^\dagger).$$

Lemma 8.3. *For all finitely generated torsion-free $\bar{G}$-modules M and all r, the map* $-\cup u$: $H_T^r(\bar{G},(C^H\otimes M)^\dagger) \longrightarrow H_T^{r+2}(\bar{G},M^\dagger)$ *is an isomorphism.*

Proof: The diagram

$$\begin{array}{ccccc} H_T^r(\bar{G},(C^H\otimes M)^\dagger) & \times & H_T^{-r-1}(\bar{G},C^H\otimes M) & \longrightarrow & H_T^{-1}(\bar{G},\mathbb{C}^\times) \subset \mathbb{C}^\times \\ \downarrow -\cup u & & \uparrow u\cup - & & \| \\ H_T^{r+2}(\bar{G},M^\dagger) & \times & H_T^{-r-3}(\bar{G},M) & \longrightarrow & H_T^{-1}(\bar{G},\mathbb{C}^\times) \subset \mathbb{C}^\times \end{array}$$

commutes because of the associativity of cup-products:

$$(a\cup u)\cup b = a\cup(u\cup b),\ a \in H_T^r(\bar{G},(C^H\otimes M)^\dagger),\ b \in H_T^{-r-3}(\bar{G},M).$$

The two pairings are nondegenerate by (8.2), and the second vertical map is an isomorphism by virtue of the Tate-Nakayama theorem (0.2). It follows that the first vertical map is an isomorphism.

Note that,

$$(C^H\otimes M)^\dagger \overset{\text{df}}{=} \text{Hom}(C^H\otimes M,\mathbb{C}^\times) = \text{Hom}(C^H,\text{Hom}(M,\mathbb{C}^\times)) = \text{Hom}(C^H,M^\dagger).$$

Therefore the isomorphism in the above lemma can also be written

$$-\cup u\colon\ H_T^r(\bar{G},\text{Hom}(C^H,M^\dagger)) \xrightarrow{\approx} H_T^{r+2}(\bar{G},M^\dagger).$$

Let

$$0 \longrightarrow C^H \longrightarrow \bar{W} \longrightarrow \bar{G} \longrightarrow 1$$

be the exact sequence of groups corresponding to the canonical class $u \in H^2(\bar{G},C^H)$. For any $\bar{W}$-module M, the Hochschild-Serre spectral sequence gives an exact sequence

$$0 \longrightarrow H^1(\bar{G},M^C) \xrightarrow{\text{Inf}} H^1(\bar{W},M) \xrightarrow{\text{Res}} H^1(C^H,M)^{\bar{G}} \xrightarrow{\tau} H^2(\bar{G},M^C).$$

The map τ (the transgression) has the following explicit description: let $a \in H^1(C^H,M)^{\bar{G}}$, and choose a 1-cocycle α representing it; extend α to a 1-cochain β on $\bar{W}$; then $d\beta$ is a 2-cocycle on $\bar{G}$, and the class it represents is $\tau(a)$.

Lemma 8.4. *If C^H acts trivially on M, then the transgression*

$$\tau\colon\ H^0(\bar{G},\text{Hom}(C^H,M)) \longrightarrow H^2(\bar{G},M)$$

is the negative of the map $-\cup u$ induced by the pairing

$$\text{Hom}(C^H,M) \times C^H \longrightarrow M.$$

Proof: Write $\bar{W} = \cup\, C^H w_g$ (disjoint union of right cosets), and let

$w_g w_{g'} = \gamma(g,g')w_{gg'}$. Then $(\gamma(g,g'))$ is a 2-cocycle representing u. Let $\alpha \in \mathrm{Hom}_{\bar{G}}(C^H,M)$, and define β by $\beta(cw_g) = \alpha(c)$, $c \in C_H$. Then

$$d\beta(g,g') \overset{df}{=} d\beta(w_g,w_{g'}) = g\beta(w_{g'}) - \beta(w_g w_{g'}) + \beta(w_g)$$
$$= 0 \quad - \alpha(\gamma(g,g')) + 0$$
$$= -\alpha(\gamma(g,g')),$$

which equals $-(\alpha\cup\gamma)(g,g')$. Therefore $\tau(\alpha) = -\alpha\cup u$.

The duality theorem

Let K be a local or global field (we could in fact work abstractly with any class formation), and let F be a finite Galois extension of K. Let M be a finitely generated torsion-free $G_{F/K}$-module. Then $M' \overset{df}{=} \mathrm{Hom}_{cts}(M,\mathbb{C}^\times) = \mathrm{Hom}(M,\mathbb{C}^\times) = M^\dagger$ are again $G_{F/K}$-modules. We shall use the notation M' when we wish to emphasize that M' has a topology. We frequently regard these groups as $W_{F/K}$-modules.

Write $W_{F/K} = \bigcup w_g C_F$ (disjoint union of left cosets). For any homomorphism $\alpha\colon C_F \longrightarrow M$, the map $\mathrm{Cor}(\alpha)\colon W_{F/K} \longrightarrow M^\dagger$ such that

$$(\mathrm{Cor}(\alpha))(w) = \sum_{g\in G} w_g\alpha(w_g^{-1}ww_{g'}),\quad ww_{g'} \equiv w_g \bmod C_F,$$

is a cocycle, and so we have a map

$$\mathrm{Cor}\colon H^1(C_F,M^\dagger) \longrightarrow H^1(W_{F/K},M^\dagger),$$

called the *corestriction* map. It is independent of the choice of coset representatives (see [Serre (1962), VII.7] or [Weiss (1969), p81]). It is clearly continuous, and so maps continuous homomorphisms to continuous cocycles.

Lemma 8.5. *The corestriction map* $\mathrm{Cor}\colon H^1(C_F,M^\dagger) \longrightarrow H^1(W_{F/K},M^\dagger)$ *factors through* $H^1(C_F,M^\dagger)_G$, $G = G_{F/K}$.

Proof: Let $\alpha \in \mathrm{Hom}(C_F, M^\dagger)$ and $h \in G$. Then $(h\alpha)(w) = w_h\alpha(w_h^{-1}ww_h)$ (this is the definition), and so $\mathrm{Cor}(h\alpha)(w) = \sum_g w_g w_h \alpha(w_h^{-1}w_g^{-1}ww_{g'}w_h)$ where g' is such that $ww_{g'} \equiv w_g \bmod C_F$. The family $(w_g w_h)_{g\in G}$ is also a set of coset representatives for C_F in $W_{F/K}$, and $w(w_{g'}w_h) \equiv (w_g w_h) \bmod C_F$. Therefore the class of $\mathrm{Cor}(h\alpha)$ is the same as that of $\mathrm{Cor}(\alpha)$, and so $\mathrm{Cor}((h-1)\alpha) = 0$ in $H^1(W_{F/K}, M^\dagger)$.

Theorem 8.6. *For any finitely generated torsion-free* $G_{F/K}$*-module* M, *the corestriction map defines an isomorphism*

$$\mathrm{Hom}_{cts}(C_F, M')_{G_{F/K}} \xrightarrow{\approx} H^1_{cts}(W_{F/K}, M').$$

Proof: Throughout the proof, we write G for $G_{F/K}$. We shall first prove that the corestriction map defines an isomorphism $\mathrm{Hom}(C_F, M^\dagger)_G \xrightarrow{\approx} H^1(W_{F/K}, M^\dagger)$ and then show (in Lemma 8.9) that it makes continuous homomorphisms correspond to continuous cocycles.

Consider the diagram

$$\begin{array}{ccccccccc}
0 \to & H_T^{-1}(G, \mathrm{Hom}(C_F, M^\dagger)) & \to & \mathrm{Hom}(C_F, M^\dagger)_G & \xrightarrow{N_G} & \mathrm{Hom}(C_F, M^\dagger)^G & \to & H_T^0(G, \mathrm{Hom}(C_F, M)^\dagger) & \\
& \downarrow \approx & & \downarrow \mathrm{Cor} & & \downarrow \mathrm{id} & & \downarrow \approx & \quad \mathbf{(8.6.1)} \\
0 \to & H^1(G, M^\dagger) & \to & H^1(W_{F/K}, M^\dagger) & \longrightarrow & H^1(C_F, M^\dagger)^G & \to & H^2(G, M^\dagger). &
\end{array}$$

The top row is the sequence defining the Tate cohomology groups of $\mathrm{Hom}(C_F, M^\dagger)$. The bottom row can be deduced from the Hochschild-Serre spectral sequence or else can be constructed in an elementary fashion. The two isomorphisms are those in Lemma 8.3. The third square (anti-) commutes because of (8.4). We shall prove in the next two lemmas that the first two squares in the diagram commute. The five-lemma will then show that Cor: $\mathrm{Hom}(C_F, M^\dagger)_G \longrightarrow H^1(W_{F/K}, M^\dagger)$ is an isomorphism. Finally Lemma 8.9 will complete the proof.

Lemma 8.7. *The first square in* (8.6.1) *commutes.*

Proof: We first show that Cor maps an element of $H_T^{-1}(G,\mathrm{Hom}(C_F,M^\dagger))$ into the subgroup $H^1(G,M^\dagger)$ of $H^1(W_{F/K},M^\dagger)$. Let α be a homomorphism $C_F \longrightarrow M^\dagger$, and let $c \in C_F$ and $w \in W$. Then

$$(\mathrm{Cor}\ \alpha)(cw) = \sum_g w_g\alpha(w_g^{-1}cw_gw_g^{-1}ww_{g'}) = \sum_g (w_g\alpha)(c) + (\mathrm{Cor}\ \alpha)(w)$$
$$= (N\alpha)(c) + (\mathrm{Cor}\ \alpha)(w).$$

Therefore, if $N\alpha = 0$ (that is, $\alpha \in H_T^{-1}(G,\mathrm{Hom}(C_F,M^\dagger)))$, then $(\mathrm{Cor}\ \alpha)(w)$ depends only on the class of w in G, and so $\mathrm{Cor}(\alpha)$ arises by inflation from an element of $H^1(G,M^\dagger)$.

It remains to show that the restriction of Cor to $H_T^{-1}(G,\mathrm{Hom}(C_F,M^\dagger))$ is $-\nu u$. Note that

$$(\mathrm{Cor}\ \alpha)(h) = \sum_g g(\alpha(w_g^{-1}w_hw_{h^{-1}g})) = \sum_g (g\alpha)(w_hw_{h^{-1}g}w_g^{-1})$$
$$= \sum_g (g\alpha)(u(h,h^{-1}g)).$$

To obtain the middle equality, we have used that

$$w_g^{-1}w_hw_{h^{-1}g} = c \Rightarrow w_hw_{h^{-1}g} = w_gc = (gc)w_g \Rightarrow w_hw_{h^{-1}g}w_g^{-1} = gc$$

and that $g(\alpha(c)) = (g\alpha)(gc)$.

It is difficult to give explicit descriptions of cup-products when both negative and positive indices are involved. We shall use the exact sequence

$$0 \longrightarrow I_G \longrightarrow \mathbb{Z}[G] \longrightarrow \mathbb{Z} \longrightarrow 0$$

to shift the problem. It remains exact when tensored with $M^\dagger$ and $\mathrm{Hom}(C_F,M^\dagger)$, and the boundary maps in the resulting cohomology sequences give the horizontal maps in the following diagram:

$$\begin{array}{ccc} H_T^{-1}(G,\mathrm{Hom}(C_F,M^\dagger)) & \xrightarrow{d^{-1}} & H_T^0(G,\mathrm{Hom}(C_F,M^\dagger)\otimes I_G) \\ \downarrow -\nu u & & \downarrow -\nu u\otimes 1 \\ H^1(G,M^\dagger) & \xrightarrow{d^1} & H^2(G,M^\dagger\otimes I_G) \end{array}.$$

Both boundary maps are isomorphisms, and $-\cup u$ is the unique map making the diagram commute. If we can show that the diagram still commutes when this map is replaced with Cor, we will have proved the lemma. This we do by an ugly cocycle calculation.

Note first that d^{-1} and d^1 have the following descriptions:

$$d^{-1}(\alpha) = N(\alpha\otimes 1) = N(\alpha\otimes 1) - (N\alpha)\otimes 1 = \sum_g g\alpha\otimes(g-1),\ \alpha \in \mathrm{Hom}(C_F, M^\dagger),\ N\alpha = 0$$

$$d^1(\beta)(g_1,g_2) = g_1\beta(g_2)\otimes(g_1-1),\quad \beta \in Z^1(G,M^\dagger),\ g_1,g_2 \in G.$$

If $\alpha \in \mathrm{Hom}(C_F, M^\dagger)$ has $N\alpha = 0$, then

$$(d^1\circ\mathrm{Cor}\ \alpha)(g_1,g_2) = \sum_{g\in G} g_1\cdot(g\alpha)(u(g_2,g_2^{-1}g))\otimes(g_1-1)$$

and

$$(d^{-1}\alpha\cup(u\otimes 1))(g_1,g_2) = \sum_g (g\alpha)(u(g_1,g_2))\otimes(g-1).$$

An element of $M^\dagger\otimes I_G$ can be written uniquely in the form $\sum m_g\otimes(g-1)$. Therefore a general element of $C^1(G,M^\dagger\otimes I_G)$ is of the form $\sum F_g\otimes(g-1)$ with F_g a map $G \longrightarrow M^\dagger$, and a coboundary in $B^2(G,M^\dagger\otimes I_G)$ can be written

$$d(\sum_g F_g\otimes(g-1))(g_1,g_2)$$
$$= \sum_g (g_1.F_{g_1^{-1}g}(g_2) - F_g(g_1g_2) + F_g(g_1))\otimes(g-1) - \sum_g g_1.F_g(g_2)\otimes(g_1-1).$$

In obtaining the second expression, we have used that

$$\begin{aligned} g_1(\sum F_g(g_2)\otimes(g-1)) &= \sum g_1.F_g(g_2)\otimes(g_1g-g_1) \\ &= \sum g_1.F_g(g_2)\otimes(g_1g-1) - \sum g_1.F_g(g_2)\otimes(g_1-1) \\ &= \sum g_1.F_{g_1^{-1}g}(g_2)\otimes(g-1) - \sum g_1.F_g(g_2)\otimes(g_1-1) \end{aligned}$$

Put $F_g(g_2) = (g\alpha)(u(g_2,g_2{}^{-1}g))$; then

$$(d\sum F_g\otimes(g-1) - (d^{-1}\alpha)\cup(u\otimes 1) + (d^1\circ\mathrm{Cor}\ \alpha)(g_1,g_2) =$$
$$\sum(g\alpha)(g_1u(g_2,g_2^{-1}g_1^{-1}g).u(g_1g_2,g_2^{-1}g_1^{-1}g)^{-1}.u(g_1,g_1^{-1}g).u(g_1,g_2)^{-1})\otimes(g-1).$$

When we put $h = g_2{}^{-1}g_1{}^{-1}g$, this becomes

$$\sum (g\alpha)(g_1u(g_2,h).u(g_1g_2,h)^{-1}.u(g_1,g_2h).u(g_1,g_2)^{-1})\otimes(g-1),$$

and each term in the sum is zero because u is a 2-cocycle. Therefore

$$d\sum(F_g \otimes (g-1)) = (d^{-1}\alpha)\cup(u\otimes 1) - (d^1 \circ \mathrm{Cor}\ \alpha),$$

which completes the proof of the lemma.

Lemma 8.8. *The composite*

$$H^1(C_F, M^\dagger) \xrightarrow{\mathrm{Cor}} H^1(W_{F/K}, M^\dagger) \xrightarrow{\mathrm{Res}} H^1(C_F, M^\dagger)$$

is equal to the norm N_G. *Hence the second square in (8.6.1) commutes.*

Proof: For $\alpha \in Z^1(C_F, M^\dagger)$ and $w \in W_{F/K}$, $\mathrm{Cor}(\alpha)(w) = \sum_g w_g \alpha(w_g^{-1} w w_{g'})$. When $w \in C_F$, this becomes $\mathrm{Cor}(\alpha)(w) = \sum_{g\in G} g\alpha(g^{-1}wg) = (N_G\alpha)(w)$.

Lemma 8.9. *Let* $\alpha \in \mathrm{Hom}(C_F, M')$; *then* $\alpha \in \mathrm{Hom}_{cts}(C_F, M')$ *if and only if* $\mathrm{Cor}(\alpha) \in Z^1_{cts}(W_{F/K}, M')$.

Proof: Clearly $\alpha \in Z^1(W_{F/K}, M')$ is continuous if and only if its restriction to C_F is continuous. Therefore (8.8) shows that it suffices to prove that a homomorphism $f\colon C_F \longrightarrow M'$ is continuous if and only if $N_G f$ is continuous. Since N_G is continuous, there is a commutative diagram

$$\begin{array}{ccccccc}
\mathrm{Hom}_{cts}(C_F, M') & \xrightarrow{N} & \mathrm{Hom}_{cts}(C_F, M')^G & \longrightarrow & H^0_T(G, \mathrm{Hom}_{cts}(C_F, M')) & \longrightarrow & 0 \\
\downarrow & & \downarrow & & \downarrow & & \\
\mathrm{Hom}(C_F, M') & \xrightarrow{N} & \mathrm{Hom}(C_F, M')^G & \longrightarrow & H^0_T(G, \mathrm{Hom}(C_F, M')) & \longrightarrow & 0,
\end{array}$$

from which it follows that it suffices to show that

$$H^0_T(G, \mathrm{Hom}_{cts}(C_F, M')) \longrightarrow H^0_T(G, \mathrm{Hom}(C_F, M'))$$

is injective. In fact, following [Labesse (1984)], we shall prove much more.

Lemma 8.10. *For all* r, *the map*

$$H_T^r(G, \mathrm{Hom}_{cts}(C_F, M')) \longrightarrow H_T^r(G, \mathrm{Hom}(C_F, M'))$$

is an isomorphism.

Proof: We consider the cases separately.

(a) K local archimedean. The only nontrivial case has $K = \mathbb{R}$ and $F = \mathbb{C}$. Here $C_K = \mathbb{C}^\times$, and we shall use the exponential sequence

$$0 \longrightarrow \mathbb{Z} \longrightarrow \mathbb{C} \longrightarrow \mathbb{C}^\times \longrightarrow 0.$$

From it we get exact sequences

$$0 \longrightarrow \mathrm{Hom}(\mathbb{C}^\times, M') \longrightarrow \mathrm{Hom}(\mathbb{C}, M') \longrightarrow \mathrm{Hom}(\mathbb{Z}, M') \longrightarrow 0$$

(because M' is divisible) and

$$0 \longrightarrow \mathrm{Hom}_{cts}(\mathbb{C}^\times, M') \longrightarrow \mathrm{Hom}_{cts}(\mathbb{C}, M') \longrightarrow \mathrm{Hom}_{cts}(\mathbb{Z}, M') \longrightarrow 0$$

(because M' is a connected commutative Lie group). The groups $\mathrm{Hom}(\mathbb{C}, M')$ and $\mathrm{Hom}_{cts}(\mathbb{C}, M')$ are uniquely divisible, and so are cohomologically trivial. Therefore, we can replace C_F in the statement of the lemma with $\mathbb{Z}$, but then it becomes obvious because $\mathbb{Z}$ is discrete.

(b) K local nonarchimedean. Here $C_F = F^\times$. From [Serre (1967a), 1.4], we know that $F^\times$ contains a cohomologically trivial open subgroup V; moreover V contains a fundamental system (V_n) of neighbourhoods of zero with each V_n an open subgroup, such that V/V_n is cohomologically trivial. (For example, when F is unramified over K, it is possible to take $V = \mathcal{O}_K^\times$.) Now, because M' is divisible, [Serre (1962), IX.6, Thm 9] shows that $\mathrm{Hom}(V, M')$ and $\mathrm{Hom}(V/V_n, M')$ are also cohomlogically trivial. As $\mathrm{Hom}_{cts}(V, M) = \varinjlim \mathrm{Hom}(V/V_n, M')$, we see that it also is cohomologically trivial. A similar argument to the above, using the sequence

$$0 \longrightarrow V \longrightarrow F^\times \longrightarrow F^\times/V \longrightarrow 0,$$

shows that it suffices to prove the lemma with C_F replaced with $F^{\times}/V$, but this group is discrete.

(c) K global. Here C_F is the idèle class group. Define $V \subset C_K$ to be $\prod V_v$ where $V_v = \hat{\mathcal{O}}_v^{\times}$ for v a nonarchimedean prime that is unramified in F and V_v is a subgroup as considered in (b) for the remaining nonarchimedean primes. This group has similar properties to the group V in (b). It therefore suffices to prove the lemma with C_F replaced with C_F/V. In the function field case this is discrete, and in the number field case it is an extension of a finite group by $\mathbb{R}^{\times}$ (with trivial action). In the first case the lemma is obvious, and in the second the exponential again shows that $\mathbb{R}^{\times}$ is the quotient of a uniquely divisible group by a discrete group.

This completes the proof of the theorem.

Corollary 8.11. *Let K be a global or local field, and let M be a finitely generated torsion-free G_K-module. There is a canonical isomorphism*

$$((C\otimes M)^{G_K})' \xrightarrow{\approx} H^1_{cts}(W_K, M')$$

where $C = \bigcup C_F$.

Proof: Let F be a finite Galois extension of K splitting M. Any continuous crossed homomorphism $f\colon W_K \longrightarrow M'$ restricts to a continuous homomorphism on W_F. Because M is commutative, f must be trivial on W_F^c and so factors through $W_K/W_F^c \overset{df}{=} W_{F/K}$. Consequently, the inflation map $H^1_{cts}(W_{F/K}, M') \longrightarrow H^1_{cts}(W_K, M')$ is bijective.

Next note that $\mathrm{Hom}(C_F, M') = (C_F\otimes M)'$. I claim that the canonical map $(C_F\otimes M)'_G \longrightarrow ((C_F\otimes M)^G)'$ is an isomorphism. Note that this is obviously so when ' is replaced with †, because $(C_F\otimes M)^G$ is the maximal subgroup of $C_F \otimes M$ on which G acts trivially, and so $((C_F\otimes M)^G)^\dagger$ is the maximal quotient group on which G acts trivially, that is, it is $((C_F\otimes M)^\dagger)_G$. The diagram

$$\begin{array}{ccccccc} 0 \longrightarrow & H^{-1}(G,(C_F\otimes M)') & \longrightarrow & (C_F\otimes M)'_G & \longrightarrow & ((C_F\otimes M)')^G \\ & \downarrow \approx & & \downarrow & & \downarrow \text{inj.} \\ 0 \longrightarrow & H^{-1}(G,(C_F\otimes M)^\dagger) & \longrightarrow & (C_F\otimes M)^\dagger_G & \longrightarrow & ((C_F\otimes M)^\dagger)^G \end{array}$$

shows that the middle vertical arrow is injective. Now the diagram

$$\begin{array}{ccc} (C_F\otimes M)'_G & \longrightarrow & ((C_F\otimes M)^G)' \\ \downarrow \text{inj} & & \downarrow \approx \\ (C_F\otimes M)^\dagger_G & \longrightarrow & ((C_F\otimes M)^G)^\dagger \end{array}$$

shows that $(C_F\otimes M)'_G \longrightarrow ((C_F\otimes M)^G)'$ is injective, which proves the claim since the map is obviously surjective. To complete the proof of the corollary, note that

$$\operatorname{Hom}(C_F,M')_{G_{F/K}} = (C_F\otimes M)'_{G_{F/K}} = ((C_F\otimes M)^{G_{F/K}})' = ((C\otimes M)^{G_K})',$$

and so the corollary simply restates the theorem.

Remark 8.12. (a) After making the obvious changes, the above arguments show that there is a canonical isomorphism

$((C\otimes M)^{G_K})^u \longrightarrow H^1_{cts}(W_K,M^u)$.

(b) Replace M in (8.11) with its linear dual. Then $H^1_{cts}(W_K,M')$ becomes $H^1_{cts}(W_K,M\otimes\mathbb{C}/\mathbb{Z})$ and $((C\otimes M)^{G_K})'$ becomes $\operatorname{Hom}_{G_K}(M,C)'$. On the other hand, (4.10) gives us an isomorphism $H^2_{cts}(G_K,M) \xrightarrow{\approx} \operatorname{Hom}_{G_K}(M,C)^*$, and $H^2_{cts}(G_K,M) = H^1_{cts}(G_K,M\otimes\mathbb{Q}/\mathbb{Z}) = H^1_{cts}(W_K,M\otimes\mathbb{Q}/\mathbb{Z})$. These results and their relations can be summarized as follows: for any finitely generated torsion-free G-module M, there is a commutative diagram

$$\begin{array}{ccccc} H^1_{cts}(W_K, M\otimes\mathbb{Q}/\mathbb{Z}) & \hookrightarrow & H^1_{cts}(W_K, M\otimes\mathbb{R}/\mathbb{Z}) & \hookrightarrow & H^1_{cts}(W_K, M\otimes\mathbb{C}/\mathbb{Z}) \\ \downarrow\approx & & \downarrow\approx & & \downarrow\approx \\ \mathrm{Hom}_{W_K}(M,C)^* & \hookrightarrow & \mathrm{Hom}_{W_K}(M,C)^u & \hookrightarrow & \mathrm{Hom}_{W_K}(M,C)', \end{array}$$

in which the horizontal maps are defined by the inclusions

$$\mathbb{Q}/\mathbb{Z} \hookrightarrow \mathbb{R}/\mathbb{Z} \hookrightarrow \mathbb{C}/\mathbb{Z}.$$

Application to tori

Let T be a torus over a field K. The *dual torus* $T^\vee$ to T is the torus such that $X_*(T^\vee)$ is the linear dual $X^*(T)$ of $X_*(T)$. When K is a global field, we say that an element of $H^1_{cts}(W_K, M')$ is locally trivial if it restricts to zero in $H^1_{cts}(W_{K_v}, M')$ for all primes v.

Theorem 8.13. *Let K be a local or global field, and let T be a torus over K.*

(a) *When K is local,* $H^1_{cts}(W_K, T^\vee(\mathbb{C}))$ *is canonically isomorphic to the group of continuous generalized characters of* T(K).

(b) *When K is a global, there is a canonical homomorphism from* $H^1_{cts}(W_K, T^\vee(\mathbb{C}))$ *onto the group of continuous generalized characters of* $T(\mathbb{A}_K)/T(K)$. *The kernel is finite and consists of the locally trivial classes.*

Proof: Take $M = X_*(T)$ in the statement of (8.11). Then $\mathrm{Hom}(M, R^\times) = T^\vee(R)$ for any ring R containing a splitting field for T. In particular, $M' = T^\vee(\mathbb{C})$, and so $H^1_{cts}(W_K, M') = H^1_{cts}(W_K, T^\vee(\mathbb{C}))$.

When K is local, $((X_*(T)\otimes C_F)^G)' = (T(F)^G)' = T(K)'$, which proves (a). In the global case, on tensoring the exact sequence

$$0 \longrightarrow F^\times \longrightarrow J_F \longrightarrow C_F \longrightarrow 0$$

with $X_*(T)$, we obtain an exact sequence

$$0 \longrightarrow T(F) \longrightarrow T(\mathbb{A}_F) \longrightarrow X_*(T)\otimes C_F \longrightarrow 0,$$

and hence an exact sequence

$$0 \longrightarrow T(K) \longrightarrow T(\mathbb{A}_K) \longrightarrow (X_*(T)\otimes C_F)^G \longrightarrow H^1(G,T(F)).$$

The last group in this sequence is finite, and so we have a surjection with finite kernel $((X_*(T)\otimes C_F)^G)' \twoheadrightarrow (T(\mathbb{A}_K)/T(K))'$. This, composed with the isomorphism $H^1_{cts}(W_{F/K},T^\vee(\mathbb{C})) \xrightarrow{\approx} ((X_*(T)\otimes C_F)^G)'$ of the theorem, gives the map.

There is a commutative diagram:

$$\begin{array}{ccc} H^1_{cts}(W_K,T^\vee(\mathbb{C})) & \longrightarrow & (T(\mathbb{A}_K)/T(K))' \\ \downarrow & & \downarrow \\ \prod H^1_{cts}(W_{K_v},T^\vee(\mathbb{C})) & \longrightarrow & \prod T(K_v)' \end{array}$$

We have just seen that the lower horizontal map is an isomorphism, and the second vertical map is injective because it is the dual of a surjective map. Therefore the kernels of the two remaining arrows are equal, as claimed by the theorem.

Re-interpretation as class field theory

Let T be a torus over K, let $M = X_*(T)$, and let $T^\vee$ be the torus such that $X^*(T^\vee) = M$. Let G_K act on $T^\vee(\mathbb{C}) = \mathrm{Hom}(M,\mathbb{C}^\times)$ through its action on M, and define ${}^LT$ to be the semi-direct product $T^\vee(\mathbb{C}) \rtimes G_K$. It is complex Lie group with identity component ${}^LT^o = T^\vee(\mathbb{C})$. A continuous homomorphism $\varphi\colon W_K \longrightarrow {}^LT$ is said to be *admissable* if it is compatible with the projections onto G_K. Two such homomorphisms φ and φ' are said to be *equivalent* if there exists a $t \in {}^LT^o$ such that $\varphi'(w) = t\varphi(w)t^{-1}$ for all w. Write $\Phi_K(T)$ for the set of equivalence classes of admissable homomorphisms, and define $\Pi_K(T)$ to be $T(K)'$ when K is local and $(T(\mathbb{A}_K)/T(K))'$ when K is global.

Theorem 8.14. *There is a canonical map* $\Phi_K(T) \longrightarrow \Pi_K(T)$; *when* K *is local, the map is an isomorphism, and when* K *is global, it is surjective with finite kernel.*

Proof: Any continuous homomorphism φ: $W_K \longrightarrow {}^LT$ can be written $\varphi = f\times\sigma$ with f and σ maps from W_K into ${}^LT^o$ and G_K respectively. One checks immediately that φ is an admissable homomorphism if and only if f is a 1-cocycle and σ is the map $W_K \longrightarrow G_K$ given as part of the structure of W_K. Moreover, every 1-cocycle arises in this way, and two φ's are equivalent if and only if the corresponding 1-cocycles are cohomologous. Thus the theorem follows immediately from (8.13).

L-series

Let K be a nonarchimedean local field. For any representation ρ of W_K on a finite-dimensional complex vector space V, the L-series

$$L(s,\rho) = (\det(1 - \rho(Fr)N(\pi)^{-s}|V^I)^{-1},$$

where Fr is an element of W_K mapping to 1 under the canonical map $W_K \longrightarrow \mathbb{Z}$, π is a local uniformizing parameter, and I is the inertia group. For a global field K and representation ρ of ${}^LT$, the Artin-Hecke L-series $L(s,\rho)$ is defined to be the product of the local L-series at the nonarchimedean primes. (It is possible also to define factors corresponding to the archimedean factors, but we shall ignore them.) For S a finite set of primes, we let $L_S(s,\rho)$ be the product of the local factors over all primes not in S.

Assume now that T splits over an unramified Galois extension F of K. On tensoring

$$0 \longrightarrow \mathcal{O}_F^\times \longrightarrow F^\times \longrightarrow \mathbb{Z} \longrightarrow 0$$

with $X_*(T)$, we obtain an exact sequence

$$0 \longrightarrow T(\mathcal{O}_F) \longrightarrow T(F) \longrightarrow X_*(T) \longrightarrow 0.$$

with $T(\mathcal{O}_F)$ a maximal compact subgroup of $T(F)$. The usual argument [Serre (1967), 1.2] shows that $H^1(G_{F/K}, T(\mathcal{O}_F)) = 0$, and so there is an exact sequence

$$0 \longrightarrow T(\mathcal{O}_K) \longrightarrow T(K) \longrightarrow X_*(T)^{G_{F/K}} \longrightarrow 0.$$

Let χ be a generalized character of $T(K)$, and assume that it is trivial on $T(\mathcal{O}_K)$ (we then say that χ is *unramified*). Such a χ gives rise to a generalized character of $X_*(T)^{G_{F/K}}$, which we can extend to a generalized character $\tilde{\chi}$ of $X_*(T)$. Because $\mathrm{Hom}(X_*(T), \mathbb{C}^\times) = X^*(T) \otimes \mathbb{C}^\times = T^\vee(C) = {}^LT^o$, we can view $\tilde{\chi}$ as an element of this last group. Let r be a representation of ${}^LT$ (as a pro-algebraic group) on a finite-dimensional complex vector space V. We define the L-series

$$L(s,\chi,r) = \det(1 - r(\tilde{\chi},\sigma)N(\omega)^{-s}|V)$$

where σ is an element of $\mathrm{Gal}(K_s/K)$ restricting to the Frobenius automorphism on F.

Now let K be a global field, and let χ be a generalized character of $T(\mathbb{A}_K)/T(K)$. By restriction, we get generalized characters χ_v of $K_v^\times$ for each v. Let F be a finite Galois extension of K splitting T, and choose a finite set of primes S of K includin all archimedean primes, all primes that ramify in F, and all primes v for which χ_v is ramified. Define the automorphic L-series

$$L_S(s,\chi,r) = \prod_{v \notin S} L(s,\chi_v,r_v)$$

where r_v is the restriction of r to the local L-group.

Theorem 8.15. (a) *Let K be a local field, and let T be a torus over K splitting over an unramified extension of K. For all $\varphi \in \Phi(T)$ and all representations r of ${}^LT$,*

$$L(s, r\circ\varphi) = L(s, \chi, r),$$

where χ *is the character of* $T(K)$ *corresponding to* φ *in* (8.14).

(b) *Let* K *be a global field, and let* T *be a torus over* K. *Let* $\varphi \in \Phi(T)$, *and let* χ *be the corresponding element of* $\Pi(T)$. *Choose a set* S *of primes of* K *containing all archimedean primes, all primes that ramify in a splitting field for* T, *and all primes* v *such that* χ_v *is ramified. Then, for all representations* r *of* ${}^LT$,

$$L_S(s, r\circ\varphi) = L_S(s, \chi, r).$$

Proof: Only (a) has to be proved, and we leave this as an exercise to the reader.

The general conjecture

Let K be a global field, and let G be a reductive group over K. Then G is determined by certain linear data (a root datum), and the group ${}^LG^o$ is defined by the dual data. The full L-group ${}^LG$ is defined to be a semi-direct product ${}^LG^o \rtimes G_K$. The set $\Phi(G)$ of equivalence classes of admissable homomorphisms $W_K \longrightarrow {}^LG$ is defined analogously to the case of a torus, but the analogue of a generalized character of $T(\mathbb{A}_K)/T(K)$ is more difficult to define. Since $G(\mathbb{A}_K)$ is neither commutative or compact, its interesting representations are infinite dimensional. The correct notion is that of an irreducible automorphic representation of G. Langlands conjectures that it is possible to associate with each $\varphi \in \Phi(G)$ a (nonempty) set of irreducible automorphic representations of G. If π is associated with φ, then the L-series of φ and π are related as in (8.15b): let r be a complex representation of ${}^LG$; corresponding to almost all primes v of K, it is possible to define a local L-series for π and r; for each of these primes v, the local L-series for π and r is equal to the corresponding factor of the Artin-Hecke L-series of $r \circ \varphi$. See [Borel (1979)].

Notes: The results in this section were proved in [Langlands (1968)] and again in [Labesse (1984)]. While the above proof of (8.6) borrows from the proofs in both papers, it is somewhat simpler than each. For applications of the theorems, see [Kottwitz (1984)], [Labesse (1984)], and [Shelstad (1986)]

§9 Other applications

We explain a few of the other applications that have been made of the duality theorems in §2 and §4.

The Hasse principle for finite modules

Let K be a global field, and let M be a finite module over G_K. We say that the *Hasse principle holds for* M if the map

$$\beta^r(K,M)\colon H^1(K,M) \longrightarrow \prod_{\text{all } v} H^1(K_v,M)$$

is injective.

Example 9.1. (a) Let F/K be a finite Galois extension of degree n such that the greatest common divisor r of local degrees $[F_w:K_v]$ is strictly less than n. (For example, let $K = \mathbb{Q}$ and $F = \mathbb{Q}(\sqrt{13}, \sqrt{17})$; then $n = 4$ and the local degrees are all 1 or 2.) Consider the exact sequence

$$0 \longrightarrow M \longrightarrow (\mathbb{Z}/n\mathbb{Z})[G] \xrightarrow{\epsilon} \mathbb{Z}/n\mathbb{Z} \longrightarrow 0$$

in which $G = \mathrm{Gal}(F/K)$ and ϵ is the augmentation map $\sum n_\sigma \sigma \mapsto \sum n_\sigma$. From its cohomology sequence, we obtain an isomorphism

$\mathbb{Z}/n\mathbb{Z} \xrightarrow{\approx} H^1(G,M)$. Let c generate $H^1(G,M)$; then rc is a nonzero element of $H^1(K,M)$ mapping to zero in all the local cohomology groups. Therefore $Ш^1(K,M) \neq 0$, and the Hasse principle does not hold for M.

(b) From the duality theorem (4.10), we see that, for M as in (a), $Ш^2(K,M) \neq 0$. For a more explicit example (based on the failure of the original form of the Grunwald theorem) see [Serre (1964), III.4.7].

In view of these examples, the theorem below is of some interest. For a module M, we write K(M) for the subfield of K_s fixed by $\mathrm{Ker}(G_K \longrightarrow \mathrm{Aut}(M))$. Thus K(M) is the smallest splitting field of M. A finite group G is said to be *ℓ-solvable* if it has a composition series whose factors of order divisible by ℓ are cyclic.

Theorem 9.2. *Let M be a finite simple G_K-module such that $\ell M = 0$ for some prime ℓ, and assume that Gal(K(M)/K) is an ℓ-solvable group.*

(a) *If S is a set of primes of K with Dirichlet density one, then the mapping*

$$\beta_S^1(K,M)\colon H^1(K,M) \longrightarrow \prod_{v\in S} H^1(K_v,M)$$

is injective.

(b) *If $\ell \neq \mathrm{char}(K)$, then the mapping*

$$\beta^2(K,M)\colon H^2(K,M) \longrightarrow \prod_{\text{all } v} H^2(K_v,M)$$

is injective.

Note that β_S^r is not quite the same as the map in §4. However the next lemma shows that $\mathrm{Ker}\,\beta_S^1(K,M) = Ш_S^1(K,M)$. For any profinite group G and G-module M, define $H_*^1(K,M)$ to be the kernel of

$$H^1(G,M) \longrightarrow \prod_Z H^1(Z,M),$$

where the product is over all closed cyclic subgroups Z of G. When $G = G_K$, we also write $H^1_*(K,M)$ for $H^1_*(G,M)$. The next result explains the significance of this notion for the theorem. As always, for each prime v of K, we choose an extension w of v to K_s.

Lemma 9.3. *Let M be a finite G_K-module, and let $F \subset K_s$ be a finite Galois extension of K containing K(M). Let S be a set of primes of K with Dirichlet density one, and let*

$$\beta^1_S(F/K,M)\colon H^1(G_{F/K},M) \longrightarrow \prod_{v\in S} H^1(G_{F_w/K_v},M)$$

be the map induced by the restriction maps. Then there is a commutative diagram

$$\begin{array}{ccc} \mathrm{Ker}(\beta^1_S(F/K,M)) & \xrightarrow{\approx} & \mathrm{Ker}(\beta^1_S(K,M)) \\ \cap & & \cap \\ H^1_*(\mathrm{Gal}(F/K),M) & \xrightarrow{\approx} & H^1_*(G_K,M). \end{array}$$

The inclusions become equalities when all the decomposition groups $\mathrm{Gal}(F_w/K_v)$ are cyclic.

Proof: There is an exact commutative diagram

$$\begin{array}{ccccccc} 0 \longrightarrow & H^1(\mathrm{Gal}(F/K),M) & \xrightarrow{\mathrm{Inf}} & H^1(K,M) & \longrightarrow & H^1(F,M) \\ & \downarrow \beta^1_S(F/K,M) & & \downarrow \beta^1_S(K,M) & & \downarrow \beta^1_S(F,M) \\ 0 \longrightarrow & \prod_{v\in S} H^1(\mathrm{Gal}(F_w/K_v),M) & \xrightarrow{\mathrm{Inf}} & \prod_{v\in S} H^1(K_v,M) & \longrightarrow & \prod_{v\in S} H^1(F_w,M) \end{array}$$

The Chebotarev density theorem shows that $\beta^1_S(F,M)$ is injective. The inflation map therefore defines an isomorphism of the kernels of the first two vertical maps, which gives us the isomorphism on the top row. The isomorphism on the bottom row can be proved by a similar argument. The Chebotarev density theorem shows that all cyclic subgroups of Gal(F/K) are of the form $\mathrm{Gal}(F_w/K_v)$ for some primes $w|v$ with $v \in S$, and so clearly $H^1_*(\mathrm{Gal}(F/K),M) \subset \mathrm{Ker}(\beta^1(F/K,M))$. The reverse inclusion holds if all the decomposition groups are cyclic.

We say that the *Hasse principle holds for a finite group* G (*and the prime* ℓ) if $H^1_*(G,M) = 0$ for all finite simple G-modules M (with $\ell M = 0$). Note that the Hasse principle obviously holds for G if all of its Sylow subgroups are cyclic.

Lemma 9.4. *Let*

$$1 \longrightarrow G' \longrightarrow G \longrightarrow G'' \longrightarrow 1$$

be an exact sequence of finite groups. If the Hasse principle holds for G' *and* G'' *relative to the prime* ℓ, *then it holds for* G *and* ℓ; *conversely, if the Hasse principal holds for* G *and to* ℓ, *then it holds for* G'' *and* ℓ.

Proof: Let M be a simple G-module such that $\ell M = 0$. As G' is normal in G, $M^{G'}$ is stable under G, and so either $M^{G'} = 0$ or $M^{G'} = M$. In the first case, there is a commutative diagram

$$\begin{array}{ccc} H^1(G,M) & \xrightarrow{\text{Res}} & H^1(G',M) \\ \downarrow & & \downarrow \\ \prod H^1(Z,M) & \xrightarrow{\text{Res}} & \prod H^1(Z \cap G',M) \end{array}$$

in which the upper restriction map has kernel $H^1(G'',M^{G'}) = 0$. When regarded as a G'-module, M is semisimple because, for any nonzero simple G'-submodule N of M, M is a sum of the simple modules gN, $g \in G$. Therefore if the Hasse principle holds for G' and ℓ, then the right hand vertical arrow is an injection. Consequently the first vertical arrow is also an injection, and this shows that the Hasse principle holds for G and ℓ.

In the case that $M^{G'} = M$, we consider the diagram

$$\begin{array}{ccccccc} 0 \longrightarrow & H^1(G'',M) & \longrightarrow & H^1(G,M) & \longrightarrow & H^1(G',M) \\ & \downarrow & & \downarrow & & \downarrow \\ 0 \longrightarrow & \prod H^1(ZG'/G',M) & \longrightarrow & \prod H^1(Z,M) & \longrightarrow & \prod H^1(Z\wedge G',M). \end{array}$$

The right hand vertical arrow is an injection because G' acts trivially on M and the groups $Z\wedge G'$ generate G'. The left hand vertical arrow has kernel $H^1_*(G'',M)$ because the groups ZG'/G' run through all cyclic subgroups of G'', and so we see that if the Hasse principle holds for G'' and ℓ then it holds also for G and ℓ.

We use the same diagram to prove the converse part of the lemma. A simple G''-module M can be regarded as a simple G-module such that $M^{G'} = M$. Therefore the diagram shows that $H^1_*(G'',M) = 0$ if $H^1_*(G,M) = 0$.

Proposition 9.5. (a) *The Hasse principle holds for a finite group (and the prime ℓ) when it holds for all the composition factors of the group (and ℓ).*

(b) *If G is ℓ-solvable, then the Hasse principle holds for G and ℓ.*

(c) *A solvable group satisfies the Hasse principle.*

Proof: Part (a) follows by induction from the lemma. Part (c) follows from (a) and the obvious fact that the Hasse principle holds for a cyclic group. Part (b) follows from (a) and (c) and the additional fact that the higher cohomology groups of a module killed by ℓ relative to a group of order prime to ℓ are all zero.

We now prove Theorem 9.2. Lemma 9.3 shows that

$$\operatorname{Ker} \beta^1_S(K,M) = \operatorname{Ker} \beta^1_S(K(M)/K,M) \subset H^1_*(G_{K(M)/K},M),$$

and (9.5b) shows that this last group is zero, which proves part (a) of the theorem. From (4.10) we know that $\mathrm{Ker}(\beta^2(K,M))$ is dual to $\mathrm{Ker}(\beta^1(K,M^D))$. Clearly M^D is simple if M is, and the extension $K(M^D)$ is ℓ-solvable if K(M) is because it is contained in $K(M)(\mu_\ell)$. Therefore part (b) of the theorem follows from part (a).

Corollary 9.6. *If* $M \approx \mathbb{Z}/\ell\mathbb{Z} \times \mathbb{Z}/\ell\mathbb{Z}$ *(as an abelian group) for some prime* ℓ *not equal to the characteristic of* K, *then* $Ш^2(K,M) = 0$.

Proof: If M is simple (or semisimple) as a G_K-module, this follows directly from the theorem. The remaining case can be proved directly.

Notes: The groups $H^r_*(G,M)$ were introduced by Tate (see [Serre (1964/71)]). Theorem 9.2 and its proof are taken from [Jannsen (1982)]. For an elementary proof of (9.6), see [Cassels (1962), §5].

The Hasse principle for algebraic groups

In this subsection, G will be a connected (not necessarily commutative) linear algebraic group over a number field K. We say that G satisfies the *Hasse principle* if

$$H^1(K,G) \longrightarrow \prod_{\text{all } v} H^1(K_v,G)$$

is injective. It is known ([Kneser (1966), (1969)] and [Harder (1965/66)]) that if G is semisimple and simply-connected without factors of type E_8, then $H^1(K_v,G) = 0$ for all nonarchimedean v, and $H^1(K,G) \xrightarrow{\approx} \prod_{v \text{ real}} H^1(K_v,G)$.

Theorem 9.7. *Let* G *be a simply connected semisimple group, and let* $\varphi\colon G \longrightarrow G'$ *be a separable isogeny. Let* M *be the kernel of* $\varphi(K_s)\colon G(K_s) \longrightarrow G'(K_s)$, *and assume that* $Ш^2(K,M) = 0$. *If the Hasse* *principle holds for* G, *then it also holds for* G'.

Lemma 9.8. *Let* M *be a finite module* G_K *whose order is not divisible by* char(K), *and assume* S *omits only finitely many primes of* K.

(a) *The cokernel of* $H^1(K,M) \longrightarrow \prod_{v\notin S} H^1(K_v,M)$ *is canonically isomorphic to the dual of* $(\operatorname{Ker}\beta^1_S(K,M^D))/(\operatorname{Ker}\beta^1(K,M^D))$.

(b) *If each* $v \notin S$ *has a cyclic decomposition group in* K(M), *then* $H^1(G_K,M) \longrightarrow \bigoplus_{v\notin S} H^1(K_v,M)$ *is surjective. In particular* $H^1(K,M) \longrightarrow \bigoplus_{v \text{ real}} H^1(K_v,M)$ *is surjective.*

Proof: (a) From (4.10) we know there is an exact sequence

$$H^1(K,M^D) \xrightarrow{\beta^1} P^1_S(K,M^D) \times \prod_{v\notin S} H^1(K_v,M^D) \xrightarrow{\gamma^1} H^1(K,M)^*.$$

Therefore the kernel-cokernel sequence (0.24) of the pair of maps

$$H^1(K,M^D) \longrightarrow P^1_S(K,M^D) \times \prod_{v\notin S} H^1(K_v,M^D) \longrightarrow P^1_S(K,M^D)$$

is an exact sequence

$$0 \longrightarrow \operatorname{Ker}\beta^1(K,M^D) \longrightarrow \operatorname{Ker}\beta^1_S(K,M^D) \longrightarrow \prod_{v\notin S} H^1(K_v,M^D) \xrightarrow{\gamma} H^1(K,M)^*.$$

The exactness at the third term says that $\operatorname{Ker}(\beta^1_S)/\operatorname{Ker}(\beta^1) = \operatorname{Ker}(\gamma)$, but this last group is the dual of the cokernel of $H^1(K,M) \longrightarrow \prod_{v\notin S} H^1(K_v,M)$.

(b) Let F be a finite Galois extension of K containing $K(M^D)$. According to the Chebotarev density theorem, for each prime $v \notin S$ having a cyclic decomposition group in Gal(F/K), there is a prime $v' \in S$ having the same decomposition group. Therefore if an element c of $H^1(G_{F/K},M^D)$ maps to zero in $H^1(G_{F_w/K_v},M^D)$ for all v in S, then it maps to zero for all v. Hence $\operatorname{Ker}\beta^1_S(F/K,M) = \operatorname{Ker}\beta^1(F/K,M)$, and Lemma 9.3 shows that this implies that $\operatorname{Ker}\beta^1_S(K,M) = \operatorname{Ker}\beta^1(K,M)$. Now (a) implies (b).

We now prove the theorem. Consider the diagram of pointed sets:

$$\begin{array}{ccccccccc}
H^1(K,M) & \longrightarrow & H^1(K,G) & \longrightarrow & H^1(K,G') & \longrightarrow & H^2(K,M) & \longrightarrow \\
\downarrow & & \downarrow \text{inj} & & \downarrow & & \downarrow \text{inj} & \\
\prod_{\text{all } v} H^1(K_v,M) & \longrightarrow & \prod_{\text{all } v} H^1(K_v,G) & \longrightarrow & \prod_{\text{all } v} H^1(K_v,G') & \longrightarrow & \prod_{\text{all } v} H^2(K_v,M) & \longrightarrow .
\end{array}$$

(See [Serre (1961), VII, Annexe].) If $c \in H^1(K,G')$ maps to zero in $H^1(K_v,G')$ for all v, then it lifts to an element $b \in H^1(K,G)$. As we observed above, $H^1(K_v,G) = 0$ for all nonarchimedean v. For each archimedean prime v, the image b_v of b in $H^1(K_v,G)$ lifts to an element a_v of $H^1(K_v,M)$. According to (9.8), there is an element $a \in H^1(K,M)$ mapping to a_v for all archimedean v. Now $b - a'$, where a' is the image of a in $H^1(K,G)$, maps to c in $H^1(K,G')$ and to 0 in $H^1(K_v,G)$ for all v. The last condition shows that $b - a'$ (hence c) is zero. This shows that the kernel of $H^1(K,G') \longrightarrow \prod H^1(K_v,G)$ is zero, and a standard twisting argument (cf. [Kneser (1969), I.1.4]) now allows one to show that the map is injective.

Corollary 9.9. *Let G be a semisimple algebraic group over K without factors of type* E_8. *Then the Hasse principle holds for G under each of the following the hypotheses:*

(a) *G has trivial centre;*

(b) *G is almost absolutely simple;*

(c) *G is split by a finite Galois extension F of K such that all Sylow subgroups of Gal(F/K) are cyclic;*

(d) *G is an inner form of a group satisfying* (a), (b), *or* (c).

Proof: An group with trivial centre is a product of groups of the form $R_{F/K}G$ with G an absolutely simple group over F, and so (a) follows from (b). An absolutely almost simple group is an inner form of a quasi-split almost simple group, and such a group is split by a extension whose Galois group is a subgroup of the group of automorphisms of its Dynkin diagram. But this automorphism group is either trivial or is $\mathbb{Z}/2\mathbb{Z}$ or S_3. Therefore (b) follows from (c) and (d). Let G be split by an extension F as in (c), and let M be the kernel of $\tilde{G}(K_s) \longrightarrow G(K_s)$ where $\tilde{G}$ is the universal covering group of G. Then M is a sum of $Gal(F_s/F)$ modules of the form μ_m, and so $Gal(F_s/F)$ acts trivially on M^D. Therefore

$$Ш^1(K,M^D) = \text{Ker } \beta^1(F/K,M^D) \subset H^1_*(Gal(F/K),M^D) = 0,$$

and so $Ш^2(K,M) = 0$. Finally (d) is obvious from the fact that the $Gal(K_s/K)$-module M is unchanged when G is replaced by an inner form.

Notes: Theorem 9.7 is proved in [Harder (1967/68), Theorem 4.3.2] and in [Kneser (1969), p 77-78]. Part (a) of Corollary 9.9 is proved in [Langlands (1983), VII.6]. All of the results in this subsection are contained in [Sansuc (1981)].

Forms of an algebraic group

The next result shows that (under certain conditions) a family of local forms of an algebraic group arises from a global form.

Theorem 9.10. *Let K be an algebraic number field, S a finite set of primes of K, and G an absolutely almost simple algebraic group over K that is either simply connected or has trivial centre. Then the canonical map*

$$H^1(K,Aut(G)) \longrightarrow \prod_{v\in S} H^1(K_v,Aut(G))$$

is surjective.

Proof: (Sketch) Let $\tilde{G}$ be the universal covering group of G, and let $M = Ker(\tilde{G}(K_s) \longrightarrow G(K_s))$. Consider the diagram

$$\begin{array}{ccccccc}
\cdots \longrightarrow & H^1(K,\tilde{G}) & \longrightarrow & H^1(K,G) & \longrightarrow & H^2(K,M) & \longrightarrow \cdots \\
 & \downarrow & & \downarrow & & \downarrow & \\
\cdots \longrightarrow & \bigoplus_{v\in S} H^1(K_v,\tilde{G}) & \longrightarrow & \bigoplus_{v\in S} H^1(K_v,G) & \longrightarrow & \bigoplus_{v\in S} H^2(K_v,M) & \longrightarrow \cdots .
\end{array}$$

Corollary 4.16 shows that the final vertical map is surjective. We have already noted that the first vertical map is surjective when G has no factors of type E_8, but in fact this condition is unnecessary. Next one shows that the map $H^1(K,G) \longrightarrow H^2(K,M)$ is surjective, and a diagram chase then shows that $H^1(K,G) \longrightarrow \bigoplus_{v\in S} H^1(K_v,G)$ is surjective. One shows that it suffices to prove the theorem for a split G, in which case Aut(G) is the semi-direct product $G \rtimes \mathrm{Aut}(D)$ of G with the automorphism group of the Dynkin diagram of G. The proof of the theorem then is completed by showing that $H^1(K,\mathrm{Aut}(D)) \longrightarrow \oplus H^1(K_v,\mathrm{Aut}(D))$ is surjective.

For the details, see [Borel and Harder (1978)], where the theorem is used to prove the existence of discrete cocompact subgroups in the groups of rational points of reductive groups over nonarchimedean local fields of characteristic zero.

The Tamagawa numbers of tori

We refer the reader to [Weil (1961)] for the definition of the Tamagawa number $\tau(G)$ of a linear algebraic group G over a global field.

Theorem 9.11. *For any torus T over a global field K*

$$\tau(G) = \frac{[H^1(K,T^\vee)]}{[Ш^1(K,T)]}$$

where $Ш^1(K,T)$ *is the kernel of* $H^1(K,T) \longrightarrow \prod_{\text{all } v} H^1(K_v,T)$ *and* $T^\vee$ *is the dual torus defined by the relation* $X^*(T) = X_*(T^\vee)$.

Proof: Let $\varphi(T) = \tau(T)\,\dfrac{[Ш^1(K,T)]}{[H^1(K,T^\vee)]}$. The proof has three main steps:

(i) φ is an additive function on the category of tori over K;

(ii) $\varphi(\mathbb{G}_m) = 1$;

(iii) for any finite separable extension F of K, $\varphi(\mathrm{Res}_{F/K}T) = \varphi(T)$.

Once these fact have been established the proof is completed as follows. The functor $T \mapsto X_*(T)$ defines an equivalence between the category of tori over K and the category $\mathbf{Rep}_{\mathbb{Z}}(G_K)$ of continuous representations of G_K on free $\mathbb{Z}$-modules of finite rank, and so we can regard φ as being defined on the latter category. Then (i) says that φ induces a homomorphism $K_0(\mathbf{Rep}_{\mathbb{Z}}(G_K)) \longrightarrow \mathbb{Q}_{>0}$. A theorem [Swan (1960)] shows that $[X] - [X']$ is a torsion element of $K_0(\mathbf{Rep}_{\mathbb{Z}}(G_K))$ if $X\otimes\mathbb{Q} = X'\otimes\mathbb{Q}$, and, as $\mathbb{Q}_{>0}$ is torsion-free, φ is zero on torsion elements of $K_0(\mathbf{Rep}_{\mathbb{Z}}(G_K))$. Therefore φ takes the equal values on isogenous tori. Artin's theorem on characters [Serre (1967b), 9.2 implies that for any torus T, there exists and integer m and finite separable extensions F_i and E_j of K such that $T^m \times \prod \mathrm{Res}_{F_i/K}\mathbb{G}_m$ is isogenous to $\prod \mathrm{Res}_{E_j/K}\mathbb{G}_m$. Now (ii) and (iii) show that $\varphi(T)^m = 1$, and therefore $\varphi(T) = 1$.

Statements (ii) and (iii) are easily proved (they follow almost directly from the definitions), and so the main point of the proof is (i). This is proved by an argument, not dissimilar to that used to prove Theorem 7.3, involving the duality theorems. See [Ono (1961), (1963)], and also [Oesterlé (1984)], which corrects errors in Ono's treatment of the function field case.

The central embedding problem

Let S be a finite set of primes of a global field K, and let G_S

be the Galois group over K of the maximal extension of K unramified outside S. Let

$$1 \longrightarrow M \longrightarrow E \longrightarrow G \longrightarrow 1$$

be an extension of finite groups with M in the centre of E, and let $\varphi: G_S \twoheadrightarrow G$ be a surjective homorphism. The *embedding problem* for E and φ is the problem of finding a surjective homomorphism $\Phi: G_S \twoheadrightarrow E$ lifting φ. Concretely this means the following: the homomorphism φ realizes G as the Galois group of an extension F of K that is unramified outside S, and the embedding problem asks for a field F' that is Galois over K with Galois group E, is also unramified outside S, contains F, and is such that the map $E \twoheadrightarrow G$ induced by the inclusion of F into F' is that in the sequence. For each v in S, let G_v be the image of $\mathrm{Gal}(K_{v,s}/K)$ in G, and let E_v be the inverse image of G_v in E. Then the (local) embedding problem asks for a homomorphism $\mathrm{Gal}(K_{v,s}/K_v) \twoheadrightarrow E_v$ lifting $\mathrm{Gal}(K_{v,s}/K_v) \twoheadrightarrow G_v$.

Let Δ be the class of the extension in $H^2(G,M)$. If the embedding problem has a solution, then Δ clearly is sent to zero by the map $H^2(G,M) \longrightarrow H^2(G_S,M)$ defined by φ. The converse is also true if $\Delta \neq 0$ and M is a simple G-module. Thus Theorem 9.2 has the following consequence.

Proposition 9.12. *Let*

$$1 \longrightarrow M \longrightarrow E \longrightarrow G \longrightarrow 1$$

be a nonsplit central extension of finite groups, and let $\varphi: G_K \twoheadrightarrow G$ *be a surjective homomorphism. If* M *is a simple* G*-module and* G *is solvable, then the embedding problem for* E *and* φ *has a solution if and only if the corresponding local problem has a solution for all* v.

Proof: The necessity of the condition is obvious. For the sufficiency, note that when the local problem has a solution, the image of Λ in $H^2(K_v,M)$ is zero for all v. According to (9.2), this implies that Λ is zero.

Unfortunately, the proposition does not lead to a proof of Shafarevich's theorem [Shafarevich (1954)]: for any number field K and finite solvable group G, there exists an extension F of K with Galois group G.

For other applications of the duality theorems to the embedding problem, see for example [Haberland (1978)], [Neumann (1977)], and [Klingen (1983)].

Abelian varieties defined over their fields of moduli

Let A be a polarized abelian variety defined over $\bar{\mathbb{Q}}$. The obstruction to A having a model over its field of moduli is a class Λ in $H^2(G,\mathrm{Aut}(A))$. In the case that Aut(A) is abelian, the duality theorems can sometimes be helpful in studying this element.

Abelian varieties and $\mathbb{Z}_p$-extensions

The duality theorems (and their generalizations to flat cohomology) have been used in the study of the behaviour of the points on an abelian variety as one progresses up a $\mathbb{Z}_p$-tower of number fields. See for example [Mazur (1972)], [Manin (1971)], [Harris (1979)], and [Rubin (1985)]

Appendix A: Class field theory for function fields.

Most of the accounts of class field theory either omit the case of a function field or make it appear harder than the number field case. In fact it is easier (at least for those knowing a little algebraic geometry). In this appendix we derive the main results of class field theory except for the existence theorem for a function field over a finite field. As a preliminary, we derive class field theory for a Henselian local field with quasi-finite residue field. We also investigate to what extent the global results hold for a function field over a quasi-finite field.

Local class field theory

A field k is *quasi-finite* if it is perfect and if the Galois group $G(k_s/k)$ is isomorphic to the profinite completion $\hat{\mathbb{Z}}$ of $\mathbb{Z}$. The main examples of quasi-finite fields are the finite fields and the power series fields $k_0((t))$ with k_0 an algebraically closed field of characteristic zero, but there are others. For example, any algebraic extension k' of a quasi-finite field k whose degree [k':k] is divisible by only a finite power of each prime number is quasi-finite. Also, given an algebraically closed field K, one can always find a quasi-finite field k having K as its algebraic closure [Serre (1962), XIII.2, Ex 3].

Whenever a quasi-finite field k is given, we shall always assume that there is also given as part of its structure a generator σ_k of $\mathrm{Gal}(k_s/k)$, or equivalently, a fixed isomorphism $\varphi_k\colon \Xi(G_k) \longrightarrow \mathbb{Q}/\mathbb{Z}$ where $\Xi(G_k)$ is the character group $\mathrm{Hom}_{cts}(G_k,\mathbb{Q}/\mathbb{Z})$ of G_k. The relation between σ and φ_k is that $\varphi_k(\chi) = \chi(\sigma)$ for all $\chi \in \Xi(G_k)$. A finite extension ℓ of a quasi-finite field k is again quasi-finite with generator $\sigma_\ell = \sigma_k^{[\ell:k]}$. When k is finite, we always take σ_k to be the Frobenius automorphism $a \mapsto a^q$, $q = [k]$.

Note that the Brauer group of a quasi-finite field is zero,

because G_k has cohomological dimension one and k_s^* is divisible.

Let R be a discrete valuation ring with residue field $k = R/m$. Write $\bar{f}$ for the reduction of an element of R or R[X] modulo m. We say that R is *Henselian* if it satisfies the conclusion of Hensel's lemma: whenever f is a monic polynomial with coefficients in R such that $\bar{f}$ factors as $\bar{f} = g_0 h_0$ with g_0 and h_0 monic and relatively prime, then f itself factors as $f = gh$ with g and h monic and such that $\bar{g} = g_0$ and $\bar{h} = h_0$. Hensel's lemma says that complete discrete valuation rings are Henselian, but not all Henselian rings are complete. For example, let v be a prime in a global field K, and let $\mathcal{O}_v$ be the ring of elements of K that are integral at v. Choose an extension w of v to K_s, let K_{dec} be the decomposition field of w in K_s, and let $\mathcal{O}_v^h$ be the ring elements of K_{dec} that are integral with respect to w. Alternatively, choose an embedding of K_s into $K_{v,s}$, and let $\mathcal{O}_v^h = K_s \cap \hat{\mathcal{O}}_v$. Then $\mathcal{O}_v^h$ is a Henselian local ring, called the *Henselization* of $\mathcal{O}_v$. See [Milne (1980), I.4].

Now let R be a Henselian discrete valuation ring with quasi-finite residue field k, and write K for its field of fractions. Many results usually stated only for complete discrete valuation rings hold in fact for Henselian discrete valution rings (often the proof uses only that the ring satisfies Hensel's lemma). For example, the valuation v on K has a unique extension to a valuation (which we shall also write v) on K_s. As usual we write K_{un} for the maximal unramified subextension of K_s over K, and R_{un} for the integral closure of R in K_{un}.

Proposition A.1. *There is a canonical isomorphism*

$\mathrm{inv}_K\colon \mathrm{Br}(K) \longrightarrow \mathbb{Q}/\mathbb{Z}$.

Proof: We first show that $\mathrm{Br}(K_{un}) = 0$. Let D be a skew field of degree n^2 over K_{un}. Because R_{un} is also a Henselian discrete valuation ring, the valuation on K_{un} has a unique extension to each commutative subfield of D, and therefore it has a unique extension to D. The usual argument in the commutative case shows that, for this extension, $n^2 = ef$. Let α in D have value 1/e; then $K_{un}[\alpha]$, being a commutative subfield of D, has degree at most n, and so $e \lesssim n$. On the other hand f = 1 because the residue field of R is algbraically closed, and it follows that n = 1.

The exact sequence

$$0 \longrightarrow H^2(\mathrm{Gal}(K_{un}/K), K_{un}^\times) \longrightarrow \mathrm{Br}(K) \longrightarrow \mathrm{Br}(K_{un})$$

shows that $\mathrm{Br}(K) = H^2(\mathrm{Gal}(K_{un}/K), K_{un}^\times)$.

Lemma A.2. *The map* $H^2(\mathrm{Gal}(K_{un}/K), K_{un}^\times) \xrightarrow{\mathrm{ord}} H^2(\mathrm{Gal}(K_{un}/K), \mathbb{Z})$ *is an isomorphism.*

Proof: As

$$0 \longrightarrow R_{un}^\times \longrightarrow K_{un}^\times \xrightarrow{\mathrm{ord}} \mathbb{Z}_{un}^\times \longrightarrow 0$$

is split as a sequence of $\mathrm{Gal}(K_{un}/K)$-modules, the map in question is surjective. Let c lie in its kernel, and let γ be a cocycle representing c. Associated with c there is a central simple algebra B over K [Herstein (1968), 4.4], and if γ is chosen to take values in $R^\times$, then the same construction that gives B gives an Azumaya algebra B_0 over R that is an order in B. The reduction $B_0 \otimes_R k$ of B_0 is a central simple algebra over k, and therefore is isomorphic to a matrix algebra. An elementary argument [Milne (1980), IV.1.6] shows then that B_0 is also isomorphic to a matrix algebra, and this implies that c = 0.

We define inv_K to be the unique map making

$$\begin{array}{ccc} H^2(\mathrm{Gal}(K_{un}/K),\mathbb{Z}) & = \mathrm{Br}(K) \xrightarrow{\mathrm{inv}_K} & \mathbb{Q}/\mathbb{Z} \\ \approx \uparrow d & & \approx \uparrow \varphi_k \\ H^1(\mathrm{Gal}(K_{un}/K),\mathbb{Q}/\mathbb{Z}) & = & \Xi(G_k) \end{array}$$

commute. It is an isomorphism. If F is a finite separable extension of K, then the integral closure R_F of R in F is again a Henselian discrete valuation ring with quasi-finite residue field, and one checks easily from the definitions that $\mathrm{inv}_F(\mathrm{Res}(a)) = [F:K]\ \mathrm{inv}_K(a)$ for all $a \in \mathrm{Br}(K)$. Therefore $(G_K,K_s^\times)$ is a class formation in the sense of §1.

We identify the cup-product pairing

$$H^0(G_K,K_s^\times) \times H^2(G_K,\mathbb{Z}) \longrightarrow H^2(G_K,K_s^\times)$$

with a pairing

$$\langle\ ,\ \rangle\colon K^\times \times \Xi(G_K) \longrightarrow \mathrm{Br}(K).$$

Theorem A.3. (Local reciprocity law). *There is a continuous homomorphism* $(-,K)\colon K^\times \longrightarrow \mathrm{Gal}(K_{ab}/K)$ *such that*

(a) *for each finite abelian extension* $F \subset K_s$ *of* K, $(-,K)$ *induces an isomorphism*

$$(-,F/K)\colon K^\times/N_{F/K}F^\times \longrightarrow \mathrm{Gal}(L/K);$$

(b) *for any* $\chi \in \Xi(G)$ *and* $a \in K^\times$, $\chi(a,K) = \mathrm{inv}_K\langle a,\chi\rangle$.

Proof: As is explained in §1, this theorem is a formal consequence of the fact that $(G_K,K_s^\times)$ is a class formation.

Corollary A.4. *Let* F_1 *and* F_2 *be extensions of* K *such that* $F_1 \cap F_2 = K$, *and let* $F = F_1F_2$. *If all three fields are finite abelian extensions of* K, *then* $NF^\times = NF_1^\times \cap NF_2^\times$ *and* $(NF_1^\times)(NF_2^\times) = K^\times$.

Proof: According the theorem, $a \in NF^{\times}$ if and only if (a,K) acts trivially on F; this is equivalent to (a,K) acting trivially on F_1 and F_2, or to it lying in $NF_1^{\times} \cap NF_2^{\times}$. The second equality can be proved similarly.

Remark A.5. When F/K is unramified, σ lifts to a unique (Frobenius) element $\tilde{\sigma}$ in $\mathrm{Gal}(F/K)$, and $(-,K)\colon K^{\times} \longrightarrow \mathrm{Gal}(F/K)$ sends an element a of $K^{\times}$ to $\tilde{\sigma}^{\mathrm{ord}(a)}$. In particular, $(a,F/K) = 1$ if $a \in R^{\times}$. When F/K is ramified, the description of $(a,F/K)$ is much more difficult (see [Serre (1967a), 3.4]).

We say that a subgroup N of $K^{\times}$ is a *norm group* if there exists a finite abelian extension F of K such that $N = N_{F/K}F^{\times}$. (The name is justified by the following result: if F/K is any finite separable extension of K, then $N_{F/K}F^{\times} = N_{L/K}L^{\times}$ where L is the maximal abelian subextension of F; see [Serre (1962), XI.4]).

Remark A.6. The reciprocity map defines an isomorphism $\varprojlim K^{\times}/N \longrightarrow G^{ab}$ (inverse limit over the norm groups in $K^{\times}$), and so to fully understand G^{ab} it is necessary to determine the norm groups. This is what the existence theorem does.

Case 1: K is complete and k is finite. This is the classical case. Here the norm groups of $K^{\times}$ are precisely the open subgroups of finite index. Every subgroup of finite index prime to $\mathrm{char}(K)$ is open. The image of the reciprocity map is the subgroup of G^{ab} of elements that act as an integral power of the Frobenius automorphism on k_s. The reciprocity map is injective, and it defines an isomorphism of the topological group $R^{\times}$ onto the inertia subgroup of G^{ab}. See [Serre (1962), XIV.6].

Case 2: K is Henselian with finite residue field. We assume that R is *excellent*. This is equivalent to the completion $\hat{K}$ of K being separable over K. From the description of it given above, it is clear that the Henselization of the local ring at a prime in a global field is excellent. Under this assumption:

(i) every finite separable extension of $\hat{K}$ is of the form $\hat{F}$ for a finite separable extension F of K with $[F:K] = [\hat{F}:\hat{K}]$;

(ii) K is algebraically closed in $\hat{K}$, and hence $\hat{K}$ is linearly disjoint from K_a over K.

To prove (i), write $\hat{F} = \hat{K}[\alpha]$, and let $F = K[\beta]$ with β a root of a polynomial in K[X] that is close to the minimal polynomial of α over K (cf. [Lang (1970), II.2]). To prove (ii), note that if K is not algebraically closed in $\hat{K}$, then there is an element α of $\hat{K}$ that is integral over R but which does not lie in R. Let f(X) be the minimal polynomial of α over K. As α is integral over $\hat{R}$, it lies in $\hat{R}$, and so f has a root in $\hat{R}$. An approximation theorem [Greenberg (1966)] now says that f has a root in R, but f was chosen to be irreducible over K. Thus K is algebraically closed in $\hat{K}$, and combined with the separability of $\hat{K}$ over K, this implies that $\hat{K}$ is linearly disjoint from K_a [Lang (1958), III.1, Thm 2].

On combining these two assertions, we find that $F \mapsto \hat{F}$ defines a degree-preserving bijection from the set of finite separable extensions F of K to the set of similar extensions of $\hat{K}$. Moreover, $NF^{\times} = N\hat{F}^{\times} \cap K^{\times}$ for each F because $NF^{\times}$ is dense in $N\hat{F}^{\times}$ and Greenberg's theorem implies that $NF^{\times}$ is open in $K^{\times}$. It follows that the norm groups of $K^{\times}$, are again precisely the open subgroups of finite index.

Case 3: K is complete with quasi-finite residue field. In this case every subgroup of K of finite index prime to char(k) is a norm group, but not every open subgroup of index a power of the characteristic of k is. In [Whaples (1952-54)], various characterizations of the norm groups are given using the proalgebraic structure on $K^{\times}$.

Case 4: K is Henselian with quasi-finite residue field. We leave this case to the reader to investigate.

Global class field theory

Let X be a complete smooth curve over a quasi-finite field (k,σ), and let $K = k(X)$. The set of closed points of X will be denoted by X^0 (thus X^0 omits only the generic point of X). To each point v of X^0, there corresponds a valuation (also written v) of K, and we write K_v for the completion of K with respect to v and R_v for the ring of integers in K_v. The residue field k(v) is a quasi-finite field of degree deg(v) over k with $\sigma^{\deg(v)}$ as the chosen generator of $\mathrm{Gal}(k(v)_s/k(v))$. Write a_v for the image of an element a of Br(K) in $\mathrm{Br}(K_v)$, and define $\mathrm{inv}_K\colon \mathrm{Br}(K) \longrightarrow \mathbb{Q}/\mathbb{Z}$ to be $a \mapsto \sum \mathrm{inv}_v(a_v)$ where inv_v is inv_{K_v}. Let $\bar{X} \overset{\mathrm{df}}{=} X \otimes_k k_s$ be X regarded as curve over k_s, and let $\bar{K} = k_s(\bar{X})$ be its function field. We write Jac_X be the Jacobian variety of X.

Theorem A.7. *There is an exact sequence*

$$0 \longrightarrow H^1(G_k, \mathrm{Jac}_X(k_s)) \longrightarrow \mathrm{Br}(K) \longrightarrow \bigoplus_{v \in X^0} \mathrm{Br}(K_v) \xrightarrow{\mathrm{inv}_K} \mathbb{Q}/\mathbb{Z} \longrightarrow 0.$$

Proof: We use the exact sequence of G_k-modules

$$0 \longrightarrow k_s^{\times} \longrightarrow \bar{K}^{\times} \longrightarrow \mathrm{Div}(\bar{X}) \longrightarrow \mathrm{Pic}(\bar{X}) \longrightarrow 0,$$

where $\mathrm{Div}(\bar{X}) = \bigoplus_{v \in X^0} \mathbb{Z}$ is the group of (Weil) divisors on $\bar{X}$. From the cohomology sequence of its truncation

$$0 \longrightarrow k_s^{\times} \longrightarrow \bar{K}^{\times} \longrightarrow Q \longrightarrow 0$$

we obtain an isomorphism $H^2(G_k, \bar{K}^{\times}) \xrightarrow{\approx} H^2(G_k, Q)$. Note that

$H^2(G_k,\bar{K}^\times) = \mathrm{Ker}(\mathrm{Br}(K) \longrightarrow \mathrm{Br}(\bar{K}))$, and recall that Tsen's theorem [Shatz (1972), Theorem 24] states that $\mathrm{Br}(\bar{K}) = 0$, and so $H^2(G_k,Q) = H^2(G_k,\bar{K}^\times) = \mathrm{Br}(K)$.

The cohomology sequence of the remaining segment of the sequence

$$0 \longrightarrow Q \longrightarrow \mathrm{Div}(\bar{X}) \longrightarrow \mathrm{Pic}(\bar{X}) \longrightarrow 0,$$

is

$$H^1(G_k,\mathrm{Div}(\bar{X})) \longrightarrow H^1(G_k,\mathrm{Pic}(\bar{X})) \longrightarrow H^2(G_k,Q)$$
$$\longrightarrow H^2(G_k,\mathrm{Div}(\bar{X})) \longrightarrow H^2(G_k,\mathrm{Pic}(\bar{X})) \rightarrow 0.$$

But $H^r(G_k,\mathrm{Div}(\bar{X})) = \bigoplus_{v\in X^0} H^r(G_k,D_v)$, where $D_v \overset{\mathrm{df}}{=} \bigoplus_{w\mapsto v} \mathbb{Z}$ is the G_k-module induced by the trivial $G_{k(v)}$-module $\mathbb{Z}$, and so $H^r(G_k,\mathrm{Div}(\bar{X})) = \bigoplus_{v\in X^0} H^r(\mathrm{Gal}(k_s/k(v)),\mathbb{Z})$. In particular, $H^1(G_k,\mathrm{Div}(\bar{X})) = 0$ and $H^2(G_k,\mathrm{Div}(\bar{X})) = \bigoplus_{v\in X^0} \Xi(G_{k(v)})$.

Almost by definition of Jac_X, there is an exact sequence

$$0 \longrightarrow \mathrm{Jac}_X(k_s) \longrightarrow \mathrm{Pic}(\bar{X}) \longrightarrow \mathbb{Z} \longrightarrow 0.$$

As $\mathrm{Jac}_X(k_s)$ is divisible [Milne (1986b), 8.2] and k has cohomological dimension one, $H^2(G_k,\mathrm{Jac}_X(k_s)) = 0$, and so $H^2(G_k,\mathrm{Pic}(\bar{X})) = H^2(G_k,\mathbb{Z}) = \Xi(G_k)$. These results allow us to identify the next sequence with the required one:

$$\begin{array}{ccccccccc}
\cdot \rightarrow H^1(G_k,\mathrm{Pic}(\bar{X})) & \longrightarrow & H^2(G_k,Q) & \longrightarrow & H^2(G_k,\mathrm{Div}(\bar{X})) & \longrightarrow & H^2(G_k,\mathrm{Pic}(\bar{X})) & \rightarrow & 0 \\
\| & & \| & & \| & & \| & & \| \\
0 \rightarrow H^1(G_k,\mathrm{Jac}_X) & \longrightarrow & \mathrm{Br}(K) & \longrightarrow & \oplus\, \Xi(G_{k(v)}) & \overset{\Sigma}{\longrightarrow} & \Xi(G_k) & \longrightarrow & 0.
\end{array}$$

The map Σ can be identified with $\mathrm{inv}_K\colon \oplus\, \mathrm{Br}(K_v) \longrightarrow \mathbb{Q}/\mathbb{Z}$.

We define the *group of idèles* J_K of K to be the subgroup of $\prod_{v\in X^0} K_v^\times$ comprising those elements $a = (a_v)$ such that $a_v \in R_v^\times$ for all but finitely many v. The quotient of J_K by $K^\times$ (embedded diagonally) is the *idèle class group* of C_K of K. We set $J = \varinjlim J_F$ and $C = \varinjlim C_F$ (limit over all finite extension F of K, $F \subset K_s$).

Corollary A.8. *If* $H^1(G_{k'}, Jac_X(k_s)) = 0$ *for all finite extensions* k' *of* k, *then it is possible to define on* (G_K, C) *a natural structure of a class formation.*

Proof: An argument similar to that in (4.13) shows that $H^r(G_K, J) = \bigoplus_{v \in X^0} H^r(G_v, K_{v,s}^\times)$, $r \geq 1$, where for each v in X^0 a choice w of an extension of v to K_s has been made in order to identify K with a subfield of $K_{v,s}$ and $G_v \overset{df}{=} G_{K_v}$ with a decomposition group in G. Consider the diagram

$$\begin{array}{ccccccccc} & 0 & \longrightarrow & Br(K) & \longrightarrow & \bigoplus_{v \in X^0} Br(K_v) & \xrightarrow{inv_K} & \mathbb{Q}/\mathbb{Z} & \longrightarrow 0 \\ & & & \| & & \| & & & \\ 0 \longrightarrow & H^1(G_K, C) & \longrightarrow & H^2(G_K, K_s^\times) & \longrightarrow & \bigoplus_{v \in X^0} H^2(G_v, K_v^\times) & \longrightarrow & H^2(G_K, C) & \longrightarrow 0, \end{array}$$

whose top row is the sequence in (A.7) and whose bottom row is the cohomology sequence of

$$0 \longrightarrow K_s^\times \longrightarrow J \longrightarrow C \longrightarrow 0.$$

The zero at lower left comes from Hilbert's theorem 90, and the zero at lower right comes from the fact that G_K has cohomological dimension $\leq cd(k) + 1 = 2$. This diagram shows that $H^1(G_K, C) = 0$ and that there is a unique isomorphism $inv_K\colon H^2(G_K, C) \xrightarrow{\approx} \mathbb{Q}/\mathbb{Z}$ making

$$\begin{array}{ccc} inv_K\colon H^2(G_K, C) & \longrightarrow & \mathbb{Q}/\mathbb{Z} \\ \uparrow & & \| \\ inv_v\colon H^2(G_v, K_{v,s}^\times) & \longrightarrow & \mathbb{Q}/\mathbb{Z} \end{array}$$

commute for all v. The same assertions are true for any finite separable extension F of K, and it obvious that the maps inv_F satisfy the conditions (1.1).

Corollary A.9. *If k is algebraic over a finite field, then* (G_K, C) *is a class formation.*

Proof: Lang's lemma shows that $H^1(G_k, A) = 0$ for any connected algebraic group A if k is finite. If k is algebraic over a finite field, then any element of $H^1(G_k, A)$ is represented by a principal homogeneous space, which is defined over a finite field and is consequently trivial by what we have just observed.

In (A.14) below, we shall see examples of fields K/k for which the conditions of (A.8) fail. We now investigate how much of class field theory continues to hold in such cases.

Fix an extension of each v to K_s, and hence embeddings $i_v\colon G_v \hookrightarrow G_K$. Define $(-,K)\colon J_K \longrightarrow \mathrm{Gal}(K_{ab}/K)$ by

$$(a,K) = \prod_{v \in X^0} i_v(a_v, K_v), \qquad a = (a_v).$$

For any finite abelian extension F of K, this induces a mapping $(-,F/K)\colon J_K/N_{F/K}J_F \longrightarrow \mathrm{Gal}(F/K)$ such that $(a,F/K) = \prod i_v(a_v, F_w/K_v)$ where F_w denotes the completion of F at the chosen prime lying over v. It follows from (A.5) above, and the fact that only finitely many primes of K ramify in F, that this last product is finite (and that the previous product converges).

Lemma A.10. *For all a in* $K^\times$, $(a,K) = 0$.

Proof: Consider the diagram:

$$\begin{array}{ccccccc}
J_K & \times & \Xi(G) & \longrightarrow & \oplus\, Br(K_v) & \xrightarrow{inv_K} & \mathbb{Q}/\mathbb{Z} \\
\| & & \| & & \uparrow & & \\
H^0(G_K, J) & \times & H^2(G_K, \mathbb{Z}) & \longrightarrow & H^2(G_K, J) & & \\
\uparrow & & \| & & \uparrow & & \\
H^0(G_K, K_s^\times) & \times & H^2(G_K, \mathbb{Z}) & \longrightarrow & H^2(G_K, K_s^\times). & &
\end{array}$$

The two lower pairings are defined by cup-product, and the top pairing sends (a,χ) to $inv_K(\sum (\langle a_v, \chi | G_v \rangle)) = \chi((a,K))$, $a = (a_v)$ (here $\langle\ ,\ \rangle$ is as in A.3). It is obvious from the various definitions that the maps are compatible with the pairings. If $a \in K^\times$, then the diagram shows that $\chi(a,K)$ lies in the image of $Br(K)$ in $\mathbb{Q}/\mathbb{Z}$, but $Br(K)$ is the kernel of inv_K, and so $\chi(a,K) = 0$ for all χ. This implies that $(a,K) = 0$.

The lemma shows that there exist maps

$$(-,K)\colon C_K \longrightarrow Gal(K_{ab}/K)$$

$$(-,F/K)\colon C_K/NC_F \longrightarrow Gal(F/K), \qquad F/K \text{ finite abelian.}$$

We shall say that the *reciprocity law holds* for K/k if, for all finite abelian extensions F/K, this last map is an isomorphism.

Unfortunately the reciprocity law does not always hold because there can exist abelian extensions F/K in which all primes of K split, that is, such that $F_w = K_v$ for all primes v. This suggests the following definition: let F be a finite abelian extension of K, and let K' be the maximal subfield of F containing K and such that all primes of K split in K'; the *reduced Galois group* $\bar{G}_{F/K}$ of F over K is the subgroup $Gal(F/K')$ of $Gal(F/K)$.

Proposition A.11. *For any finite abelian extension* F *of* K, *the map* (-,F/K) *induces an isomorphism* $C_K/NC_F \xrightarrow{\approx} \bar{G}_{F/K}$.

Proof: For each prime v, the image of $Gal(F_w/K_v)$ is contained in $\bar{G}_{F/K}$, and so the image of J in Gal(F/K) is also contained in $\bar{G}_{F/K}$.

It clearly suffices to prove the surjectivity of (,F/K) in the case that F/K' is cyclic of prime order. Then there exists a prime v such that $F_w \neq K_v$, and $K_v^\times \longrightarrow Gal(F_w/K_v) \approx \bar{G}_{F/K}$ is surjective local class field theory.

To prove the injectivity, we count. If F_1 and F_2 are finite abelian extensions of F such that $F_1 \cap F_2 = K$ and $F_1F_2 = F$, then it follows from (A.4) that $NC_{F_1} \cap NC_{F_2} = NC_F$ and $(NC_{F_1})(NC_{F_2}) = NC_F$. As $\bar{G}_{F_1/K} \cap \bar{G}_{F_2/K} = 1$ and $\bar{G}_{F_1/K} \cdot \bar{G}_{F_2/K} = \bar{G}_{F/K}$, it suffices to prove that C_K/NC_F and $\bar{G}_{F/K}$ have the same order for F/K cyclic of prime power order.

Let F/K be cyclic of prime power, and consider the diagram

$$\begin{array}{ccccc} Br(F/K) & \longrightarrow & \bigoplus_{v \in X^0} Br(F_w/K_v) & \xrightarrow{inv_K} & \mathbb{Q}/\mathbb{Z} \\ \uparrow \approx & & \uparrow \approx & & \\ K^\times/NF^\times & \longrightarrow & J_K/NJ_F & \longrightarrow & C_K/NC_F \longrightarrow 0. \end{array}$$

The top row is part of the sequence in (A.7), and the bottom row is part of the Tate cohomology sequence of

$$0 \longrightarrow F^\times \longrightarrow J_F \longrightarrow C_F \longrightarrow 0.$$

The first two vertical arrows are the isomorphisms given by the periodicity of the cohomology of cyclic groups. The order of the image of inv_K is the maximum of the orders of the $Br(F_w/K_v)$, and the order of $Br(F_w/K_v)$ is $[F_w:K_v]$. Thus the order of the image equals the order of $\bar{G}(L/K)$. From the diagram, we see that it is also the order of C_K/NC_F.

For any curve Y over a quasi-finite field, we define the *Brauer group* Br(Y) of Y to be the kernel of $\mathrm{Br}(F) \longrightarrow \bigoplus_{v\in Y^0}\mathrm{Br}(F_v)$, where F is the function field of Y. This definition will be justified in A.15 below. Note that Theorem A.7 shows that $\mathrm{Br}(Y) = H^1(G_k, \mathrm{Jac}_Y(k_s))$.

Proposition A.12. *The following statements are equivalent:*

(a) *the reciprocity law holds for K/k;*

(b) *for all finite cyclic extensions F/K, the sequence*

$$\mathrm{Br}(F/K) \longrightarrow \bigoplus_{v\in X^0} \mathrm{Br}(F_w/K_v) \longrightarrow [F:K]^{-1}\mathbb{Z}/\mathbb{Z} \longrightarrow 0$$

is exact;

(c) *for all finite cyclic extensions F/K,* $H^1(\mathrm{Gal}(F/K),\mathrm{Br}(Y)) = 0$, *where* Br(Y) *is the Brauer group of the projective smooth curve with function field* F.

Proof: It follows from (A.11) that the reciprocity law holds for K/k if and only if $\bar{G}_{F/K} = G_{F/K}$ for all finite abelian extensions F/K, and it suffices to check this for cyclic extensions. But, as we saw in the above proof, for such an extension the order of $\bar{G}(F/K)$ is the order of the cokernel of $\mathrm{Br}(F/K) \longrightarrow \bigoplus \mathrm{Br}(F_w/K_v)$. The equivalence of (a) and (b) is now clear.

Consider the exact commutative diagram

$$\begin{array}{ccccccccc}
\mathrm{Br}(X) & \longrightarrow & \mathrm{Br}(K) & \longrightarrow & \bigoplus \mathrm{Br}(K_w) & \longrightarrow & \mathbb{Q}/\mathbb{Z} & \longrightarrow & 0 \\
\downarrow & & \downarrow \alpha & & \downarrow & & \downarrow n & & \\
0 & \longrightarrow & (\mathrm{Br}(F)/\mathrm{Br}(Y))^{\mathrm{Gal}(F/K)} & \longrightarrow & (\bigoplus \mathrm{Br}(F_w))^{\mathrm{Gal}(F_w/K_v)} & \longrightarrow & \mathbb{Q}/\mathbb{Z} & &
\end{array}$$

where $n = [F:K]$. From the Hochschild-Serre spectral sequences, we get exact sequences

$$0 \longrightarrow \mathrm{Br}(F/K) \longrightarrow \mathrm{Br}(K) \longrightarrow \mathrm{Br}(F)^{\mathrm{Gal}(F/K)} \longrightarrow H^3(\mathrm{Gal}(F/K),F^{\times})$$

$$0 \longrightarrow \mathrm{Br}(F_w/K_v) \longrightarrow \mathrm{Br}(K_v) \longrightarrow \mathrm{Br}(F_w)^{\mathrm{Gal}(F_w/K_v)} \longrightarrow H^3(\mathrm{Gal}(F_w/K_v),F_w^{\times}),$$

and from the periodicity of the cohomology of cyclic groups, we see that

$$H^3(\mathrm{Gal}(F/K),F^\times) = H^1(\mathrm{Gal}(F/K),F^\times) = 0,$$

$$H^3(\mathrm{Gal}(F_w/K_v),F_v^\times) = H^1(\mathrm{Gal}(F_w/K_v),F_w^\times) = 0.$$

Thus the preceding diagram gives an exact sequence of kernels and cokernels,

$$\mathrm{Br}(F/K) \longrightarrow \oplus\, \mathrm{Br}(F_w/K_v) \longrightarrow n^{-1}\mathbb{Z}/\mathbb{Z} \longrightarrow \mathrm{coker}(\alpha) \longrightarrow 0.$$

But, as $\mathrm{Br}(K) \longrightarrow \mathrm{Br}(F)^{\mathrm{Gal}(F/K)}$ is surjective,

$$\mathrm{Coker}(\alpha) = \mathrm{Coker}(\mathrm{Br}(F)^{\mathrm{Gal}(F/K)} \longrightarrow (\mathrm{Br}(F)/\mathrm{Br}(Y))^{\mathrm{Gal}(F/K)}),$$

which equals $H^1(\mathrm{Gal}(F/K),\mathrm{Br}(Y))$ because $H^1(\mathrm{Gal}(F/K),\mathrm{Br}(F)) = 0$ (look at the Hochschild-Serre spectral sequence). Thus (b) is equivalent to (c).

We say that the *Hasse principle* holds for K/k if the map $K^\times/NF^\times \longrightarrow \oplus\, K_v^\times/NF_w^\times$ is injective for all finite cyclic field extensions F of K.

Proposition A.13. *The following are equivalent:*

(a) *the Hasse principle holds for* K;

(b) $H^1(G_k,\mathrm{Jac}_X(k_s)) = 0$;

(c) $\mathrm{Br}(X) = 0$.

In particular, the Hasse principle holds for K/k *if* k *is algebraic over a finite field.*

Proof: As $K^\times/NF^\times \approx \mathrm{Br}(F/K)$ for F/K finite and cyclic, we see that that the Hasse principle holds for K/k if and only if $\mathrm{Br}(F/K) \longrightarrow \oplus\, \mathrm{Br}(F_w/K_v)$ is injective for all F/K finite and cyclic. As $\mathrm{Br}(\bar{K}) = 0$, $\mathrm{Br}(K) = \cup\, \mathrm{Br}(F/K)$ where the union runs over all finite cyclic extensions. Thus the Hasse principle holds for K if and only if $Br(K) \longrightarrow \oplus Br(K_v)$ is injective, but the kernel of this map is $Br(X) = H^1(G_k, Jac_X(k_s))$.

Remark A.14. Let k_0 be an algebraically closed field of characteristic zero, and let k be the quasi-finite field $k_0((t))$. In this case there exist elliptic curves E over k with $H^1(k,E) \neq 0$, and there exist function fields K over k with finite extensions F linearly disjoint from k_s such that every prime of K splits completely in F (see [Rim and Whaples (1966)]). In lectures in 1966, Rim asked (rather pessimistically) whether the following conditions on a quasi-finite field k are equivalent:

(a) k is algebraic over a finite field;

(b) $H^1(G_k, A) = 0$ for all connected commutative algebraic group varieties over k;

(c) the reciprocity law holds for all K/k;

(d) the Hasse principle holds for all K/k.

We have seen that (a) $\Rightarrow$ (b) $\Rightarrow$ (c), (d), but (b) does not imply (a). In fact Jordan has shown [Jordan (1972), (1974)] that if k is finitely generated over $\mathbb{Q}$, then for almost all $\sigma \in Gal(k_s/k)$, the fixed field $k(\sigma)$ of σ is quasi-finite and has the property that every absolutely irreducible variety over it has a rational point; thus (b) holds for $k(\sigma)$.

Remark A.15. We use étale cohomology to show that the group $Ker(Br(K) \longrightarrow \oplus Br(K_v))$ is indeed the Brauer group of X. Let $\pi: X \longrightarrow Spec(k)$ be the structure morphism, and consider the exact sequence of sheaves

$$0 \longrightarrow \mathbb{G}_m \longrightarrow g_*\mathbb{G}_m \longrightarrow \mathrm{Div}_X \longrightarrow 0$$

on X_{et} (see [Milne (1980), II.3.9]); here g is the inclusion of the generic point into X and Div_X is the sheaf of Weil divisors). On applying the right derived functors of π_*, we get a long exact sequence of sheaves on $\mathrm{Spec}(k)_{et}$ which we can regard as G_k-modules. The sequence is

$$0 \longrightarrow k_s^\times \longrightarrow \bar{K}^\times \longrightarrow \mathrm{Div}(\bar{X}) \longrightarrow \mathrm{Pic}(\bar{X}) \longrightarrow 0,$$

which is exactly the sequence considered in the proof of Theorem A.7. Here it tells us that $\pi_*\mathbb{G}_m = k_s^\times$, $R^1\pi_*\mathbb{G}_m = \mathrm{Pic}(\bar{X})$, and $R^r\pi_*\mathbb{G}_m = 0$ for $r \geq 2$. Therefore the Leray spectral sequence for π reduces to a long exact sequence

$$\cdots \longrightarrow H^r(G_k, k_s^\times) \longrightarrow H^r(X_{et}, \mathbb{G}_m) \longrightarrow H^{r-1}(G_k, \mathrm{Pic}(\bar{X})) \longrightarrow \cdots .$$

From this we can read off that $H^2(X_{et}, \mathbb{G}_m) = H^1(G_k, \mathrm{Pic}(\bar{X}))$, which proves what we want because $H^2(X_{et}, \mathbb{G}_m)$ is equal to the Brauer group of X.

Exercise A.16. Investigate to what extent the results in the second section remain true when the fields K_v are replaced by the Henselizations of K at its primes.

Notes: Class field theory for complete fields with quasi-local residue fields was first developed in [Whaples (1952/54)] (see also [Serre (1962)]). The same theory for function fields over quasi-finite fields was investigated in [Rim and Whaples (1966)].

CHAPTER II

ETALE COHOMOLOGY

In §1 we prove a duality theorem for $\mathbb{Z}$-constructible sheaves on the spectrum of a Henselian discrete valuation ring with finite residue field. The result is obtained by combining the duality theorems for modules over the Galois groups of the finite residue field and the field of fractions. After making some preliminary calculations in §2, we prove in §3 a generalization of the duality theorem of Artin and Verdier to $\mathbb{Z}$-constructible sheaves on the spectrum of the ring of integers in a number field or on curves over finite fields. In the following section, the theorems are extended to certain nonconstructible sheaves and to tori; also the relation between the duality theorems in this and the preceding chapter is examined. Section 5 treats duality theorems for abelian schemes, §6 considers singular schemes, and in §7 the duality theorems are extended to schemes of dimension greater than one.

In this chapter, the reader is assumed to be familiar with the more elementary parts of étale cohomology, for example, with Chapters II and III of [Milne (1980)]. All schemes are endowed with the étale topology.

§0 Preliminaries

We begin by reviewing parts of [Milne (1980)].

Cohomology with support on a closed subscheme

([Milne (1980), p73-78, p91-95])

Consider a diagram

$$Z \xrightarrow{i} X \xleftarrow{j} U$$

in which i and j are closed and open immersions respectively, and X is the disjoint union of i(Z) and j(U). There are the following functors between the categories of sheaves:

$$S(Z_{et}) \;\underset{\xleftarrow{\;i^!\;}}{\overset{\xleftarrow{\;i^*\;}}{\xrightarrow{\;i_*\;}}}\; S(X_{et}) \;\underset{\xleftarrow{\;j_*\;}}{\overset{\xleftarrow{\;j_!\;}}{\xrightarrow{\;j^*\;}}}\; S(U_{et}).$$

Each functor is left adjoint to the one listed below it; for example, $\mathrm{Hom}_Z(i^*F,F') \approx \mathrm{Hom}_X(F,i_*F')$. The functors i^*, i_*, $j_!$, and j^* are exact, and $i^!$ and j_* are left exact. The functors i_*, $i^!$, j_*, and j^* map injective sheaves to injective sheaves. For any sheaf F on X, there is a canonical exact sequence

$$0 \longrightarrow j_!j^*F \longrightarrow F \longrightarrow i_*i^*F \longrightarrow 0.$$

Proposition 0.1. (a) *For any sheaves* F *on* U *and* F' *on* X, *there is a canonical isomorphism* $\mathrm{Ext}^r_X(j_!F,F') \xrightarrow{\approx} \mathrm{Ext}^r_U(F,j^*F')$, *all* $r \geq 0$; *in particular*, $\mathrm{Ext}^r_X(j_!\mathbb{Z},F') = H^r(U,F'|U)$.

(b) *For any sheaves* F *on* X *and* F' *on* U, *there is a spectral sequence* $\mathrm{Ext}^r_X(F,R^sj_*F') \Rightarrow \mathrm{Ext}^{r+s}_U(F|U,F')$.

(c) *For any sheaves* F *on* X *and* F' *on* Z, *there is a canonical isomorphism* $\mathrm{Ext}^r_X(F,i_*F') \xrightarrow{\approx} \mathrm{Ext}^r_Z(i^*F,F')$, *all* $r \geq 0$.

(d) *For any sheaf* F *on* X, *there is a canonical isomorphism* $H^r_Z(X,F) \xrightarrow{\approx} \mathrm{Ext}^r_X(i_*\mathbb{Z},F)$, *all* $r \geq 0$; *consequently, for any sheaf* F *on* Z, $H^r_Z(X,i_*F) = H^r(Z,F)$, *all* $r \geq 0$.

(e) *For any sheaves* F *on* Z *and* F' *on* X, *there is a spectral sequence*

$$\mathrm{Ext}_Z^r(F, R^s i^! F') \Rightarrow \mathrm{Ext}_X^{r+s}(i_* F, F').$$

(f) *For any sheaf* F *on* X, *there is a long exact sequence*

$$\dots \longrightarrow H_Z^r(X,F) \longrightarrow H^r(X,F) \longrightarrow H^r(U,F) \longrightarrow \dots .$$

Proof: (a) As j^* is exact and preserves injectives, on forming the derived functors of $\mathrm{Hom}_X(j_!F,-) = \mathrm{Hom}_U(F, j^*(-))$, we obtain canonical isomorphisms $\mathrm{Ext}_X^r(j_!F,-) \approx \mathrm{Ext}_U^r(F, j^*(-))$.

(b) As j_* is left exact and preserves injectives, and $\mathrm{Hom}_X(F,-)\circ j_* = \mathrm{Hom}_U(j_!F,-)$, this is the spectral sequence of a composite of functors.

(c) As i^* is exact and preserves injectives, on forming the derived functors of $\mathrm{Hom}_X(F, i_*(-)) = \mathrm{Hom}_Z(i^*F,-)$, we obtain canonical isomorphisms $\mathrm{Ext}_X^r(F, i_*(-)) \approx \mathrm{Ext}_Z^r(i^*F,-)$.

(d) From the exact sequence

$$0 \longrightarrow \mathrm{Hom}_X(i_*\mathbb{Z}, F) \longrightarrow \mathrm{Hom}_X(\mathbb{Z}, F) \longrightarrow \mathrm{Hom}_X(j_!\mathbb{Z}, F)$$

we see that $\mathrm{Hom}_X(i_*\mathbb{Z}, F) = \mathrm{Ker}(\Gamma(X,F) \longrightarrow \Gamma(U,F))$. By definition, this kernel is $\Gamma_Z(X,F)$. Hence $\mathrm{Hom}_X(i_*\mathbb{Z},-) = \Gamma_Z(X,-)$, and on passing to the derived functors we obtain the required canonical isomorphism. The second statement can be obtained by combining the first with (c).

(e) As $i^!$ is left exact and preserves injectives, and $\mathrm{Hom}_Z(F,-)\circ i^! = \mathrm{Hom}_U(i_*F,-)$, this is the spectral sequence of a composite of functors.

(f) This can most simply be constructed as the $\mathrm{Ext}_X(-,F)$-sequence arising from

$$0 \longrightarrow j_!\mathbb{Z} \longrightarrow \mathbb{Z} \longrightarrow i_*\mathbb{Z} \longrightarrow 0.$$

The exact sequence in (0.1f) is referred to as the *cohomology sequence of the pair* $X \supset U$.

Extensions of sheaves

We generalize Theorem 0.3 of Chapter I to the étale topology. Recall that if Y is a Galois covering of a scheme X with Galois group G, then for any G-module M, there is a unique locally constant sheaf F_M on X such that $\Gamma(Y,F_M) = M$ (as a G-module). In the next theorem we use the same letter for M and F_M.

Theorem 0.2. *Let Y be a finite Galois covering of X with Galois group G, and let N and P be sheaves on* X_{et}*. Then, for any G-module M such that* $\mathcal{T}or_r^{\mathbb{Z}}(M,N) = 0$ *for* $r > 0$*, there is a spectral sequence*

$$\mathrm{Ext}_G^r(M,\mathrm{Ext}_Y^s(N,P)) \Rightarrow \mathrm{Ext}_X^{r+s}(M\otimes_{\mathbb{Z}}N,P).$$

In particular, there is a spectral sequence

$$H^r(G,\mathrm{Ext}_Y^s(N,P)) \Rightarrow \mathrm{Ext}_X^{r+s}(N,P).$$

Proof: The second spectral sequence is obtained from the first by taking $M = Z$. After a few preliminaries, the first will be shown to be the spectral sequence of a composite of functors.

Lemma 0.3. *For any sheaves N and P on X and G-module M, there is a canonical isomorphism* $\mathrm{Hom}_G(M,\mathrm{Hom}_Y(N,P)) \xrightarrow{\approx} \mathrm{Hom}_X(M\otimes_{\mathbb{Z}}N,P)$.

Proof: Almost by definition of tensor products, there is a canonical isomorphism $\mathrm{Hom}_Y(M,\mathcal{H}om_Y(N,P)) \xrightarrow{\approx} \mathrm{Hom}_Y(M\otimes_{\mathbb{Z}}N,P)$. Because M becomes the constant sheaf on Y, $\mathrm{Hom}_Y(M,\mathcal{H}om_Y(N,P)) = \mathrm{Hom}(M,\mathrm{Hom}_Y(N,P))$ (homomorphisms of abelian groups). On taking G-invariants, we get the required isomorphism.

Lemma 0.4. *If* I *is an injective sheaf on* X *and* F *is a flat sheaf on* X, *then* $\mathrm{Hom}_Y(F,I)$ *is an injective* G-*module*.

Proof: We have to check that $\mathrm{Hom}_G(-,\mathrm{Hom}_Y(F,I))\colon \mathbf{S}(X_{et}) \longrightarrow \mathbf{Ab}$ is exact, but (0.3) expresses it as the composite of two exact functors $-\otimes_{\mathbb{Z}}F$ and $\mathrm{Hom}_X(-,I)$.

Lemma 0.5. *Let* N *and* I *be sheaves on* X *with* I *injective, and let* M *be a* G-*module. If* $\mathcal{T}or_r^{\mathbb{Z}}(M,N) = 0$ *for* $r > 0$, *then* $\mathrm{Ext}_G^r(M,\mathrm{Hom}_Y(N,I)) = 0$ *for* $r > 0$.

Proof: Let $F^{\cdot} \longrightarrow N$ be a flat resolution of N. The assumption on M implies that $M\otimes F^{\cdot} \longrightarrow M\otimes N$ is a resolution of $M\otimes N$, and it follows from the injectivity of I that $\mathrm{Hom}_X(M\otimes N,I) \longrightarrow \mathrm{Hom}_X(M\otimes F^{\cdot},I)$ is then a resolution of $\mathrm{Hom}_X(M\otimes N,I)$.

Regard $\mathbb{Z}[G]$ as a sheaf on X. Then

$$\mathrm{Hom}_X(\mathbb{Z}[G]\otimes_{\mathbb{Z}}N,I) \longrightarrow \mathrm{Hom}_X(\mathbb{Z}[G]\otimes_{\mathbb{Z}}F^{\cdot},I)$$

is a resolution of $\mathrm{Hom}_X(\mathbb{Z}[G]\otimes_{\mathbb{Z}}N,I)$. But

$$\mathrm{Hom}_X(\mathbb{Z}[G]\otimes_{\mathbb{Z}}F,I) = \mathrm{Hom}_G(\mathbb{Z}[G],\mathrm{Hom}_Y(F,I)) = \mathrm{Hom}_Y(F,I)$$

for any sheaf F, and so the resolution can be regarded as an injective resolution $\mathrm{Hom}_Y(N,I) \longrightarrow \mathrm{Hom}_Y(F^{\cdot},I)$ of the G-module $\mathrm{Hom}_Y(N,I)$. We use this to compute $\mathrm{Ext}_G^r(M,\mathrm{Hom}_Y(N,I))$. From (0.3) we know that $\mathrm{Hom}_G(M,\mathrm{Hom}_Y(F^{\cdot},I)) = \mathrm{Hom}_X(M\otimes F^{\cdot},I)$, and we saw in the above paragraph that this last complex is exact except at the first step. Consequently $\mathrm{Ext}_G^r(M,\mathrm{Hom}_Y(N,I)) = 0$ for $r > 0$.

We now prove the theorem. Lemma 0.3 shows that $\mathrm{Hom}_X(M\otimes_{\mathbb{Z}}N,-)$ is the composite of the functors $\mathrm{Hom}_Y(N,-)$ and $\mathrm{Hom}_G(M,-)$, and Lemma 0.5 shows that the first of these sends injective objects I to objects that are acyclic for the second functor. The spectral sequence therefore arises in a standard way from a composite of functors.

Corollary 0.6. *If* Y *is a finite Galois covering of* X *and* M *and* N *are sheaves on* X, *then* $\mathrm{Ext}^r_Y(M,N) = \mathrm{Ext}^r_X(\pi_*\pi^*M,N)$.

Proof: On applying the theorem with $M = \mathbb{Z}[G]$, we find that

$$\mathrm{Hom}_G(\mathbb{Z}[G],\mathrm{Ext}^r_Y(M,N)) = \mathrm{Ext}^r_X(\mathbb{Z}[G]\otimes M,N),$$

but the first group is $\mathrm{Ext}^r_Y(M,N)$, and, in the second, $\mathbb{Z}[G]\otimes M = \pi_*\pi^*M$.

Pairings

For any sheaves M, N, and P on X, there are canonical pairings

$$\mathrm{Ext}^r_X(N,P) \times \mathrm{Ext}^s_X(M,N) \longrightarrow \mathrm{Ext}^{r+s}_X(M,P),$$

which can be defined in the same way as the pairings in (I.0). Also, if X is quasi-projective over an affine scheme (as all our schemes will be), then we can identify the cohomology groups with the Čech groups and use the standard formulas (see [Milne (1980), V.1.19]) to define cup-product pairings

$$H^r(X,M) \times H^s(X,N) \longrightarrow H^{r+s}(X,M\otimes N).$$

Recall also [Milne (1980), III.1.22] that there is a spectral sequence

$$H^r(X,\mathcal{E}xt^s_X(M,N)) \Rightarrow \mathrm{Ext}^{r+s}_X(M,N)$$

whose edge morphisms are maps $H^r(X,\mathcal{H}om_X(M,N)) \longrightarrow \mathrm{Ext}^r_X(M,N)$. As we have already noted (see the proof of (0.3)), a pairing $M \times N \longrightarrow P$ corresponds to a map $M \longrightarrow \mathcal{H}om_X(N,P)$.

Proposition 0.7. *Let* $M \times N \longrightarrow P$ *be a pairing of sheaves on* X, *and consider the composed map*

$$H^r(X,M) \longrightarrow H^r(X,\mathcal{H}om_X(N,P)) \longrightarrow \mathrm{Ext}^r_X(M,N).$$

Then the diagram

$$\begin{array}{ccccc} H^r(X,M) & \times & H^s(X,N) & \longrightarrow & H^{r+s}(X,P) \\ \downarrow & & \downarrow & & \downarrow \\ \mathrm{Ext}^r_X(N,P) & \times & H^s(X,N) & \longrightarrow & H^{r+s}(X,P) \end{array}$$

(*cup-product pairing*)

(*Ext pairing*)

commutes.

Proof: [Milne (1980), V.1.20].

The Čech complex

Let F be a sheaf on X_{et}. For any U étale over X, let $C^{\cdot}(V/U,F)$ be the Čech complex corresponding to a covering $(V \longrightarrow U)$, and define $\mathcal{C}^{\cdot}(F)(U)$ to be $\varinjlim C^{\cdot}(V/U,F)$ (direct limit over the étale coverings $(V \longrightarrow U)$). Then $\mathcal{C}^{\cdot}(F)$ is a complex of presheaves on X, and we let $C^{\cdot}(U,F) = \Gamma(U,\mathcal{C}^{\cdot}(F))$ be the complex of its sections over U. In the next proposition, we write $\mathcal{H}^r(F)$ for the presheaf $U \mapsto H^r(U,F)$.

Proposition 0.8. *Assume that* X *is quasi-projective over an affine scheme.*

(a) *For any sheaf* F *on* X, $H^r(\mathcal{C}^{\cdot}(F)) = \mathcal{H}^r(F)$ *and* $H^r(C^{\cdot}(X,F)) = H^r(X,F)$.

(b) *For any morphism* $f\colon Y \longrightarrow X$, *there is a canonical map* $f^*\mathcal{C}^{\cdot}(F) \longrightarrow \mathcal{C}^{\cdot}(f^*F)$, *which is a quasi-isomorphism if* f *is étale.*

(c) *For any pair of sheaves* F *and* F', *there is a canonical pairing* $\mathcal{C}^{\cdot}(F) \times \mathcal{C}^{\cdot}(F') \longrightarrow \mathcal{C}^{\cdot}(F\otimes F')$ *inducing the cup-product on cohomology.*

(d) *Let* X *be the spectrum of a field* K, *and let* F *be the sheaf on* X *corresponding to the* G_K*-module* M. *Then* $C^{\cdot}(X,F)$ *is the standard resolution of* M *(defined using inhomogeneous cochains).*

Proof: (a) By definition, $H^r(C^{\cdot}(V/U,F)) = \check{H}^r(V/U,F)$, and therefore

$$H^r(\mathcal{C}^{\cdot}(F)(U)) = \varinjlim H^r(C^{\cdot}(V/U,F)) = \varinjlim \check{H}^r(V/U,F) = \check{H}^r(U,F).$$

Under our assumptions, the Čech groups agree with derived-functor groups, and so this says that $H^r(\Gamma(U,\mathcal{C}^{\cdot}(F))) = H^r(U,F) \overset{\text{df}}{=} \Gamma(U,\mathcal{H}^r(F))$, which proves both the equalities.

(b) For any étale map $V \longrightarrow X$, there is a canonical map $\Gamma(V,F) \longrightarrow \Gamma(V_{(Y)},f^*(F))$. In particular, when U is étale over X and $(V \longrightarrow U)$ is an étale covering of U, then there is a canonical map $\Gamma(V^r,F) \longrightarrow \Gamma(V^r_{(Y)},f^*(F))$ (here V^r denotes $V\times_U V\times_U \cdots$), all r. On passing to the limit over V, we obtain a map

$$\Gamma(U,\mathcal{C}^r(F)) \longrightarrow \Gamma(U_{(Y)},\mathcal{C}^r(f^*F))$$

for all r, and these maps give a map of complexes $\mathcal{C}^{\cdot}(F) \longrightarrow f_*\mathcal{C}^{\cdot}(f^*F)$. By adjointness, we get a map $f^*\mathcal{C}^{\cdot}(F) \longrightarrow \mathcal{C}^{\cdot}(f^*F)$. The last part of the statement is obvious because, when f is étale,

$$H^r(f^*\mathcal{C}^{\cdot}(F)) = \mathcal{H}^r(F)|U = \mathcal{H}^r(F|U) = H^r(\mathcal{C}^{\cdot}(f^*F)).$$

(c) For each $U \longrightarrow X$, the standard formulas define a pairing of complexes

$$\Gamma(U,\mathcal{C}^{\cdot}(F)) \times \Gamma(U,\mathcal{C}^{\cdot}(F')) \longrightarrow \Gamma(U,\mathcal{C}^{\cdot}(F\otimes F')),$$

and these pairings are compatible with the restriction maps.

(d) If U is a finite Galois covering of X with Galois group G, then it is shown in [Milne (1980), III.2.6] that $C^{\cdot}(U/X,F)$ is the standard complex for the G-module F(U). The result follows by passing to the limit.

Constructible sheaves

Let X be a scheme of Krull dimension one. A sheaf F on such a scheme is *constructible* if there is a dense open subset U of X such that

(a) for some finite étale covering $U' \longrightarrow U$, the restriction of F to U' is the constant sheaf defined by a finite group;

(b) for all $x \notin U$, the stalk $F_{\bar{x}}$ of F is finite.

It is said to be *ℤ-constructible* if its restriction to some such U' is the constant sheaf defined by a finitely generated group and the stalks $F_{\bar{x}}$ are finitely generated. Note that a constructible sheaf is ℤ-constructible and a ℤ-constructible sheaf is constructible if and only if it is torsion.

The constructible sheaves form an abelian subcategory of $\mathbf{S}(X_{et})$ and if

$$0 \longrightarrow F' \longrightarrow F \longrightarrow F'' \longrightarrow 0$$

is exact, then F is constructible if and only if F' and F" are constructible. When π is a morphism that is locally of finite type, π^* carries constructible sheaves to constructible sheaves, and when π is finite, π_* has the same property. Similar statements hold for ℤ-constructible sheaves.

Proposition 0.9. *If X is quasi-compact, then every sheaf on X is a filtered direct limit of ℤ-constructible sheaves. Therefore, every torsion sheaf is a filtered direct limit of constructible sheaves.*

Proof: Let F be a sheaf on X, and consider all pairs $(g: U \longrightarrow X, s)$ with g étale, U affine, and s a section of F over U. For each such pair, we have a map $\mathbb{Z} \longrightarrow F|U$ sending 1 to s. This induces a map $g_*\mathbb{Z} \longrightarrow F$ and $g_*\mathbb{Z}$ is ℤ-constructible. Therefore the image of $g_*\mathbb{Z}$ in F is ℤ-constructible, and it is clear that the union of all subsheaves of this form is F. When F is torsion, then each of the subsheaves, being torsion and ℤ-constructible, is constructible.

Mapping cones

For a complex $A^{\cdot}$, $A^{\cdot}[1]$ denotes the complex with $(A^{\cdot}[1])^r = A^{r+1}$ and the differential $d^r = -d_A^{r+1}$. Let $u: A^{\cdot} \longrightarrow B^{\cdot}$ be a map of complexes. The *mapping cone* $C^{\cdot}(u)$ corresponding to u is the complex $A^{\cdot}[1] \oplus B^{\cdot}$ with the differential $d^r = -d_A^{r+1} + u^{r+1} + d_{B^{\cdot}}^r$. Thus $C^r(u) = A^{r+1} \oplus B^r$, and the differential is $(a,b) \mapsto (-da, ua + db)$. There is an obvious injection $i: B^{\cdot} \longrightarrow C^{\cdot}(u)$ and an obvious projection $p: C^{\cdot}(u)[-1] \longrightarrow A^{\cdot}$, and the distinguished triangle corresponding to u is

$$C^{\cdot}(u)[-1] \xrightarrow{-p} A^{\cdot} \xrightarrow{u} B^{\cdot} \xrightarrow{i} C^{\cdot}(u).$$

By definition, every distinguished triangle is isomorphic to one of this form.

A short exact sequence

$$0 \longrightarrow A^{\cdot} \xrightarrow{u} B^{\cdot} \xrightarrow{v} C^{\cdot} \longrightarrow 0$$

gives rise to a distinguished triangle

$$C^{\cdot}[-1] \xrightarrow{w} A^{\cdot} \xrightarrow{u} B^{\cdot} \xrightarrow{v} C^{\cdot}$$

in which w is defined as follows: let $q: C^{\cdot}(u) \longrightarrow C^{\cdot}$ be v on $B^{\cdot}$ and zero on $A^{\cdot}[1]$; then q is a quasi-isomorphism, and so we can define w to be $(-p)\circ q^{-1}[-1]$. A distinguished triangle of complexes of sheaves on X_{et},

$$C^{\cdot}[-1] \xrightarrow{p} A^{\cdot} \xrightarrow{u} B^{\cdot} \xrightarrow{i} C^{\cdot},$$

gives rise to a long exact sequence of hypercohomology groups

$$\dots \longrightarrow \mathbb{H}^r(X,A^{\cdot}) \longrightarrow \mathbb{H}^r(X,B^{\cdot}) \longrightarrow \mathbb{H}^r(X,C^{\cdot}) \longrightarrow \mathbb{H}^{r+1}(X,A^{\cdot}) \longrightarrow \dots .$$

Proposition 0.10. (a) *A morphism of exact sequences of complexes*

$$\begin{array}{ccccccccc} 0 & \longrightarrow & A^{\cdot} & \longrightarrow & B^{\cdot} & \longrightarrow & C^{\cdot} & \longrightarrow & 0 \\ & & \downarrow a & & \downarrow b & & \downarrow c & & \\ 0 & \longrightarrow & D^{\cdot} & \longrightarrow & E^{\cdot} & \longrightarrow & F^{\cdot} & \longrightarrow & 0 \end{array}$$

defines a distinguished triangle

$$C^{\cdot}(c)[-1] \longrightarrow C^{\cdot}(a) \longrightarrow C^{\cdot}(b) \longrightarrow C^{\cdot}(c).$$

(b) *Assume that the rows of the diagram*

$$\begin{array}{ccccccc} C^{\cdot}[-1] & \longrightarrow & A^{\cdot} & \longrightarrow & B^{\cdot} & \longrightarrow & C^{\cdot} \\ & & \downarrow & & \downarrow & & \\ F^{\cdot}[-1] & \longrightarrow & D^{\cdot} & \longrightarrow & E^{\cdot} & \longrightarrow & F^{\cdot} \end{array}$$

are distinguished triangles and that the diagram commutes; then the diagram can be completed to a morphism of distinguished triangles.

(c) *For any maps* $u: A^{\cdot} \longrightarrow B^{\cdot}$, $v: B^{\cdot} \longrightarrow C^{\cdot}$ *of complexes, there is a distinguished triangle*

$$C^{\cdot}(v)[-1] \longrightarrow C^{\cdot}(u) \longrightarrow C^{\cdot}(v \circ u) \longrightarrow C^{\cdot}(v).$$

Proof: The statements are all easy to verify. (Note that (b) and (c) are special cases of the axioms (TR2) and (TR3) [Hartshorne (1966), I.1] for a triangulated category; also that the distinguished triangle in (c) is the analogue for complexes of the kernel-cokernel sequence of a pair of maps.)

§1 Local results

Except when stated otherwise, X will be the spectrum of an excellent Henselian discrete valuation ring R with field of fractions K and residue field k. For example, R could be a complete discrete valuation ring or the Henselization of the local ring at a prime in a global field. We shall use the following notations:

$$\begin{array}{ccccc} & & & & K_s \\ & & & & | \\ k_s & — & R_{un} & — & K_{un} \\ | & & | & & | \\ k & — & R & — & K \end{array} \qquad \begin{array}{l} I = \mathrm{Gal}(K_s/K_{un}) \\ G = \mathrm{Gal}(K_s/K) \\ g = \mathrm{Gal}(k_s/k) = G/I \end{array}$$

$$\mathrm{Spec}\ k = x \xrightarrow{i} X \xleftarrow{j} u = \mathrm{Spec}\ K.$$

Preliminary calculations

We compute the cohomology groups of $\mathbb{Z}$ and $\mathbb{G}_m$.

Proposition 1.1. (a) *Let* F *be a sheaf on* u; *then* $H^r(X, j_!F) = 0$ *for all* $r \geq 0$, *and consequently there is a canonical isomorphism* $H^r(u,F) \xrightarrow{\approx} H_x^{r+1}(X, j_!F)$ *for all* $r \geq 0$.

(b) *For any sheaf* F *on* X, *the map* $H^r(X,F) \longrightarrow H^r(x, i^*F)$ *is an isomorphism all* $r \geq 0$.

Proof: (a) The cohomology sequence of the pair $X \supset u$

$$\cdots \longrightarrow H_x^r(X, j_!F) \longrightarrow H^r(X, j_!F) \longrightarrow H^r(u, j_!F|u) \longrightarrow \cdots$$

shows that the first part of the statement implies the second.

Let M be the stalk $F_{\bar{u}}$ of F at u regarded as a G-module. The functor $F \mapsto i^*j_*F$ can be identified with

$$M \mapsto M^I : \mathbf{Mod}_G \longrightarrow \mathbf{Mod}_g.$$

The equality $\mathrm{Hom}_g(N, M^I) = \mathrm{Hom}_G(N,M)$ for N a g-module shows that i^*j_* has an exact left adjoint, namely, "regard the g-module as a G-module", and so i^*j_* preserves injectives. Consider the exact sequence (see §0)

$$0 \longrightarrow j_!F \longrightarrow j_*F \longrightarrow i_*i^*j_*F \longrightarrow 0.$$

If F is injective, this is an injective resolution of $j_!F$ because j_*

and i_* preserve injectives. As $H^0(X, j_*F) \longrightarrow H^0(X, i_*i^*j_*F)$ is the isomorphism $M^G \xrightarrow{\approx} (M^I)^g$, the cohomology sequence of the sequence shows that $H^0(X, j_!F) = 0$ for all F and that $H^r(X,F) = 0$ for all $r \geq 0$ if F is injective. In particular, we see that if F is injective, then $j_!F$ is acyclic for $\Gamma(X,-)$.

Let $F \longrightarrow I^{\cdot}$ be an injective resolution of F. Then $j_!F \longrightarrow j_!I^{\cdot}$ is an acyclic resolution of $j_!F$, and so $H^r(X, j_!F) = H^r(\Gamma(X, j_!I^{\cdot}))$. But $H^r(\Gamma(X, j_!I^{\cdot})) = (R^rf)(F)$ where f is the functor $F \mapsto \Gamma(X, j_!F) = 0$, and so $H^r(X, j_!F) = 0$ for all r.

(b) The cohomology sequence of

$$0 \longrightarrow j_!j^*F \longrightarrow F \longrightarrow i_*i^*F \longrightarrow 0$$

yields the required isomorphisms.

Corollary 1.2. *For all* r, $H^r(X,\mathbb{Z}) = H^r(g,\mathbb{Z})$; *in particular, when k is finite,* $H^0(X,\mathbb{Z}) = \mathbb{Z}$, $H^1(X,\mathbb{Z}) = 0$, $H^2(X,\mathbb{Z}) = \mathbb{Q}/\mathbb{Z}$, *and* $H^r(X,\mathbb{Z}) = 0$ *for* $r \geq 3$.

Proof: This follows immediately from part (b) of the proposition.

Lemma 1.3. *If k is algebraically closed, then* $H^r(K,\mathbb{G}_m) = 0$ *for all* $r \geq 1$.

Proof: The assumption that R is excellent means that $\hat{K}$ is separable over K, and we have seen in (I.A.6) that K is algebraically closed in $\hat{K}$. Therefore $\hat{K}$ is a regular extension of K, and so we can apply [Shatz (1972), Theorem 27, p116] to obtain that K is a C_1 field. It follows that K has cohomological dimension at most 1, and so $H^r(K,\mathbb{G}_m) = 0$ for $r \geq 2$ [Serre (1964), II.3].

Lemma 1.4. *If k is perfect, then* $R^rj_*\mathbb{G}_m = 0$ *for all* $r > 0$; *therefore* $\mathrm{Ext}^r_X(F, j_*\mathbb{G}_m) = \mathrm{Ext}^r_U(F|U, \mathbb{G}_m)$ *for all sheaves F on X and all r.*

Proof: The stalks of $R^r j_* \mathbb{G}_m$ are (see [Milne (1980), III.1.15])

$$(R^r j_* \mathbb{G}_m)_{\bar{x}} = H^r(K_{un}, \mathbb{G}_m) = 0, \ r > 0 \qquad \text{(by 1.3)}$$

$$(R^r j_* \mathbb{G}_m)_{\bar{u}} = H^r(K_s, \mathbb{G}_m) = 0, \ r > 0.$$

This proves the first assertion, and the second follows from (0.1b).

Proposition 1.5. *Assume that* k *is finite.*

(a) *For all* $r > 0$, $H^r(X, \mathbb{G}_m) = 0$.

(b) *The groups* $H^0_x(X, \mathbb{G}_m) = 0$, $H^1_x(X, \mathbb{G}_m) = \mathbb{Z}$, $H^2_x(X, \mathbb{G}_m) = 0$, $H^3_x(X, \mathbb{G}_m) = \mathbb{Q}/\mathbb{Z}$, *and* $H^r_x(X, \mathbb{G}_m) = 0$ *for* $r > 3$.

Proof: (a) As $R^r j_* \mathbb{G}_m = 0$ for $r \geq 0$, $H^r(X, j_* \mathbb{G}_m) = H^r(K, \mathbb{G}_m)$ all r. Clearly $H^r(X, i_* \mathbb{Z}) = H^r(x, \mathbb{Z})$ for all r, and so the exact sequence

$$0 \longrightarrow \mathbb{G}_m \longrightarrow j_* \mathbb{G}_m \xrightarrow{\text{ord}} i_* \mathbb{Z} \longrightarrow 0$$

gives rise to an exact sequence

$$0 \longrightarrow H^0(X, \mathbb{G}_m) \longrightarrow K^\times \xrightarrow{\text{ord}} \mathbb{Z} \longrightarrow H^1(X, \mathbb{G}_m) \longrightarrow 0 \longrightarrow 0$$

$$\longrightarrow H^2(X, \mathbb{G}_m) \longrightarrow H^2(K, \mathbb{G}_m) \xrightarrow{\approx} H^2(k, \mathbb{Z}) \longrightarrow 0 \longrightarrow H^3(X, \mathbb{G}_m) \longrightarrow 0 \longrightarrow \dots,$$

which yields the result.

(b) Consider the cohomology sequence of the pair $X \supset u$

$$\begin{array}{ccccccccccc} 0 \longrightarrow H^0_x(X, \mathbb{G}_m) & \longrightarrow & H^0(X, \mathbb{G}_m) & \longrightarrow & H^0(K, \mathbb{G}_m) & \longrightarrow & H^1_x(X, \mathbb{G}_m) & \longrightarrow & H^1(X, \mathbb{G}_m) & \longrightarrow & \dots \\ & & \| & & \| & & & & \| & & \\ & & R^\times & \xrightarrow{\text{ord}} & K^\times & & & & 0 . & & \end{array}$$

From the part we have displayed, it is clear that $H^0_x(X, \mathbb{G}_m) = 0$ and $H^1_x(X, \mathbb{G}_m) = \mathbb{Z}$. The remainder of the sequence gives isomorphisms $H^r(K, \mathbb{G}_m) \xrightarrow{\approx} H^{r+1}_x(X, \mathbb{G}_m)$ for $r \geq 1$, from which the values of $H^r_x(X, \mathbb{G}_m)$, $r \geq 2$, can be read off.

Corollary 1.6. *Assume that* k *is finite. If* n *is prime to* char(k), *then* $H^r_x(X, \mu_n) = \mathbb{Z}/n\mathbb{Z}$ *for* $r = 1, 2$, *and* $H^r_x(X, \mu_n) = 0$ *otherwise.*

Proof: As n is prime to char(k), the sequence

$$0 \longrightarrow \mu_n \longrightarrow \mathbb{G}_m \xrightarrow{n} \mathbb{G}_m \longrightarrow 0$$

is exact, and the result follows immediately from (1.5).

Remark 1.7. (a) Part (a) of (1.5) can also be obtained as a consequence of the following more general result [Milne (1980), III.3.11]: if G is a smooth commutative group scheme over the spectrum of a Henselian local ring, then $H^r(X,G) = H^r(x,G_0)$ for $r \geq 1$, where x is the closed point of X, and G_0 is the closed fibre of G/X.

Alternatively, (1.1b) shows that $H^r(X,\mathbb{G}_m) = H^r(x,i^*\mathbb{G}_m)$. Obviously $i^*\mathbb{G}_m$ corresponds to the g-module $R_{un}^{\times}$, and it is not difficult show that $H^r(g,R_{un}^{\times}) = 0$ for $r > 0$ (see A.2).

(b) There is an alternative way of computing the groups $H_x^r(X,\mathbb{G}_m)$. Whenever k is perfect, (1.4) shows that $H^r(X,j_*\mathbb{G}_m) \longrightarrow H^r(u,\mathbb{G}_m)$ is an isomorphism, and it follows immediately that $H_x^r(X,j_*\mathbb{G}_m) = 0$ for all r. Therefore the exact sequence

$$0 \longrightarrow \mathbb{G}_m \longrightarrow j_*\mathbb{G}_m \xrightarrow{\text{ord}} i_*\mathbb{Z} \longrightarrow 0$$

leads to isomorphisms $H_x^r(X,i_*\mathbb{Z}) \xrightarrow{\approx} H_x^{r+1}(X,\mathbb{G}_m)$, and we have seen in (0.1d) that $H_x^r(X,i_*\mathbb{Z}) = H^r(x,\mathbb{Z})$. Therefore $H_x^r(X,\mathbb{G}_m) = H^{r-1}(x,\mathbb{Z})$ for all $r \geq 1$.

This argument works whenever k is perfect. In particular, when k is algebraically closed, it shows that

$$H_x^r(X,\mathbb{G}_m) = 0,\ \mathbb{Z},\ 0,\ \ldots \text{ for } r = 0,\ 1,\ 2,\ \ldots$$

from which it follows that, if n is prime to char(k), then

$$H_x^r(X,\mu_n) = 0,\ 0,\ \mathbb{Z}/n\mathbb{Z},\ 0,\ \ldots \text{ for } r = 0,\ 1,\ 2,\ \ldots .$$

These values of $H^r(X,\mu_n)$ are those predicted by the purity conjecture [Artin, Grothendieck, and Verdier (1972/73), XIX].

Since $R^r i^! \mathbb{G}_m$ is the sheaf on z asssociated with the presheaf $z' \mapsto H^r_{z'}(X', \mathbb{G}_m)$ (here $X' \longrightarrow X$ is the étale covering of X corresponding to $z' \longrightarrow z$), we see that if k is any perfect field, then $R^1 i^! \mathbb{G}_m = \mathbb{Z}$ and $R^r i^! \mathbb{G}_m = 0$ for $r \neq 1$, and consequently that if $(n, \mathrm{char}(k)) = 1$, then $R^2 i^! \mu_n = \mathbb{Z}/n\mathbb{Z}$ and $R^r i^! \mu_n = 0$ for $r \neq 2$. For this last statement concerning μ_n it is not even necesssary to assume that k is perfect.

The duality theorem

We now assume that the residue field k is finite. There are two natural candidates for a trace map $H^3_x(X, \mathbb{G}_m) \xrightarrow{\approx} \mathbb{Q}/\mathbb{Z}$. The first is that in (1.5), namely, the composite of the inverse of $H^2(u, \mathbb{G}_m) \xrightarrow{\approx} H^3_x(X, \mathbb{G}_m)$ with $H^2(u, \mathbb{G}_m) = H^2(G_K, K_s^\times) \xrightarrow{\mathrm{inv}_K} \mathbb{Q}/\mathbb{Z}$. The second is that in (1.7b), namely, the composite of the inverse of $H^2(x, \mathbb{Z}) = H^2_x(X, \mathbb{Z}) \xrightarrow{\approx} H^3_x(X, \mathbb{G}_m)$ with $H^2(x, \mathbb{Z}) = H^2(G_k, \mathbb{Z}) \xrightarrow{\mathrm{inv}_k} \mathbb{Q}/\mathbb{Z}$. From the definition of inv_K (see I.1.6) it is clear that the two methods lead to the same map.

Recall (0.1d) that for any sheaf F on X, $H^r_x(X, F) = \mathrm{Ext}^r_X(i_* \mathbb{Z}, F)$, and so there is a canonical pairing

$$\mathrm{Ext}^r_X(F, \mathbb{G}_m) \times H^{3-r}_x(X, F) \longrightarrow H^3_x(X, \mathbb{G}_m).$$

On combining the pairing with the trace map, we obtain a map

$$\alpha^r(X, F): \mathrm{Ext}^r_X(F, \mathbb{G}_m) \longrightarrow H^{3-r}_x(X, F)^*.$$

Before we can state the theorem, we need to endow $\mathrm{Hom}_X(F, \mathbb{G}_m)$ with a topology. We shall see below that the restriction map

$$\mathrm{Hom}_X(F, \mathbb{G}_m) \longrightarrow \mathrm{Hom}_u(F|u, \mathbb{G}_m) = \mathrm{Hom}_G(F_{\bar{u}}, K_s^\times)$$

is injective. The last group inherits a topology from that on $K_s^\times$, and we give $\mathrm{Hom}_X(F, \mathbb{G}_m)$ the subspace topology. When F is $\mathbb{Z}$-constructible, all subgroups of $\mathrm{Hom}_X(F,\mathbb{G}_m)$ of finite index prime to char(K) are open.

Theorem 1.8. (a) *Let* F *be a* $\mathbb{Z}$*-constructible sheaf on* X *without* p*-torsion if* K *has characteristic* $p \neq 0$. *For* $r \geq 2$, *the groups* $\mathrm{Ext}^r_X(F,\mathbb{G}_m)$ *are torsion of cofinite-type, and* $\alpha^r(X,F)$ *is an isomorphism. For* $r \leq 1$, $\alpha^r(X,F)$ *defines an isomorphism*

$$\mathrm{Ext}^r_X(F,\mathbb{G}_m)^\wedge \longrightarrow H^{3-r}(X,F)^*,$$

where ^ *denotes the completion for the topology of subgroups of finite index when* $r = 1$ *and the completion for the topology of open subgroups of finite index when* $r = 0$. *The group* $\mathrm{Ext}^1_X(F,\mathbb{G}_m)$ *is finitely generated.*

(b) *Let* F *be a constructible sheaf on* X, *and assume that* K *is complete or that* $pF = F$ *for* $p = \mathrm{char}\, K$. *Then the pairing*

$$\mathrm{Ext}^r_X(F,\mathbb{G}_m) \times H^{3-r}_x(X,F) \longrightarrow H^3_x(X,\mathbb{G}_m) = \mathbb{Q}/\mathbb{Z}.$$

is nondegenerate; if $pF = F$, *then all the groups are finite.*

Proof: We first consider a sheaf of the form i_*F, F a $\mathbb{Z}$-constructible sheaf on x. Recall (0.1d), that for such a sheaf $H^r_x(X,i_*F) = H^r(x,F)$, all r.

Lemma 1.9. *For any sheaf* F *on* x, *there is a canonical isomorphism* $\mathrm{Ext}^{r-1}_x(F,\mathbb{Z}) \xrightarrow{\approx} \mathrm{Ext}^r_X(i_*F,\mathbb{G}_m)$, *all* $r \geq 1$.

Proof: From the exact sequence

$$0 \longrightarrow \mathbb{G}_m \longrightarrow j_*\mathbb{G}_m \longrightarrow i_*\mathbb{Z} \longrightarrow 0$$

we obtain an exact sequence

$$\dots \longrightarrow \mathrm{Ext}^r_X(i_*F,\mathbb{G}_m) \longrightarrow \mathrm{Ext}^r_X(i_*F,j_*\mathbb{G}_m) \longrightarrow \mathrm{Ext}^r_X(i_*F,i_*\mathbb{Z}) \longrightarrow \dots .$$

But $\mathrm{Ext}^r_X(i_*F, j_*\mathbb{G}_m) = \mathrm{Ext}^r_u(i_*F|u, \mathbb{G}_m)$ in view of (1.4) and (0.1b), and the second group is zero because $i_*F|u = 0$. Also $\mathrm{Ext}^r_X(i_*F, i_*\mathbb{Z}) = \mathrm{Ext}^r_x(F, \mathbb{Z})$ by (0.1c), and so the sequence gives the required isomorphisms.

It is obvious from the various definitions that the diagram

$$\begin{array}{ccccc} \mathrm{Ext}^r_X(i_*F, \mathbb{G}_m) & \times \; H^{3-r}_x(X, i_*F) & \longrightarrow & H^3_x(X, \mathbb{G}_m) & = \mathbb{Q}/\mathbb{Z} \\ \uparrow \approx & \| & & \uparrow \approx & \\ \mathrm{Ext}^{r-1}_x(F, \mathbb{Z}) & \times \; H^{3-r}(x, F) & \longrightarrow & H^2(x, \mathbb{Z}) & = \mathbb{Q}/\mathbb{Z} \end{array}$$

commutes. The lower pairing can be identified with the pairing in (I.1.10). We deduce: $\mathrm{Ext}^1(i_*F, \mathbb{G}_m)$ is finitely generated and $\alpha^1(X, i_*F)$ defines an isomorphism $\mathrm{Ext}^1(i_*F, \mathbb{G}_m)^\wedge \longrightarrow H^2_x(X, i_*F)^*$ (completion for the profinite topology); $\alpha^2(X, i_*F)$ is an isomorphism of finite groups; $\alpha^3(X, i_*F)$ is an isomorphism of torsion groups of cofinite-type; for all other values of r, the groups are zero. When F is constructible, it corresponds to a finite g-module, and so all the groups are finite (and discrete). This completes the proof of the theorem for a sheaf of the form i_*F.

We next consider a sheaf of the form $j_!F$, with F a $\mathbb{Z}$-constructible sheaf on u without p-torsion. Consider the diagram

$$\begin{array}{ccccc} \mathrm{Ext}^r_X(j_!F, \mathbb{G}_m) & \times \; H^{3-r}_x(X, j_!F) & \longrightarrow & H^3_x(X, \mathbb{G}_m) & = \mathbb{Q}/\mathbb{Z} \\ \downarrow \approx & \uparrow \approx & & \uparrow \approx & \\ \mathrm{Ext}^r_u(F, \mathbb{G}_m) & \times \; H^{2-r}(u, F) & \longrightarrow & H^2(u, \mathbb{G}_m) & = \mathbb{Q}/\mathbb{Z} \end{array}$$

in which the first isomorphism is restriction from X to u (see 0.1a), and the two remaining isomorphisms are boundary maps in the cohomology sequence of the pair $X \supset u$ (see (1.1) and (1.5)). It is again clear from the various definitions that the diagram commutes. The lower pairing can be identified with that in (I.2.1). We deduce:

$\mathrm{Hom}_X(j_!F, \mathbb{G}_m)$ is finitely generated and $\alpha^0(X, j_!F)$ defines an isomorphism $\mathrm{Hom}_X(j_!F, \mathbb{G}_m)^\wedge \longrightarrow H^3_x(X, j_!F)^*$ (completion for the topology of open subgroups of finite index); $\alpha^1(X, j_!F)$ is an isomorphism of finite groups; $\alpha^2(X, i_*F)$ is an isomorphism of torsion groups of cofinite-type; for all other values of r, the groups are zero. When F is constructible, it corresponds to a finite G-module, and all the groups are finite (and discrete).

This completes the proof of the theorem when $F \approx i_*i^*F$ or $F \approx j_!j^*F$ and F is without p-torsion. For a general $\mathbb{Z}$-constructible sheaf F without p-torsion, we use the exact sequence

$$0 \longrightarrow j_!(F|U) \longrightarrow F \longrightarrow i_*i^*F \longrightarrow 0,$$

and apply the five-lemma to the diagram

$$\begin{array}{ccccccccc} \cdot \longrightarrow & \mathrm{Ext}^r_X(i_*i^*F, \mathbb{G}_m) & \longrightarrow & \mathrm{Ext}^r_X(F, \mathbb{G}_m) & \longrightarrow & \mathrm{Ext}^r_X(j_!(F|U), \mathbb{G}_m) & \longrightarrow & \cdot \\ \downarrow\approx & \downarrow\approx & & \downarrow & & \downarrow\approx & & \downarrow\approx \\ \cdot \longrightarrow & H^{3-r}_x(X, i_*i^*F)^* & \longrightarrow & H^{3-r}_x(X, F)^* & \longrightarrow & H^{3-r}_x(X, j_!(F|U))^* & \longrightarrow & \cdot \end{array}.$$

This leads immediately to a proof of (a) of the theorem for $r \geq 2$. For $r < 2$, one only has to replace the first four terms in the top row of the diagram with their completions:

$$\begin{array}{ccccccc} 0 \longrightarrow & \mathrm{Hom}(F, \mathbb{G}_m)^\wedge & \longrightarrow & \mathrm{Hom}(j_!(F|U), \mathbb{G}_m)^\wedge & \longrightarrow & \mathrm{Ext}^1(i_*i^*F, \mathbb{G}_m)^\wedge & \longrightarrow \dots \\ & \downarrow & & \downarrow\approx & & \downarrow\approx & \\ 0 \longrightarrow & H^3_x(X, F)^* & \longrightarrow & H^3_x(X, j_!(F|U))^* & \longrightarrow & H^2_x(X, i_*i^*F)^* & \longrightarrow \dots \end{array}.$$

Note that the top row is exact by virtue of (I.0.20a).

The remaining case, where K is complete and F is constructible with p torsion, can be treated similarly.

Corollary 1.10. *Let* p = char k.

(a) *Let* F *be a locally constant constructible sheaf on* X *such that*

$pF = F$, *and let* $F^D = \mathcal{H}om(F, \mathbb{G}_m)$. *Then there is a canonical non-degenerate pairing of finite groups*

$$H^r(X, F^D) \times H_x^{3-r}(X, F) \longrightarrow H_x^3(X, \mathbb{G}_m) = \mathbb{Q}/\mathbb{Z}.$$

(b) *Let* F *be a constructible sheaf on* u *such that* $pF = F$, *and let* $F^D = \mathcal{H}om_u(F, \mathbb{G}_m)$. *The pairing* $F^D \times F \longrightarrow \mathbb{G}_m$ *extends to a pairing* $j_*F^D \times j_*F \longrightarrow \mathbb{G}_m$, *and the resulting pairing*

$$H^r(X, j_*F^D) \times H_x^{3-r}(X, j_*F) \longrightarrow H_x^3(X, \mathbb{G}_m) = \mathbb{Q}/\mathbb{Z}$$

is nondegenerate.

Proof: (a) We shall use the spectral sequence [Milne (1980), III.1.22]

$$H^r(X, \mathcal{E}xt_X^s(F, \mathbb{G}_m)) \Rightarrow \mathrm{Ext}_X^{r+s}(F, \mathbb{G}_m)$$

to show that the term $\mathrm{Ext}_X^r(F, \mathbb{G}_m)$ in the theorem can be replaced with $H^r(X, F^D)$. According to [Milne (1980), III.1.31], the stalk of $\mathcal{E}xt_X^s(F, \mathbb{G}_m)$ at $\bar{x}$ is $\mathrm{Ext}^s(F_{\bar{x}}, R_{un}^\times)$ (Ext as abelian groups). This group is zero for $s > 0$ because $R_{un}^\times$ is divisible by all primes dividing the order of $F_{\bar{x}}$. The stalk of $\mathcal{E}xt_X^s(F, \mathbb{G}_m)$ at $\bar{u}$ is $\mathrm{Ext}^s(F_{\bar{x}}, K_s^\times)$, which is zero for $s > 0$ by the same argument. Therefore $\mathcal{E}xt_X^s(F, \mathbb{G}_m)$ is zero for $s > 0$, and the spectral sequence collapses to give isomorphisms $H^r(X, F^D) \approx \mathrm{Ext}_X^r(F, \mathbb{G}_m)$.

(b) On applying j_* to the isomorphism $F^D \xrightarrow{\approx} \mathcal{H}om_u(F, \mathbb{G}_m)$, we obtain an isomorphism $j_*F^D \xrightarrow{\approx} j_*\mathcal{H}om_u(F, \mathbb{G}_m)$. But

$$j_*\mathcal{H}om_u(F, \mathbb{G}_m) = j_*\mathcal{H}om(j^*j_*F, \mathbb{G}_m) = \mathcal{H}om_X(j_*F, j_*\mathbb{G}_m)$$

(see [Milne (1980), II.3.21]), and from the Ext sequence of

$$0 \longrightarrow \mathbb{G}_m \longrightarrow j_*\mathbb{G}_m \longrightarrow i_*\mathbb{Z} \longrightarrow 0$$

and the vanishing of $\mathrm{Hom}_X(j_*F, i_*\mathbb{Z}) = \mathrm{Hom}_x(i^*j_*F, \mathbb{Z})$ we find that $\mathcal{H}om_X(j_*F, j_*\mathbb{G}_m) = \mathcal{H}om_X(j_*F, \mathbb{G}_m)$. Thus $j_*F^D = \mathcal{H}om_X(j_*F, \mathbb{G}_m)$ and the existence of the pairing $j_*F^D \times j_*F \longrightarrow \mathbb{G}_m$ is obvious.

Next we shall show that $\mathcal{E}xt^r_X(j_*F,\mathbb{G}_m) = 0$ for $r > 0$. Let $M = F_{\bar{u}}$. If $M^I = M$, then j_*F is locally constant, and we showed in the proof of part (a) of the corollary that the higher $\mathcal{E}xt$'s vanish for such sheaves. If $M^I = 0$, then $j_*F = j_!F$, and $\mathcal{E}xt^r_X(j_!F,\mathbb{G}_m) = j_*\mathcal{E}xt^r_u(F,\mathbb{G}_m)$. The stalk of $j_*\mathcal{E}xt^r_u(F,\mathbb{G}_m)$ at x is $H^r(K_{un},M^D)$. Because M has order prime to p, $H^r(I,M^D) = H^r(I/I_p,M^D)$, where I_p is the p-Sylow subgroup of I. But $H^r(I/I_p,M^D)$ is zero for $r > 1$, and $H^1(I/I_p,M^D)$ is dual to $H^0(I/I_p,M)$, which equals M^I (cf. the proof of I.2.18). By assumption, this is zero. Because I is normal in G_K, every G_K-module has a composition series whose quotients Q are such that either $Q^I = Q$ or $Q^I = 0$. Our arguments therefore show that $\mathcal{E}xt^r_X(j_*F,\mathbb{G}_m) = 0$ for $r > 0$.

The spectral sequence

$$H^r(X,\mathcal{E}xt^s_X(j_*F,\mathbb{G}_m)) \Rightarrow \mathrm{Ext}^{r+s}_X(j_*F,\mathbb{G}_m)$$

therefore reduces to a family of isomorphisms $H^r(X,j_*F^D) \xrightarrow{\approx} \mathrm{Ext}^r_X(j_*F,\mathbb{G}_m)$, and the corollary follows from the theorem.

Remark 1.11. (a) Part (a) of the theorem is true without the condition that F has no p torsion, p = char K, provided one endows $\mathrm{Ext}^1_X(F,\mathbb{G}_m)$ with a topology deduced from that on K and defines $\mathrm{Ext}^1_X(F,\mathbb{G}_m)^\wedge$ to be the completion with respect to the topology of open subgroups of finite index. Note that $\alpha^1(X,i_*\mathbb{Z})$ is the natural inclusion $\mathbb{Z} \hookrightarrow \hat{\mathbb{Z}}$, and that $\alpha^1(X,j_!\mathbb{Z}/p\mathbb{Z})$ for p = char K is an isomorphism of infinite compact groups $K^\times/K^{\times p} \xrightarrow{\approx} (K/\wp K)^*$ when K is complete. The first example shows that it is necessary to complete $\mathrm{Ext}^1_X(F,\mathbb{G}_m)$ in order to obtain an isomorphism, and the second shows that it is necessary to endow $\mathrm{Ext}^1_X(F,\mathbb{G}_m)$ with a topology coming from K because not all subgroups of finite index in $K^\times/K^{\times p}$ are open.

(b) By using derived categories, it is possible to restate (1.8) in the form of (1.10) for any constructible sheaf F such that pF = F. Simply set $F^D = \mathbf{R}\mathcal{H}om(F,\mathbb{G}_m)$ (an object in the derived category of the category of constructible sheaves on X), and note that $\mathbb{H}^r(X,F^D) = \mathrm{Ext}^r_X(F,\mathbb{G}_m)$. (The point of the proof of (1.10) is to show that $H^r(\mathbf{R}\mathcal{H}om(F,\mathbb{G}_m)) = 0$ for $r > 0$ when F is locally constant.)

Singular schemes

We now let X = Spec R with R the Henselization of an excellent integral local ring of dimension 1 with finite residue field k. Then R is again excellent, but it need not be reduced. Let u = $\{u_1,\dots,u_m\}$ be the set of points of X of dimension 0. Then $\mathcal{O}_{X,u_i}$ is a field K_i, and the normalization $\tilde{R}$ of R is a product of excellent Henselian discrete valuation rings R_i such that R_i has field of fractions K_i (see [Raynaud (1970), IX]). We have a diagram

$$\begin{array}{ccccc} \{x_1,\dots,x_m\} & \xrightarrow{\tilde{i}} & \tilde{X} & \xleftarrow{\tilde{j}} & u, \\ \downarrow & & \downarrow \pi & \swarrow j & \\ x & \xrightarrow{i} & X & & \end{array}$$

with $\tilde{X} = \mathrm{Spec}\,\tilde{R}$ and x and x_i the closed points of X and Spec R_i respectively. For h in the total ring of fractions of R, define ord(h) = $\sum\,[k(x_i):k(x)]\,\mathrm{ord}_i(h)$ where ord_i is the valuation on K_i. On can define a similar map for any U étals over X, and so obtain a homomorphism ord: $j_*\mathbb{G}_m \longrightarrow i_*\mathbb{Z}$. Define $\mathbb{G}$ to be the complex of sheaves $j_*\mathbb{G}_m \longrightarrow i_*\mathbb{Z}$ on X.

Lemma 1.12. (a) *For all* $r > 0$, $R^r j_*\mathbb{G}_m = 0$; *therefore* $\mathrm{Ext}^r_X(F,j_*\mathbb{G}_m) =$

$\mathrm{Ext}^r_u(F|u,\mathbb{G}_m)$ *for all sheaves* F *on* X *and all* r.

(b) *For all* r, *there is a canonical isomorphism* $H^{r-1}(x,\mathbb{Z}) \longrightarrow H^r_x(X,\mathbb{G})$.

Proof: (a) The map π is finite, and therefore π_* is exact. As $j_* = \pi_*\tilde{j}_*$, this shows that $R^r j_*\mathbb{G}_m = \pi_* R^r\tilde{j}_*\mathbb{G}_m$, which is zero by (1.4).

(b) From (a) and (0.1) we see that

$$H^r_x(X,j_*\mathbb{G}_m) = \mathrm{Ext}^r_X(i_*\mathbb{Z},j_*\mathbb{G}_m) = \mathrm{Ext}^r_u(i_*\mathbb{Z}|u,\mathbb{G}_m) = 0,$$

all r. The exact sequence

$$\cdots \longrightarrow H^r_x(X,\mathbb{G}) \longrightarrow H^r_x(X,j_*\mathbb{G}_m) \longrightarrow H^r_x(X,i_*\mathbb{Z}) \longrightarrow \cdots$$

now leads immediately to the isomorphism.

We define the trace map $H^3_x(X,\mathbb{G}) \xrightarrow{\approx} \mathbb{Q}/\mathbb{Z}$ to be the composite of the inverse of $H^2(x,\mathbb{Z}) \xrightarrow{\approx} H^3_x(X,\mathbb{G})$ and $\mathrm{inv}_k\colon H^2(g,\mathbb{Z}) \xrightarrow{\approx} \mathbb{Q}/\mathbb{Z}$.

Theorem 1.13. *For any constructible sheaf* F *on* X,

$$\mathrm{Ext}^r_X(F,\mathbb{G}) \times H^{3-r}_x(X,F) \longrightarrow H^3_x(X,\mathbb{G}) = \mathbb{Q}/\mathbb{Z}.$$

is a nondegenerate pairing of finite groups.

Proof: As in the case that X is regular, it suffices to prove this for sheaves of the form i_*F and $j_!F$.

Lemma 1.14. *For all* r, $\mathrm{Ext}^r_X(i_*F,j_*\mathbb{G}_m) = 0$; *therefore the boundary maps* $\mathrm{Ext}^{r-1}_x(F,\mathbb{Z}) \longrightarrow \mathrm{Ext}^r_X(i_*F,\mathbb{G})$ *are isomorphisms.*

Proof: The proof is the same as that of (1.9).

The theorem for i_*F now follows from the diagram:

$$\begin{array}{ccccc}
\mathrm{Ext}^r_X(i_*F,\mathbb{G}) & \times H^{3-r}_x(X,i_*F) & \longrightarrow H^3_x(X,\mathbb{G}) & = \mathbb{Q}/\mathbb{Z} \\
\uparrow\approx & \| & \uparrow\approx & \\
\mathrm{Ext}^{r-1}_x(F,\mathbb{Z}) & \times H^{3-r}(x,F) & \longrightarrow H^2(x,\mathbb{Z}) & = \mathbb{Q}/\mathbb{Z}.
\end{array}$$

We next consider a sheaf of the form $j_!F$.

Lemma 1.15. *For all* r, $H^r(X,j_!F) = 0$; *therefore the maps*

$$H^{r-1}(u,F) \longrightarrow H^r_x(X,j_!F)$$

are isomorphisms.

Proof: Consider the diagram

$$\begin{array}{ccccccc}
0 \longrightarrow & \pi_*\tilde{j}_!F \longrightarrow & \pi_*\tilde{j}_*F & & \\
 & & \| & & \\
0 \longrightarrow & j_!F \longrightarrow & j_*F \longrightarrow & i_*i^*j_*F \longrightarrow 0.
\end{array}$$

Because $(\pi_*\tilde{j}_!F)_{\bar{x}} = 0$, the map of $\pi_*\tilde{j}_!F$ into $i_*i^*j_*F$ is zero, and therefore the image of $\pi_*\tilde{j}_!F$ is contained in $j_!F$. The resulting map $\pi_*\tilde{j}_!F \longrightarrow j_!F$ induces isomorphisms on the stalks and therefore is itself an isomorphism. The first assertion now follows from (1.1), and the second is an immediate consequence of the first.

The theorem for $j_!F$ now follows from the diagram:

$$\begin{array}{ccccc}
\mathrm{Ext}^r_X(j_!F,\mathbb{G}) \times & H^{3-r}_x(X,j_!F) \longrightarrow & H^3_x(X,\mathbb{G}) & = & \mathbb{Q}/\mathbb{Z} \\
\downarrow\approx & \uparrow\approx & \uparrow & & \uparrow\Sigma \\
\bigoplus_i \mathrm{Ext}^r_{u_i}(F,\mathbb{G}_m) \times & \bigoplus_i H^{2-r}(u_i,F_i) \longrightarrow & \bigoplus_i H^2(u_i,\mathbb{G}_m) & = & (\mathbb{Q}/\mathbb{Z})^m.
\end{array}$$

Higher dimensional schemes

We obtain a partial generalization of (1.8) to d-local fields. Recall from (I.2) that a 0-local field is a finite field, and that a d-local field is a field that is complete with repect to a discrete valuation and has a (d-1)-local field as residue field. If p is either 1 or a prime and M is a torsion group or sheaf, we write M(non-p) for $\varinjlim M_m$ where the limit is over all integers prime to p. We also write $\mu_\infty(r)$ for the sheaf $\varinjlim_m \mu_m^{\otimes r}$ on X_{et} (limit over all integers m).

Theorem 1.16. *Let* K *be a* d*-local field with* $d \geq 2$, *and let* $p = \mathrm{char}(K_1)$ *where* K_1 *is the* 1*-local field in the inductive definition of* K. *Let* X *be* Spec R *with* R *the discrete valuation ring in* K, *and let* x *and* u *be the closed and open points* Spec k *and* Spec K *of* X.

(a) *There is a canonical isomorphism*

$$H_x^{d+2}(X,\mu_\infty(d))(\text{non-}p) \xrightarrow{\approx} \mathbb{Q}/\mathbb{Z}(\text{non-}p).$$

(b) *For any constructible sheaf* F *on* X *such that* $pF = F$,

$$\mathrm{Ext}_X^r(F,\mu_\infty(d)) \times H_x^{d+2-r}(X,F) \longrightarrow H_x^{d+2}(X,\mu_\infty(d)) \longrightarrow \mathbb{Q}/\mathbb{Z}$$

is a nondegenerate pairing of finite groups, all r.

Proof: (a) Let i and j be the inclusions of x and u respectively into X. As we observed in (1.7), $R^r i^! \mu_m = \mathbb{Z}/m\mathbb{Z}$ for $r = 2$ and is zero otherwise. On tensoring both sides with $\mu_m^{\otimes d-1}$, we find that $R^r i^! \mu_m^{\otimes d} = \mu_m^{\otimes d-1}$ for $r = 2$ and is zero otherwise. Next, on passing to the direct limit, we find that $R^r i^! \mu_\infty(d)(\text{non-}p) = \mu_\infty(d-1)(\text{non-}p)$ for $r = 2$ and is zero otherwise. Now the spectral sequence

$$H^r(x,R^s i^! \mu_\infty(d)) \Rightarrow H_x^{r+s}(X,\mu_\infty(d))$$

shows that $H_x^{d+2}(X,\mu_\infty(d))(\text{non-}p) = H^d(x,\mu_\infty(d-1))(\text{non-}p)$, which equals $(\mathbb{Q}/\mathbb{Z})(\text{non-}p)$ by (I.2.17).

We give a second derivation of this trace map. Note that (1.1) implies that $H^{d+1}(u,\mu_\infty(d)) \xrightarrow{\approx} H_x^{d+2}(X,j_!\mu_\infty(d))$. Moreover

$H^{d+2}_x(X, j_!\mu_\infty(d)) \longrightarrow H^{d+2}_x(X, \mu_\infty(d))$ is an isomorphism because k has cohomological dimension d (this is implied by (I.2.17) applied to k). Therefore we have a trace map

$$H^{d+2}_x(X, \mu_\infty(d))(\text{non-}p) \approx H^{d+1}(u, \mu_\infty(d))(\text{non-}p) \approx (\mathbb{Q}/\mathbb{Z})(\text{non-}p).$$

The inductive approach we adopted to define the trace map in (I.2.17) shows that the two definitions give the same trace map are equal.

(b) As in the previous cases, it suffices to prove this for sheaves of the form i_*F and $j_!F$.

Lemma 1.17. *For all* r, *there are canonical isomorphisms*

$$\mathrm{Ext}^{r-2}_x(F, \mu_\infty(d-1)) \longrightarrow \mathrm{Ext}^r_X(i_*F, \mu_\infty(d)).$$

Proof: Because $R^r i^! \mu_\infty(d) = 0$ for $r \neq 2$ and $R^2 i^! \mu_\infty(d) = \mu_\infty(d-1)$, the spectral sequence,

$$\mathrm{Ext}^r_x(F, R^s i^! \mu_\infty(d)) \Rightarrow \mathrm{Ext}^{r+s}_X(i_*F, \mu_\infty(d)),$$

collapses to give the required isomorphisms.

The theorem for i_*F now follows from the diagram,

$$\begin{array}{ccccccc} \mathrm{Ext}^r_X(i_*F, \mu_\infty(d)) & \times & H^{d+2-r}_x(X, i_*F) & \longrightarrow & H^{d+2}_x(X, \mu_\infty(d)) & \longrightarrow & \mathbb{Q}/\mathbb{Z} \\ \uparrow \approx & & \| & & \uparrow \approx & & \\ \mathrm{Ext}^{r-2}_x(F, \mu_\infty(d-1)) & \times & H^{d+2-r}(x, F) & \longrightarrow & H^d(x, \mu_\infty(d-1)) & \longrightarrow & \mathbb{Q}/\mathbb{Z}, \end{array}$$

and (I.2.17) applied to k.

Let F be a sheaf on u. The theorem for $j_!F$ follows from the diagram,

$$\begin{array}{ccccc} \mathrm{Ext}^r_X(j_!F, \mu_\infty(d)) & \times H^{d+2-r}_x(X, j_!F) & \longrightarrow & H^{d+2}_x(X, \mu_\infty(d) & = \mathbb{Q}/\mathbb{Z} \\ \downarrow \approx & \uparrow \approx & & \uparrow \approx & \\ \mathrm{Ext}^r_u(F, \mu_\infty(d)) & \times H^{d+1-r}(u, F) & \longrightarrow & H^{d+1}(u, \mu_\infty(d) & = \mathbb{Q}/\mathbb{Z} \end{array}$$

and (I.2.17) applied to K. This completes the proof of Theorem 1.16.

For any ring A we write K_rA for the rth Quillen K-group of A, and for any scheme X, we write $\mathcal{K}_r$ for the sheaf on X_{et} associated with the presheaf $U \mapsto K_r\Gamma(U,\mathcal{O}_U)$.

Theorem 1.18. *Let* K *and* p *be as in* (1.16).

(a) *There is a canonical isomorphism*

$$H_x^{d+2}(X,\mathcal{K}_{2d-1})(\text{non-p}) \xrightarrow{\approx} (\mathbb{Q}/\mathbb{Z})(\text{non-p}).$$

(b) *For any constructible sheaf* F *on* X *such that* pF = F,

$$\mathrm{Ext}_X^r(F,\mathcal{K}_{2d-1}) \times H_x^{d+2-r}(X,F) \longrightarrow H_x^{d+2}(X,\mathcal{K}_{2d-1}) \longrightarrow \mathbb{Q}/\mathbb{Z}$$

is nondegenerate pairing of finite groups.

Proof: The main part of the proof is contained in the next lemma. Recall [Browder (1977)] that for any ring A, there are K-groups with coefficients $K_r(A,\mathbb{Z}/m\mathbb{Z})$ fitting into exact sequences

$$0 \longrightarrow K_r(A)^{(m)} \longrightarrow K_r(A,\mathbb{Z}/m\mathbb{Z}) \longrightarrow K_{r-1}(A)_m \longrightarrow 0.$$

Also that for any ring A and integer m that is invertible in A, there is a canonical map $\mu_m \longrightarrow K_2(A,\mathbb{Z}/m\mathbb{Z})$. Using the product structure on the groups $K_r(A,\mathbb{Z}/m\mathbb{Z})$, we obtain a canonical map $\mu_m(r) \longrightarrow K_{2r}(A,\mathbb{Z}/m\mathbb{Z}) \longrightarrow K_{2r-1}(A)$.

Lemma 1.19. *Let* X *be any scheme. If* m *is invertible on* X, *then there is an exact sequence*

$$0 \longrightarrow \mu_m(d) \longrightarrow \mathcal{K}_{2d-1} \xrightarrow{m} \mathcal{K}_{2d-1} \longrightarrow 0$$

of sheaves on X_{et}.

Proof: This is fairly direct consequence of the following two theorems.

(i) Let k be an algebraically closed field; then $K_{2r}k$ is uniquely divisible for all r, and $K_{2r-1}k$ is divisible with torsion subgroup equal to $\mu_\infty(r)$ [Suslin (1983b), (1984)].

(ii) If R is a Henselian local ring with residue field k and m is invertible in R, then $K_r(R,\mathbb{Z}/m\mathbb{Z}) \xrightarrow{\approx} K_r(k,\mathbb{Z}/m\mathbb{Z})$ for all r ([Gabber (1983)]; see also [Suslin (1984)]).

For any field extension L/k of degree p^n, there are maps

$$f_*: K_r(k) \longrightarrow K_r(L), \qquad f^*: K_r(L) \longrightarrow K_r(k)$$

such that $f_*\circ f^* = p^n = f^*\circ f_*$. Therefore $K_r(k)(\text{non-}p) \longrightarrow K_r(L)(\text{non-}p)$ is an isomorphism. This remark, together with (i) and a direct limit argument, implies that for a separably closed field k with char(k) = p, $(K_{2r}k)(\text{non-}p)$ is uniquely divisible for all r and $(K_{2r-1}k)(\text{non-}p)$ is divisible with torsion subgroup equal to $\mu_\infty(r)(\text{non-}p)$. In terms of K-groups with coefficients, this says that $K_{2r}(k,\mathbb{Z}/m\mathbb{Z}) = K_{2r-1}(k)_m = \mu_m(r)$ and $K_{2r-1}(k,\mathbb{Z}/m\mathbb{Z}) = 0$ for all m prime to p.

Let R be a strictly Henselian local ring. From the diagram

$$\begin{array}{ccccccccc}
0 \longrightarrow & K_{2r}(R)^{(m)} & \longrightarrow & K_{2r}(R,\mathbb{Z}/m\mathbb{Z}) & \longrightarrow & K_{2r-1}(R)_m & \longrightarrow 0 \\
& \downarrow & & \downarrow\approx & & \downarrow & \\
0 \longrightarrow & K_{2r}(k)^{(m)} & \longrightarrow & K_{2r}(k,\mathbb{Z}/m\mathbb{Z}) & \xrightarrow{\approx} & K_{2r-1}(k)_m & \longrightarrow 0,
\end{array}$$

we see that $K_{2r-1}(R)_m \xrightarrow{\approx} K_{2r-1}(k)_m$, and therefore that the map $\mu_m(r) \longrightarrow K_{2r-1}(R)_m$ is an isomorphism. As $K_{2r-1}(R,\mathbb{Z}/m\mathbb{Z}) = K_{2r-1}(k,\mathbb{Z}/m\mathbb{Z}) = 0$, we know that $K_{2r-1}(R)^{(m)} = 0$, and therefore that the sequence

$$0 \longrightarrow \mu_m(r) \longrightarrow K_{2r-1}(R) \xrightarrow{m} K_{2r-1}(R) \longrightarrow 0.$$

is exact. This implies the lemma because the exactness of a sequence of sheaves can be checked on the stalks.

Lemma 1.20. *Let X be the spectrum of a Henselian discrete valuation ring with closed point x; then for any sheaf F on X, $H^r_x(X,F)$ is torsion for $r \geq 2$.*

Proof: Let u be the open point of X, and consider the exact sequence

$$\cdots \longrightarrow H^1(u,F) \longrightarrow H^2_x(X,F) \longrightarrow H^2(X,F) \longrightarrow \cdots .$$

From (1.1) we know that $H^r(X,F) \xrightarrow{\approx} H^r(x,i^*F)$ (excellence is not used in the proof of (1.1)), and $H^r(x,i^*F)$ and $H^r(u,F|u)$ are both torsion for $r > 0$ because they are Galois cohomology groups. The lemma follows.

We now complete the proof of (1.18). The exact sequence in the lemma leads to an exact sequence (ignoring p-torsion)

$$0 \longrightarrow H^{d+1}_x(X,\mathcal{K}_{2d-1})\otimes\mathbb{Q}/\mathbb{Z} \longrightarrow H^{d+2}_x(X,\mu_\infty(d)) \longrightarrow H^{d+2}_x(X,\mathcal{K}_{2d-1}) \longrightarrow 0,$$

which shows that $H^{d+2}_x(X,\mu_\infty)(\text{non-}p) \longrightarrow H^{d+2}_x(X,\mathcal{K}_{2d-1})(\text{non-}p)$ because the first term in the sequence is zero. Similarly, $\mathrm{Ext}^r_X(F,\mu_\infty) \xrightarrow{\approx} \mathrm{Ext}^r_X(F,\mathcal{K}_{2d-1})$ for all r, and so the (1.16) implies (1.18).

Remark 1.21. The corollary is a satisfactory generalization of (1.8) in the case that K_1 has characteristic zero. The general case, where the characteristic jumps from p to zero at some later stage, is not yet understood. For a discussion of what the best result should be, see §7 below.

Notes: Part (b) of Theorem 1.8 is usually referred to as the local form of the duality theorem of Artin and Verdier, although [Artin and Verdier (1964)] only discusses global results. In [Deninger (1986c)] it is pointed out that the result extends to singular schemes when $\mathbb{G}_m$

is replaced by $\mathbb{G}$. The extension to higher dimensional schemes in (Theorems 1.16 and 1.18) is taken from [Deninger and Wingberg, (1986)]. The key lemma 1.19 has probably been proved by several people.

§2 Global results: preliminary calculations

Throughout this section, K will be a global field. When K is a number field, X denotes the spectrum of the ring of integers in K, and when K is a function field, k denotes the field of constants of K and X denotes the unique connected smooth complete curve over k having K as its function field. The inclusion of the generic point into X is denoted by g: Spec K $\hookrightarrow$ X, and we sometimes write η for Spec K. For any open subset U of X, U^0 is the set of closed points of U, often regarded as the set of primes of K corresponding to points of U. The residue field at a nonarchi-medean prime v is denoted by k(v). The field K_v is the completion of K at v if v is archimedean, and it is the field of fractions of the Henselization $\mathcal{O}_v^h$ of $\mathcal{O}_v$ otherwise. For a sheaf F on X or an open sub-set of X, we sometimes write F_v for the sheaf on Spec K_v obtained by pulling back relative to the obvious map

$$f_v\colon \operatorname{Spec} K_v \longrightarrow \operatorname{Spec} K \longrightarrow X.$$

Note that when v is nonarchimedean, there is a commutative diagram

$$\begin{array}{ccc} \operatorname{Spec} K_v & \longrightarrow & \operatorname{Spec} K \\ \downarrow & & \downarrow \\ \operatorname{Spec} \mathcal{O}_v^h & \longrightarrow & X \end{array} .$$

For an archimedean prime v of K and a sheaf F on Spec(K_v), we set $H^r(K_v, F) = H_T^r(G_v, M)$ (notation as in I.0) where M is the G_v-module corresponding to F and $G_v = \mathrm{Gal}(K_{v,s}/K_v)$. Therefore $H^r(K,F)$ is zero for all integers r when v is complex, and $H^r(K_v,F)$ is isomorphic to $H^0_T(G_v,M)$ or $H^1(G_v,M)$ according as r is even or odd when v is real. We let $g_v = \mathrm{Gal}(k(v)_s/k(v))$.

The cohomology of $\mathbb{G}_m$

Proposition 2.1. *Let U be an open subset of X, and let S denote the set of all primes of K (including the archimedean primes) not corresponding to a point of U. Then*

$$H^0(U,\mathbb{G}_m) = \Gamma(U,\mathcal{O}_U)^\times,$$
$$H^1(U,\mathbb{G}_m) = \mathrm{Pic}(U),$$

there is an exact sequence

$$0 \longrightarrow H^2(U,\mathbb{G}_m) \longrightarrow \bigoplus_{v\in S} \mathrm{Br}(K_v) \xrightarrow{\sum \mathrm{inv}_v} \mathbb{Q}/\mathbb{Z} \longrightarrow H^3(U,\mathbb{G}_m) \longrightarrow 0,$$

and

$$H^r(U,\mathbb{G}_m) = \bigoplus_{v \text{ real}} H^r(K_v,\mathbb{G}_m), \quad r \geq 4.$$

Proof: Let $g\colon \eta \hookrightarrow U$ be the inclusion of the generic point of U. There is an exact sequence [Milne (1980), II.3.9]

$$0 \longrightarrow \mathbb{G}_m \longrightarrow g_*\mathbb{G}_{m,\eta} \longrightarrow \mathrm{Div}_U \longrightarrow 0$$

with $\mathrm{Div}_U = \bigoplus_{v\in U^0} i_{v*}\mathbb{Z}$ the sheaf of Weil divisors on U. The same argument as in (1.4) shows that $R^s g_*\mathbb{G}_m = 0$ for $s > 0$, and so the Leray spectral sequence for g reduces to a family of isomorphisms

$$H^r(U,g_*\mathbb{G}_{m,\eta}) \xrightarrow{\approx} H^r(K,\mathbb{G}_m), \quad r \geq 0.$$

Clearly,

$$H^r(U,\mathrm{Div}_U) = \bigoplus H^r(v,i_*\mathbb{Z}) = \bigoplus H^r(k(v),\mathbb{Z}),$$

and we know from (I.A.2) that $H^r(k(v),\mathbb{Z})$ can be identified with $\mathbb{Z}$, 0, $\mathrm{Br}(K_v)$, 0 ... for r = 0, 1, 2, 3 ... . The cohomology sequence of the above short exact sequence therefore gives exact sequences

$$0 \longrightarrow H^0(U,\mathbb{G}_m) \longrightarrow K^{\times} \longrightarrow \bigoplus_{v \in U^0} \mathbb{Z} \longrightarrow H^1(U,\mathbb{G}_m) \longrightarrow 0$$

$$0 \longrightarrow H^2(U,\mathbb{G}_m) \longrightarrow \mathrm{Br}(K) \longrightarrow \bigoplus_{v \in U^0} \mathrm{Br}(K_v) \longrightarrow H^3(U,\mathbb{G}_m) \longrightarrow H^3(K,\mathbb{G}_m) \longrightarrow 0$$

$$H^r(U,\mathbb{G}_m) \xrightarrow{\approx} H^r(K,\mathbb{G}_m), \quad r \geq 4.$$

The first sequence shows that $H^0(U,\mathbb{G}_m)$ and $H^1(U,\mathbb{G}_m)$ have the values claimed in the statement of the proposition. Global class field theory ([Tate (1967a)] and (I.A.7)) provides an exact sequence

$$0 \longrightarrow \mathrm{Br}(K) \longrightarrow \bigoplus_{\text{all } v} \mathrm{Br}(K_v) \longrightarrow \mathbb{Q}/\mathbb{Z} \longrightarrow 0.$$

From this and the second of the above sequences, we see that the kernel-cokernel sequence of the pair of maps

$$\mathrm{Br}(K) \longrightarrow \bigoplus_{\text{all } v} \mathrm{Br}(K_v) \longrightarrow \bigoplus_{v \in U^0} \mathrm{Br}(K_v)$$

is the required exact sequence

$$0 \longrightarrow H^2(U,\mathbb{G}_m) \longrightarrow \bigoplus_{v \in S} \mathrm{Br}(K_v) \xrightarrow{\sum} \mathbb{Q}/\mathbb{Z} \longrightarrow H^3(U,\mathbb{G}_m) \longrightarrow H^3(K,\mathbb{G}_m) \longrightarrow 0.$$

To complete the proof in the number field case we apply (I.4.18) (or (I.4.21)), which tells us that the map $H^r(K,\mathbb{G}_m) \longrightarrow \bigoplus_{v \text{ real}} H^r(K_v,\mathbb{G}_m)$ is an isomorphism for $r \geq 3$, and note that $H^r(\mathbb{R},\mathbb{G}_m) = 0$ for $r = 3$. In the function field case, we know that K has cohomological dimension ≤ 2, which implies that $H^r(K,\mathbb{G}_m) = 0$ for $r \geq 3$.

Remark 2.2. (a) If S contains at least one nonarchimedean prime, then the map $\sum \mathrm{inv}_v\colon \bigoplus_{v \in S} \mathrm{Br}(K_v) \longrightarrow \mathbb{Q}/\mathbb{Z}$ is surjective, and so in this case we get an exact sequence

$$0 \longrightarrow H^2(U,\mathbb{G}_m) \longrightarrow \bigoplus_{v \in S} \mathrm{Br}(K_v) \longrightarrow \mathbb{Q}/\mathbb{Z} \longrightarrow 0$$

and isomorphisms

$$H^r(U,\mathbb{G}_m) \xrightarrow{\approx} \bigoplus_{v \text{ real}} H^r(K_v,\mathbb{G}_m), \quad r \geq 3, \text{ K a number field},$$

$H^r(U,\mathbb{G}_m) = 0$, $r \geq 3$, K a function field.

(b) If K has no real primes, then $H^2(X,\mathbb{G}_m) = 0$, $H^3(X,\mathbb{G}_m) = \mathbb{Q}/\mathbb{Z}$, and $H^r(X,\mathbb{G}_m) = 0$ for $r \geq 4$.

Cohomology with compact support

We shall define *cohomology groups with compact support* $H^r_c(U,F)$, $r \in \mathbb{Z}$, that take into account the real primes. They will fit into an exact sequence

$$\cdots \longrightarrow H^r_c(U,F) \longrightarrow H^r(U,F) \longrightarrow \bigoplus_{\text{all } v} H^r(K_v,F_v) \longrightarrow H^{r+1}_c(U,F) \longrightarrow \cdots .$$

In particular, $H^r_c(U,F) \xrightarrow{\approx} \bigoplus_{v \text{ real}} H^{r-1}(K_v,F_v)$ for $r < 0$. Except in the case that K is a real number field, $H^r_c(U,F) = H^r(X,j_!F)$, and in that case $H^r_c(U,F)$ differs from $H^r(X,j_!F)$ by a group killed by 2. Therefore the reader who is prepared to ignore the prime 2 can skip this subsection.

Let F be a sheaf on U, and write $\mathcal{C}^{\cdot}(F)$ for the canonical Čech complex of F defined in §0. Thus $\Gamma(U,\mathcal{C}^{\cdot}(F)) = C^{\cdot}(U,F)$, and $H^r(C^{\cdot}(U,F)) = H^r(U,F)$. For each prime v, there is a canonical map $f_v^*\mathcal{C}^{\cdot}(F) \longrightarrow \mathcal{C}^{\cdot}(F_v)$ and therefore also a map $C^{\cdot}(U,F) \longrightarrow C^{\cdot}(K_v,F_v)$. As we noted in (0.8), $C^{\cdot}(K_v,F_v)$ can be identified with the standard (inhomogeneous) resolution $C^{\cdot}(M_v)$ of the G_v-module M_v associated with F_v. When v is real, we write $S^{\cdot}(M_v)$ (or $S^{\cdot}(K_v,F_v)$) for a standard complete resolution of M_v, and otherwise we set $S^{\cdot}(M_v) = C^{\cdot}(M_v)$. In either case there is a canonical map $C^{\cdot}(M_v) \longrightarrow S^{\cdot}(M_v)$. On combining this with the previous maps, we obtain a canonical morphism of complexes

$$u\colon C^{\cdot}(U,F) \longrightarrow \bigoplus_{v \notin U} S^{\cdot}(M_v)$$

(sum over all primes of K not in U). We define $H_c(X,F)$ to be the translate $C^{\cdot}(u)[-1]$ of the mapping cone of u, and we define $H_c^r(X,F) = H^r(H_c(X,F))$.

Proposition 2.3. (a) *For any sheaf* F *on an open subscheme* $U \subset X$, *there is an exact sequence*

$$\dots \longrightarrow H_c^r(U,F) \longrightarrow H^r(U,F) \longrightarrow \bigoplus_{v \notin U} H^r(K_v,F_v) \longrightarrow \dots .$$

(The sum is over all primes of K, *including the archimedean primes, not in* U.*)*

(b) *For any short exact sequence*

$$0 \longrightarrow F' \longrightarrow F \longrightarrow F'' \longrightarrow 0$$

of sheaves on U, *there is an exact sequence of cohomology groups*

$$\dots \longrightarrow H_c^r(U,F') \longrightarrow H_c^r(U,F) \longrightarrow H_c^r(U,F'') \longrightarrow \dots .$$

(c) *If* $i: Z \hookrightarrow U$ *is a closed immersion and* F *is a sheaf on* Z, *then* $H_c^r(U,i_*F) = H^r(Z,F)$.

(d) *If* $j: V \hookrightarrow U$ *is an open immersion and* F *is a sheaf on* V, *then* $H_c^r(U,j_!F) = H_c^r(V,F)$. *Therefore, for any sheaf* F *on* U, *there is an exact sequence*

$$\dots \longrightarrow H_c^r(V,F|V) \longrightarrow H_c^r(U,F) \longrightarrow \bigoplus_{v \in U-V} H^r(v,i_v^*F) \longrightarrow \dots .$$

(e) *For any finite map* $\pi: U' \longrightarrow U$ *and sheaf* F *on* U', *there is a canonical isomorphism* $H_c^r(U,\pi_*F) \xrightarrow{\approx} H_c^r(U',F)$.

Proof: (a) This is obvious from the definition of $H_c(U,F)$ and the properties of mapping cones (see §0).

(b) From the morphism

$$\begin{array}{ccccccccc} 0 \longrightarrow & C^{\cdot}(U,F') & \longrightarrow & C^{\cdot}(U,F) & \longrightarrow & C^{\cdot}(U,F'') & \longrightarrow 0 \\ & \downarrow u' & & \downarrow u & & \downarrow u'' & \\ 0 \longrightarrow & \bigoplus S^{\cdot}(K_v,F'_v) & \longrightarrow & \bigoplus S^{\cdot}(K_v,F_v) & \longrightarrow & \bigoplus S^{\cdot}(K_v,F''_v) & \longrightarrow 0, \end{array}$$

of short exact sequences of complexes, we obtain a distinguished triangle

$$H_c(U,F'')[-1] \longrightarrow H_c(U,F') \longrightarrow H_c(U,F) \longrightarrow H_c(U,F'').$$

(see 0.10a). This yields the required exact sequence.

(c) Since the stalk of i_*F at the generic point is zero, $H^r(K_v,(i_*F)_v) = 0$ for all v. Therefore $H^r_c(U,i_*F) \xrightarrow{\approx} H^r(U,i_*F)$, and the second group is isomorphic to $H^r(Z,F)$.

(d) We first need a lemma.

Lemma 2.4. *Under the hypotheses of* (d), *there is a long exact sequence*

$$\cdots \longrightarrow H^r(U,j_!F) \longrightarrow H^r(V,F) \longrightarrow \bigoplus_{v\in U-V} H^r(K_v,F) \longrightarrow \cdots .$$

Proof: The cohomology sequence of the pair $U \supset V$ is

$$\cdots \longrightarrow H^r_{U-V}(U,j_!F) \longrightarrow H^r(U,j_!F) \longrightarrow H^r(V,F) \longrightarrow \cdots .$$

By excision [Milne (1980), III.1.28], $H^r_{U-V}(U,j_!F) = \oplus H^r_v(U_v,j_!F)$ where $U_v = \operatorname{Spec} \mathcal{O}^h_v$, and according to (1.1), $H^r_v(U_v,j_!F) = H^{r-1}(K_v,F)$. The lemma is now obvious.

On carrying out the proof of the lemma on the level of complexes, we find that the mapping cone of $C^\cdot(U,j_!F) \longrightarrow C^\cdot(V,j_!F|V) = C^\cdot(V,F)$ is quasi-isomorphic to $\bigoplus_{v\in U-V} C^\cdot(K_v,F_v)$. The cokernel of $\bigoplus_{v\notin U} S^\cdot(K_v,F_v) \longrightarrow \bigoplus_{v\notin V} S^\cdot(K_v,F_v)$ is also $\bigoplus_{v\in U-V} C^\cdot(K_v,F)$, and the mapping cone of $H_c(U,j_!F) \longrightarrow H_c(V,F)$ is therefore quasi-isomorphic to the mapping cone of a map $\bigoplus_{v\in U-V} C^\cdot(K_v,F) \longrightarrow \bigoplus_{v\in U-V} C^\cdot(K_v,F)$. The map is the identity, and therefore $H_c(U,j_!F) \longrightarrow H_c(V,F)$ is a quasi-isomorphism. This completes the proof of the first part of (d), and to deduce the second part one only has to replace $H^r_c(U,j_!F)$ with $H^r_c(V,F)$ in the cohomology sequence of

$$0 \longrightarrow j_! j^* F \longrightarrow F \longrightarrow i^* i_* F \longrightarrow 0.$$

Finally, (e) of the proposition follows easily from the existence of isomorphisms $H^r(U, \pi_* F) \xrightarrow{\approx} H^r(U', F)$ and $H^r(K_v, \pi_* F) \xrightarrow{\approx} H^r(K'_v, F)$.

Proposition 2.5. (a) *For any sheaves* F *and* F' *on* $U \subset X$, *there is a canonical pairing*

$$\langle\ ,\ \rangle\colon \mathrm{Ext}^r_U(F, F') \times H^s_c(U, F) \longrightarrow H^{r+s}_c(U, F').$$

(b) *For any pairing* $F \times F' \longrightarrow F''$ *of sheaves on* $U \subset X$, *there is a natural cup-product pairing*

$$H^r(U, F) \times H^s_c(U, F') \longrightarrow H^{r+s}_c(U, F'').$$

(c) *The following diagram commutes*:

$$\begin{array}{ccccc} H^r(U, \mathcal{H}om(F, F')) & \times H^s_c(U, F) & \longrightarrow & H^{r+s}_c(U, F') & (\textit{Ext pairing}) \\ \downarrow & \| & & \| & \\ \mathrm{Ext}^r_U(F, F') & \times H^s_c(U, F) & \longrightarrow & H^{r+s}_c(U, F') & (\textit{cup-product}). \end{array}$$

Proof: (a) For example, represent an element of $\mathrm{Ext}^r_U(F, F')$ by an r-fold extension, and take $H^r_c(U, F) \longrightarrow H^{r+s}_c(U, F')$ to be the corresponding r-fold boundary map.

(b) The cup-product pairing on the Čech complexes [Milne (1980), V.1.19]

$$C^{\cdot}(U, F) \times C^{\cdot}(U, F') \longrightarrow C^{\cdot}(U, F'')$$

combined with the cup-product pairing on the standard complexes [Cartan and Eilenberg (1956), XII]

$$S^{\cdot}(K_v, F_v) \times S^{\cdot}(K_v, F'_v) \longrightarrow S^{\cdot}(K_v, F''_v)$$

give a natural pairing

$$H_c(U,F) \times H_c(U,F') \longrightarrow H_c(U,F'').$$

(c) Combine (I.0.14) with (0.7).

The cohomology of $\mathbb{G}_m$ with compact support

Proposition 2.6. *Let* U *be an open subscheme of* X. *Then* $H_c^2(U,\mathbb{G}_m) = 0$, $H_c^3(U,\mathbb{G}_m) = \mathbb{Q}/\mathbb{Z}$, *and* $H_c^r(U,\mathbb{G}_m) = 0$, $r > 3$.

Proof: Part (a) of (2.3) gives an exact sequence:

$$0 \longrightarrow H_c^2(U,\mathbb{G}_m) \longrightarrow H^2(U,\mathbb{G}_m) \longrightarrow \bigoplus_{v\in S} Br(K_v)$$
$$\longrightarrow H_c^3(U,\mathbb{G}_m) \longrightarrow H^3(U,\mathbb{G}_m) \longrightarrow \bigoplus_{v \text{ real}} H^3(K_v,\mathbb{G}_m) \longrightarrow \cdots .$$

In the case that $U \neq X$, this and (2.2a) immediately give the proposition. Part (d) of (2.3) and the next lemma show that $H_c^r(U,\mathbb{G}_m) = H_c^r(X,\mathbb{G}_m)$ for $r \geq 2$.

Lemma 2.7. *For any closed immersion* $i: Z \hookrightarrow U$ *with* $i(Z) \neq U$, $H^r(Z,i^*\mathbb{G}_m) = 0$ *all* $r \geq 1$.

Proof: It suffices to prove this with Z equal to a single point v of U. Then $i^*\mathbb{G}_m$ corresponds to the g_v-module $\mathcal{O}_{v,un}^\times$, and so $H^r(Z,i^*\mathbb{G}_m) = H^r(g_v,\mathcal{O}_{v,un}^\times)$. The sequence

$$0 \longrightarrow \mathcal{O}_{v,un}^\times \longrightarrow K_{v,un}^\times \xrightarrow{\text{ord}} \mathbb{Z} \longrightarrow 0$$

is split as a sequence of g_v-modules, and so $H^r(g_v,\mathcal{O}_{v,un}^\times)$ is a direct summand of $H^r(g_v,K_{v,un}^\times)$. Therefore Hilbert's theorem 90 shows that $H^1(g_v,\mathcal{O}_{v,un}^\times) = 0$, and we know from (I.A.2) that $H^2(g_v,\mathcal{O}_{v,un}^\times) = 0$. As g_v has strict cohomological dimension 2, this completes the proof.

Remark 2.8. (a) Let K be a number field, and let R be the ring of integers in K. Then there is an exact sequence

$$0 \longrightarrow H^0_c(X,\mathbb{G}_m) \longrightarrow R^\times \longrightarrow \bigoplus_{v \text{ real}} K_v^\times/K_v^{\times 2} \longrightarrow H^1_c(X,\mathbb{G}_m) \longrightarrow \mathrm{Pic}(R) \longrightarrow 0,$$

where Pic(R) is the ideal class group of R. In particular,

$$H^0_c(X,\mathbb{G}_m) = \{a \in R^\times \mid \mathrm{sign}(a_v) > 0 \text{ all real } v\}$$
$$= \text{group of totally positive units in } K.$$

The cohomology sequence with compact support of

$$0 \longrightarrow \mathbb{G}_m \longrightarrow g_*\mathbb{G}_m \longrightarrow \bigoplus_{v \in X^0} i_{v*}\mathbb{Z} \longrightarrow 0$$

is

$$H^0_c(X,g_*\mathbb{G}_m) \longrightarrow \bigoplus_{v \text{ nonarch}} \mathbb{Z} \longrightarrow H^1_c(X,\mathbb{G}_m) \longrightarrow H^1_c(X,g_*\mathbb{G}_m).$$

The exact sequence given by (2.3a)

$$0 \longrightarrow H^0_c(X,g_*\mathbb{G}_m) \longrightarrow K^\times \longrightarrow \bigoplus_{v \text{ real}} K_v^\times/K_v^{\times 2} \longrightarrow H^1_c(X,g_*\mathbb{G}_m) \longrightarrow 0$$

shows that $H^0_c(X,g_*\mathbb{G}_m)$ is the group of totally positive elements of $K^\times$ and $H^1_c(X,g_*\mathbb{G}_m) = 0$. Let Id(R) be the group of ideals in R. Then

$$H^1_c(X,\mathbb{G}_m) = \mathrm{Id}(R)/\{(a) \mid a \in K^\times,\ \mathrm{sign}(a_v) > 0 \text{ all real } v\}$$
$$= \text{group of ideal classes of } K \text{ in the narrow sense}$$

(see [Narkiewicz (1974), III, §2, §3]).

(b) Unfortunately $H^1_c(X,\mathbb{G}_m)$ is not equal to the group of isometry classes of Hermitian invertible sheaves on X (the "compactified Picard group of R" in the sense of Arakelov theory; see [Szpiro (1985), §1]). I have no idea if there is a reasonable definition of the étale cohomology groups of an Arakelov variety. Our definition of the cohomology groups with compact support has been chosen so as to lead to good duality theorems.

Cohomology of locally constant sheaves

Let U be an affine open subset of X, and let S be the set of primes of K not corresponding to a point of U. Since to be affine in the function field case simply means that $U \neq X$, S will satisfy the conditions in the first paragraph of I.4. With the notations of (I.4), $G_S = \pi_1(U,\bar{\eta})$, and the functor $F \mapsto F_{\bar{\eta}}$ defines an equivalence between the category of locally constant $\mathbb{Z}$-constructible sheaves on U and the category of finitely generated G_S-modules. We write $\tilde{U}$ for the normalization of U in K_S; thus $\tilde{U} = \operatorname{Spec} R_S$ where, as in (I.4), R_S is the integral closure of $R_{K,S}$ in K_S.

Proposition 2.9. *Let* F *be a locally constant* $\mathbb{Z}$*-constructible sheaf on an open affine subscheme* U *of* X, *and let* $M = F_{\bar{\eta}}$. *Then* $H^r(U,F)$ *is a torsion group for all* $r \geq 1$, *and* $H^r(U,F)(\ell) = H^r(G_S,M)(\ell)$ *if* ℓ *is invertible on* U *or* $\ell = \operatorname{char}(K)$.

Proof: The Hochschild-Serre spectral sequence for $\tilde{U}/U$ is

$$H^r(G_S,H^s(\tilde{U},F|\tilde{U})) \Longrightarrow H^{r+s}(\tilde{U},F).$$

As $H^0(\tilde{U},F) = M$, we have to show that $H^s(\tilde{U},F|\tilde{U})$ is torsion for $s > 0$ and that $H^s(\tilde{U},F|\tilde{U})(\ell) = 0$ if ℓ is invertible on U or equals $\operatorname{char}(K)$. By assumption $F|\tilde{U}$ is constant, and so there are three cases to consider: $F|\tilde{U} = \mathbb{Z}/\ell\mathbb{Z}$ with ℓ invertible on U, $F|\tilde{U} = \mathbb{Z}/p\mathbb{Z}$ with $p = \operatorname{char}(K)$, and $F|\tilde{U} = \mathbb{Z}$.

The first cohomology group can be disposed off immediately, because

$$H^1(\tilde{U},F) = \operatorname{Hom}(\pi_1(\tilde{U},\bar{\eta}),F(\tilde{U})),$$

and $\pi_1(\tilde{U},\bar{\eta})$ is zero.

Now let $F|\tilde{U} = \mathbb{Z}/\ell\mathbb{Z}$ with ℓ a prime that is invertible in R_S. Then $\mathbb{Z}/\ell\mathbb{Z} \approx \mu_\ell$ on $\tilde{U}$, and the remark just made shows that the cohomology sequence of

$$0 \longrightarrow \mu_\ell \longrightarrow \mathbb{G}_m \xrightarrow{\ell} \mathbb{G}_m \longrightarrow 0$$

is

$$0 \longrightarrow \mathrm{Pic}(\tilde{U}) \xrightarrow{\ell} \mathrm{Pic}(\tilde{U}) \longrightarrow H^2(\tilde{U},\mathbb{Z}/\ell\mathbb{Z}) \longrightarrow \mathrm{Br}(\tilde{U})_\ell \longrightarrow 0.$$

The Picard group of $\tilde{U}$ is the direct limit of the Picard groups of the finite étale coverings U' of U and so is torsion. The sequence shows that $\mathrm{Pic}(\tilde{U})(\ell) = 0$, and so $H^2(\tilde{U},\mathbb{Z}/\ell\mathbb{Z})$ injects into $\mathrm{Br}(\tilde{U})$. Let $L \subset K_S$ be a finite extension of K containing the ℓth roots of 1, and consider the exact sequence (see 2.2a)

$$0 \longrightarrow \mathrm{Br}(R_{L,S}) \longrightarrow \bigoplus_{w\in S_L} \mathrm{Br}(L_w) \longrightarrow \mathbb{Q}/\mathbb{Z} \longrightarrow 0.$$

Let L' be a finite extension of L; it is clear from the sequence and local class field theory that an element a of $\mathrm{Br}(R_{L,S})_\ell$ maps to zero in $\mathrm{Br}(R_{L',S})$ if ℓ divides the local degree of L'/L at all w in S_L. Let H be the Hilbert class field of L. Then the prime ideal corresponding to w becomes principal in H with generator c_w say. The field L' generated over H by the elements $c_w^{1/\ell}$, $w \in S_L$, splits a. As L' is contained in K_S, this argument shows that $\varinjlim \mathrm{Br}(L)_\ell = 0$, and therefore that $\mathrm{Br}(\tilde{U})(\ell) = 0$. Hence $H^2(\tilde{U},\mathbb{Z}/\ell\mathbb{Z}) = 0$. Finally, (2.2) shows that $H^r(U_L,\mathbb{G}_m) = 0$ for $r > 2$, where $U_L = \mathrm{Spec}\, R_{L,S}$, because L has no real primes, and so $H^r(\tilde{U},\mathbb{G}_m)(\ell) = 0$ for all $r > 2$.

In the case $F = \mathbb{Z}/p\mathbb{Z}$, $p = \mathrm{char}(K)$, we replace the Kummer sequence with the Artin-Schreier sequence:

$$0 \longrightarrow \mathbb{Z}/p\mathbb{Z} \longrightarrow \mathcal{O}_{\tilde{U}} \xrightarrow{\wp} \mathcal{O}_{\tilde{U}} \longrightarrow 0, \qquad \wp(a) = a^p - a.$$

As $H^r(\tilde{U}_{et},\mathcal{O}) = H^r(\tilde{U}_{Zar},\mathcal{O})$, and the latter group is zero for $r \geq 1$, we see immediately that $H^r(\tilde{U},\mathbb{Z}/p\mathbb{Z}) = 0$ for $r \geq 2$.

Finally consider $\mathbb{Z}$. The next lemma shows that $H^r(\tilde{U},\mathbb{Z})$ is torsion for $r > 0$, and so from the cohomology sequence of

$$0 \longrightarrow \mathbb{Z} \xrightarrow{\ell} \mathbb{Z} \longrightarrow \mathbb{Z}/\ell\mathbb{Z} \longrightarrow 0$$

and the results in the preceding three paragraphs, we can deduce that $H^r(\tilde{U},\mathbb{Z})(\ell) = 0$ for $r > 0$ if $\ell = \mathrm{char}(K)$ or ℓ is invertible on U.

Lemma 2.10. *For any normal Noetherian scheme Y and constant sheaf F on Y, the cohomology groups* $H^r(Y,F)$ *are torsion for all* $r > 0$. *In particular,* $H^r(Y,F) = 0$ *for* $r > 0$ *if F is constant and uniquely divisible.*

Proof: We may assume that Y is connected. Let $g\colon \eta \hookrightarrow Y$ be its generic point. Then $g_*g^*F = F$ and the stalks of $R^rg_*(g^*F)$, being Galois cohomology groups, are torsion (see [Milne (1980), II.3.7, III.1.15]). Therefore if F is uniquely divisible, then $R^rg_*(g^*F) = 0$ for $r > 0$, and $H^r(Y,F) = H^r(\eta,g^*F) = 0$ for $r > 0$, which proves the lemma in this case. Next note that it suffices to prove the lemma for a torsion-free constant F. For such a sheaf, the cohomology sequence of

$$0 \longrightarrow F \longrightarrow F\otimes\mathbb{Q} \longrightarrow (F\otimes\mathbb{Q})/F \longrightarrow 0$$

shows that $H^{r-1}(Y,(F\otimes\mathbb{Q})/F)$ maps onto $H^r(Y,F)$ for $r \geq 1$ because $F\otimes\mathbb{Q}$ is uniquely divisible. This completes the proof as $(F\otimes\mathbb{Q})/F$ is torsion.

Corollary 2.11. *Let U be an open subscheme of X, and let S denote the set of primes of K not corresponding to a point of U.*

(a) *For all* $r < 0$, $H^r_c(U,\mathbb{Z}) \approx \bigoplus_{v \text{ real}} H^{r-1}(K_v,\mathbb{Z})$; *in particular,* $H^r_c(U,\mathbb{Z}) = 0$ *if r is even and* < 0.

(b) *There is an exact sequence*

$$0 \longrightarrow H^0_c(U,\mathbb{Z}) \longrightarrow \mathbb{Z} \longrightarrow \bigoplus_{v\in S} H^0(K_v,\mathbb{Z}) \longrightarrow H^1_c(X,\mathbb{Z}) \longrightarrow 0.$$

If S contains at least one nonarchimedean prime, then $H^0_c(U,\mathbb{Z}) = 0$.

(c) *There is an exact sequence*

$$0 \longrightarrow H^2_c(U,\mathbb{Z}) \longrightarrow H^2(U,\mathbb{Z}) \longrightarrow \bigoplus_{v\in S} H^2(K_v,\mathbb{Z}) \longrightarrow H^3_c(U,\mathbb{Z}) \longrightarrow H^3(U,\mathbb{Z}) \longrightarrow 0.$$

For all primes ℓ *that are invertible on U or equal the characteristic of K, there is an exact sequence*

$$0 \longrightarrow H^2_c(U,\mathbb{Z})(\ell) \longrightarrow H^2(G_S,\mathbb{Z})(\ell) \longrightarrow \bigoplus_{v\in S} H^2(K_v,\mathbb{Z})(\ell)$$
$$\longrightarrow H^3_c(U,\mathbb{Z})(\ell) \longrightarrow H^3(G_S,\mathbb{Z})(\ell) \longrightarrow 0.$$

(d) *For all* $r \geq 4$, $H^r_c(U,\mathbb{Z}) = 0$.

Proof: All the statements follow from the exact sequence

$$\cdots \longrightarrow H^r_c(U,\mathbb{Z}) \longrightarrow H^r(U,\mathbb{Z}) \longrightarrow \bigoplus_{v\in S} H^r(K_v,\mathbb{Z}) \longrightarrow \cdots .$$

For $r < 0$, $H^r(U,\mathbb{Z}) = 0$, and so the sequence gives isomorphisms

$$\bigoplus_{y\in S} H^{r-1}(K_v,\mathbb{Z}) \xrightarrow{\approx} H^r_c(U,\mathbb{Z}).$$

For $r \neq 0,1,2$, $\bigoplus_{v\in S} H^r(K_v,\mathbb{Z}) = \bigoplus_{v \text{ real}} H^r(K_v,\mathbb{Z})$. Since $H^r(\mathbb{R},\mathbb{Z}) = 0$ for odd r, these calculations prove (a).

As $H^0(U,\mathbb{Z}) = \mathbb{Z}$ and $H^1(U,\mathbb{Z}) = \mathrm{Hom}_{cts}(\pi_1(U,\eta),\mathbb{Z}) = 0$, we have an exact sequence

$$0 \longrightarrow H^0_c(U,\mathbb{Z}) \longrightarrow \mathbb{Z} \longrightarrow \bigoplus_{v\in S} H^0(K_v,\mathbb{Z}) \longrightarrow H^1_c(U,\mathbb{Z}) \longrightarrow 0.$$

When S contains a nonarchimedean prime, the middle map is injective, and so this proves (b).

The first part of (c) is obvious from the fact that $H^1(K_v,\mathbb{Z}) = 0 = H^3(K_v,\mathbb{Z})$ for all v, and the proposition allows us to replace $H^r(U,\mathbb{Z})(\ell)$ with $H^r(G_S,\mathbb{Z})(\ell)$ for the particular ℓ.

For (d), we begin by showing that $H^r_c(U,\mathbb{Z})(\ell) = 0$ for $r \geq 4$ when ℓ is a prime that is invertible on U. Consider the diagram

$$\begin{array}{ccc} H^{r-1}(G_S,\mathbb{Q}/\mathbb{Z}) & \longrightarrow & \bigoplus_{v \text{ real}} H^{r-1}(K_v,\mathbb{Q}/\mathbb{Z}) \\ \downarrow & & \downarrow \\ H^r(G_S,\mathbb{Z}) & \longrightarrow & \bigoplus_{v \text{ real}} H^r(K_v,\mathbb{Z}). \end{array}$$

The vertical arrows (boundary maps) are isomorphisms for $r \geq 2$ because $\mathbb{Q}$ is uniquely divisible, and Theorem I.4.10c shows that the top arrow is an isomorphism on the ℓ-primary components for $r \geq 4$ if ℓ is invertible on U. Therefore the maps

$$H^r(G_S,\mathbb{Z})(\ell) \longrightarrow \bigoplus_{v \text{ real}} H^r(K_v,\mathbb{Z})(\ell)$$

are isomorphisms for $r \geq 4$ and all ℓ that are invertible on U. As $H^3(\mathbb{R},\mathbb{Z}) = 0$, this proves that $H^r_c(U,\mathbb{Z})(\ell) = 0$ for $r \geq 4$ and such ℓ. Since $H^r_c(U,F) = H^r_c(U[1/\ell],F)$ for any $r \geq 4$, this completes the proof except for the p-primary component in characteristic p. We may assume that U is affine. The cohomology sequence of

$$0 \longrightarrow \mathbb{Z}/p\mathbb{Z} \longrightarrow \mathcal{O}_U \xrightarrow{\wp} \mathcal{O}_U \longrightarrow 0$$

shows that $H^r(U,\mathbb{Z}/p\mathbb{Z}) = 0$ for $r \geq 2$. Therefore $H^r(U,\mathbb{Z})(p) = 0$ for $r \geq 3$, and this implies that $H^r_c(U,\mathbb{Z})(p) = 0$ for $r \geq 4$.

Remark 2.12. It has been conjectured that $\operatorname{scd}_\ell(G_S) = 2$ for all primes ℓ that are invertible in $R_{K,S}$. This would imply that the map $\bigoplus_{v\in S} H^2(K_v,\mathbb{Z})(\ell) \longrightarrow H^3_c(U,\mathbb{Z})(\ell)$ in (2.11c) is surjective on the ℓ-primary components for such ℓ.

Euler-Poincaré characteristics

Let F be a constructible sheaf on U such that mF = 0 for some m that is invertible on U. We shall see that the groups $H^r(U,F)$ and $H^r_c(U,F)$ are all finite, and so it makes sense to define

$$\chi(U,F) = \frac{[H^0(U,F)][H^2(U,F)]}{[H^1(U,F)][H^3(U,F)]}, \qquad \chi_c(U,F) = \frac{[H^0_c(U,F)][H^2_c(U,F)]}{[H^1_c(U,F)][H^3_c(U,F)]}.$$

Theorem 2.13. *Let F be a constructible sheaf on U such that mF = 0 for some m that is invertible on U.*

(a) *The groups $H^r(U,F)$ are finite, and*

$$\chi(U,F) = \prod_{v \text{ arch}} \frac{[F(K_v)]}{[H^0(K_v,F)]|[F(K_s)]|_v}.$$

(b) *The groups* $H_c^r(U,F)$ *are finite, and* $\chi_c(U,F) = \prod_{v \text{ arch}} [F(K_v)]$.

Proof: (a) Choose an open affine subscheme V of U such that $F|V$ is locally constant. Theorem 2.9 shows that $H^r(V,F) = H^r(G_S,M)$ for all r, where S is the set of primes of K not in V and M is the G_S module corresponding to $F|V$. Therefore Theorem (I.5.1) shows that

$$\chi(V,F|V)[H^3(V,F|V)] = \chi(G_S,M) = \prod_{v \text{ arch}} [H^0(G_v,M)]/|[M]|_v.$$

As $H^3(V,F|V) \xrightarrow{\approx} \prod H^3(K_v,M)$ (by 4.10c), and the groups $H^r(K_v,M)$ for a fixed nonarchimedean prime v all have the same order (recall that they are Tate cohomology groups), this proves the result for $F|V$, and it remains to show that $\chi(U,F) = \chi(V,F|V)$. The sequence

$$\cdots \longrightarrow \bigoplus_{v\in U-V} H_v^r(\mathcal{O}_v^h,F) \longrightarrow H^r(U,F) \longrightarrow H^r(V,F) \longrightarrow \cdots$$

shows that $\chi(U,F) = \chi(V,F|V) \times \prod \chi_v(\mathcal{O}_v^h,F)$, and the sequence

$$\cdots \longrightarrow H_v^r(\mathcal{O}_v^h,F) \longrightarrow H^r(\mathcal{O}_v^h,F) \longrightarrow H^r(K_v,F) \longrightarrow \cdots$$

shows that $\chi_v(\mathcal{O}_v^h,F) = \chi(\mathcal{O}_v^h,F)\chi(K_v,F)^{-1}$. But $F(K_{v,s})$ has order prime to the residue characteristic of K, and so (I.2.8) shows that $\chi(K_v,F) = 1$. Moreover (see 1.1) $H^r(\mathcal{O}_v^h,F) = H^r(g_v,F(\mathcal{O}_v^h))$ and $F(\mathcal{O}_v^h)$ is finite, and so it is obvious that $\chi(\mathcal{O}_v^h,F) = 1$ (see [Serre (1962), XIII.1]).

(b) The sequence (2.3d)

$$\cdots \longrightarrow H_c^r(V,F|V) \longrightarrow H_c^r(U,F) \longrightarrow \bigoplus_{v\in U-V} H^r(v,i^*F) \longrightarrow \cdots$$

shows that $\chi_c(U,F) = \chi_c(V,F|V)$ for any open subscheme V of U, and so we can assume that $U \neq X$ and that F is locally constant. There is an exact sequence

$$0 \longrightarrow \prod_{v \text{ arch}} H^{-1}(K_v,F) \longrightarrow H^0_c(U,F) \longrightarrow H^0(U,F) \longrightarrow \prod_{v \notin U} H^0(K_v,F) \longrightarrow \cdots$$

$$\cdots \longrightarrow H^3_c(U,F) \longrightarrow H^3(U,F) \longrightarrow \prod_{v \text{ arch}} H^3(K_v,F) \longrightarrow 0$$

because (see I.4.10c) $H^3(U,F) \longrightarrow \prod_{v \text{ arch}} H^3(K_v,F)$ is surjective (in fact, an isomorphism). As the groups $H^r(K_v,F)$ for v archimedean all have the same order,

$$\chi_c(U,F) = \chi(U,F) \times \prod_{v \in X-U} \chi(K_v,F)^{-1} \times \prod_{v \text{ arch}} [H^0(K_v,F)].$$

According to (I.2.8), $\chi(K_v,F) = |[F(K_s)]|_v$, and so

$$\chi_c(U,F) = \prod_{v \text{ arch}} [F(K_v)] \prod_{v \notin U} |[F(K_s)]|_v^{-1}.$$

But $|[F(K_s)]|_v = 1$ for $v \in U$, and so $\prod_{v \notin U} |[F(K_s)]|_v = 1$ in virtue of the product formula, and so we obtain the formula.

Remark 2.14. (a) Let F be a locally constant sheaf on U with $mF = 0$ for some m that is invertible on U. In the next section, we shall show that $H^r(U,F)$ is dual to $H^{3-r}_c(U,F^D)$ for all r. This implies that $\chi(U,F)\chi_c(U,F^D) = 1$. If we let M be the G_S-module module corresponding to F, then (2.13) shows that

$$\chi(U,F)\chi_c(U,F^D) = \prod_{v \text{ arch}} \frac{[H^0(G_v,M)]\ [H^0(G_v,M^D)]}{|[M]|_v \quad [H^0(K_v,M)]},$$

which (I.2.13c) shows to be one. Thus our results are consistent.

(b) Assume that $U \neq X$ and that F is locally constant. Then $H^0(U,F) \hookrightarrow \prod H^0(K_v,F)$ and, of course, $H^{-1}(U,F) = 0$. Therefore $\prod H^{-1}(K_v,F) \xrightarrow{\approx} H^0_c(U,F)$, and so (2.13b) becomes in this case

$$\frac{[H^2_c(U,F)]}{[H^1_c(U,F)][H^3_c(U,F)]} = \prod_{v \text{ arch}} \frac{[F(K_v)]}{[H^0(K_v,F)]}.$$

Notes: So far as I know, K. Kato was the first to suggest defining cohomology groups "with compact support" fitting into an exact sequence

$$\dots \longrightarrow H^r_c(X,F) \longrightarrow H^r(X,F) \longrightarrow \bigoplus_{v \text{ real}} H^r(K_v,F_v) \longrightarrow \dots$$

(letter to Tate, about 1973). Our definition differs from, but is equivalent to, his.

§3 Global results: the main theorem

We continue with the notations of the last section. From (2.6) (and its proof) we know that there are trace maps $H^3_c(U,\mathbb{G}_m) \xrightarrow{\approx} \mathbb{Q}/\mathbb{Z}$ such that

(a) for any $V \subset U$,

$$\begin{array}{ccc} H^3_c(V,\mathbb{G}_m) & \xrightarrow{\approx} & \mathbb{Q}/\mathbb{Z} \\ \downarrow & & \| \\ H^3_c(U,\mathbb{G}_m) & \xrightarrow{\approx} & \mathbb{Q}/\mathbb{Z} \end{array}$$

commutes;

(b) for any $v \notin U$,

$$\begin{array}{ccc} Br(K_v) & \xrightarrow{\text{inv}} & \mathbb{Q}/\mathbb{Z} \\ \downarrow & & \| \\ H^3_c(U,\mathbb{G}_m) & \xrightarrow{\approx} & \mathbb{Q}/\mathbb{Z} \end{array}$$

commutes.

On combining the pairings

$$Ext^r_U(F,\mathbb{G}_m) \times H^{3-r}_c(U,F) \longrightarrow H^3_c(U,\mathbb{G}_m)$$

with this trace map, we obtain maps

$$\alpha^r(U,F)\colon Ext^r_U(F,\mathbb{G}_m) \longrightarrow H^{3-r}_c(U,F)^*.$$

Theorem 3.1. *Let F be a $\mathbb{Z}$-constructible sheaf on an open subscheme U of X.*

(a) *For $r \neq 0,1$, $Ext^r_U(F,\mathbb{G}_m)$ is a torsion group of cofinite-type, and $\alpha^r(U,F)$ is an isomorphism. For $r = 0,1$, $Ext^r_U(F,\mathbb{G}_m)$ is finitely generated and $\alpha^r(U,F)$ defines isomorphisms*

$$\mathrm{Ext}^r_U(F,\mathbb{G}_m)^\wedge \longrightarrow H^{3-r}_c(U,F)^*$$

where ^ denotes the completion for the topology of subgroups of finite index.

(b) *If* F *is constructible, then* $\alpha^r(U,F)$ *is an isomorphism of finite groups for all* $r \in \mathbb{Z}$.

Note that (b) implies that $H^r(U,F)$ is finite if F is a constructible sheaf on U without char(K)-torsion, because (2.3a) and (I.2.1) show that then $H^r(U,F)$ differs from $H^r_c(U,F)$ by a finite group.

Corollary 3.2. *For any constructible sheaf* F *on an open subscheme* $j\colon U \longrightarrow X$ *of* X *and prime number* ℓ, *there is a nondegenerate pairing of finite groups*

$$\mathrm{Ext}^r_U(F,\mathbb{G}_m)(\ell) \times H^{3-r}(X,j_!F)(\ell) \longrightarrow H^3(X,j_!\mathbb{G}_m)(\ell) = (\mathbb{Q}/\mathbb{Z})(\ell)$$

except when $\ell = 2$ *and* K *is number field with a real prime.*

Proof: As $H^r_c(U,F) = H^r_c(X,j_!F)$ (see 2.3d), it is clear from (2.3a) that $H^r_c(U,F)$ differs from $H^r(X,j_!F)$ by at most a 2-torsion subgroup, and that it differs not at all unless K is a number field having at least one real prime.

Corollary 3.3. *Let* F *be a constructible sheaf on* U *such that* $mF = 0$ *for some* m *that is invertible on* U, *and let* $F^D = R\mathcal{H}om(F,\mathbb{G}_m)$ *(an object of the derived category of* $S(U_{et})$*).*

(a) *If* F *is locally constant, then* $H^r(F^D) = 0$ *for* $r > 0$; *thus in this case* F^D *can be identified with the sheaf* $\mathcal{H}om(F,\mathbb{G}_m)$.

(b) *There is a canonical nondegenerate pairing of finite groups*

$$H^r(U,F^D) \times H^{3-r}_c(U,F) \longrightarrow H^3(U,\mathbb{G}_m) = \mathbb{Q}/\mathbb{Z}.$$

Proof: Part (a) can be proved by the same argument as in (1.10a). Part (b) is obvious, because $\mathbb{H}^r(U,F^D) = \mathrm{Ext}^r_U(F,\mathbb{G}_m)$.

Define $D^r(U,F) = \mathrm{Im}(H^r_c(U,F) \longrightarrow H^r(U,F))$.

Corollary 3.4. *Let* F *be a locally constant constructible sheaf on* U *such that* mF = F *for some* m *invertible on* U. *Then there is a non-degenerate pairing of finite groups*

$$D^r(U,F) \times D^{3-r}(U,F^D) \longrightarrow \mathbb{Q}/\mathbb{Z}$$

for all $r \in \mathbb{Z}$.

Proof: After (3.3), the dual of the sequence

$$0 \longrightarrow D^r(U,F) \longrightarrow H^r(U,F) \longrightarrow \bigoplus_{v \notin S} H^r(K_v,F)$$

is a sequence

$$\bigoplus_{v \notin S} H^{2-r}(K_v,F^D) \longrightarrow H^{3-r}_c(U,F^D) \longrightarrow D^r(U,F)^* \longrightarrow 0.$$

But this second sequence identifies $D^r(U,F)^*$ with $D^{3-r}(U,F^D)$.

The proof of the theorem is rather long and intricate. In (3.5 - 3.8) we show that it suffices to prove the theorem with U replaced by an open subset. Proposition 3.9 and Corollary 3.10 relate the theorem on U to the theorem on U' for some finite covering of U. In Lemma 3.12 it is shown that the groups vanish for large r when K has no real primes, and hence proves the theorem for such K and r. Lemma 3.13 allows us to assume that K has no real primes. In (3.14 - 3.17) the theorem is proved by an induction argument for constructible sheaves, and we then deduce it for all $\mathbb{Z}$-constructible sheaves.

Throughout the proof, U will be an open subscheme of the scheme X. We set $\hat{\alpha}^r(U,F)$ equal to the map $\mathrm{Ext}^r_U(F,\mathbb{G}_m)^\wedge \longrightarrow H^{3-r}(U,F)^*$ induced by $\alpha^r(U,F)$ when $r = 0,1$, and equal to $\alpha^r(U,F)$ otherwise.

Lemma 3.5. *Theorem 3.1 is true if F has support on a proper closed subset of U.*

Proof: We can assume that our sheaf is of the form i_*F where i is the inclusion of a single closed point v into U. According to (2.3c), $H^r_c(U,i_*F) = H^r(v,F)$. From the exact sequence

$$0 \longrightarrow \mathbb{G}_m \longrightarrow g_*\mathbb{G}_m \longrightarrow \bigoplus_{u\in U^0} i_{u*}\mathbb{Z} \longrightarrow 0$$

we obtain an exact sequence

$$\dots \longrightarrow \mathrm{Ext}^r_U(i_*F,\mathbb{G}_m) \longrightarrow \mathrm{Ext}^r_U(i_*F,g_*\mathbb{G}_m) \longrightarrow \oplus\, \mathrm{Ext}^r_U(i_*F,i_{u*}\mathbb{Z}) \longrightarrow \dots .$$

As we observed in the proof of (2.1), $R^s g_*\mathbb{G}_m = 0$ for $s \geq 1$, and so $\mathrm{Ext}^r_U(i_*F,g_*\mathbb{G}_m) = \mathrm{Ext}^r_\eta(i_*F|\eta,\mathbb{G}_m)$, which is zero for all r because $i_*F|\eta = 0$. Moreover (0.1c) shows that $\mathrm{Ext}^r_U(i_*F,i_{u*}\mathbb{Z}) = \mathrm{Ext}^r_u(i_u^* i_{v*}F,\mathbb{Z})$, which equals 0 unless $u = v$, in which case it equals $\mathrm{Ext}^r_v(F,\mathbb{Z})$. Therefore the sequence gives isomorphisms $\mathrm{Ext}^{r-1}_v(F,\mathbb{Z}) \xrightarrow{\approx} \mathrm{Ext}^r_U(i_*F,\mathbb{G}_m)$ for all r. Let M be the g_v-module corresponding to F. Then we have a commutative diagram

$$\begin{array}{ccccc}
\mathrm{Ext}^r_U(i_*F,\mathbb{G}_m) & \times & H^{3-r}_c(U,i_*F) & \longrightarrow & H^3(U,\mathbb{G}_m) \\
\uparrow \approx & & \| & & \uparrow \approx \\
\mathrm{Ext}^{r-1}_{g_v}(M,\mathbb{Z}) & \times & H^{3-r}(g_v,M) & \longrightarrow & H^2(g_v,\mathbb{Z}),
\end{array}$$

and so the theorem follows in this case from (I.1.10).

[There is an alternative approach. One can show (as in 1.7b) that $R^r i^! \mathbb{G}_m = \mathbb{Z}$ for $r = 1$ and is zero otherwise. The spectral sequence (0.1e)

$$\mathrm{Ext}^r_v(F,R^s i^! \mathbb{G}_m) \Rightarrow \mathrm{Ext}^{r+s}_U(i_*F,\mathbb{G}_m)$$

now yields isomorphisms $\mathrm{Ext}^{r-1}_v(F,\mathbb{Z}) \xrightarrow{\approx} \mathrm{Ext}^r_U(i_*F,\mathbb{G}_m)$.]

Lemma 3.6. *For any $\mathbb{Z}$-constructible sheaf on U, the groups $\mathrm{Ext}^r_U(F,\mathbb{G}_m)$ are finitely generated for r = 0,1, torsion of cofinite-type for r = 2,3, and finite for all other values of r. Consequently, if F is constructible, all the groups are finite.*

Proof: We first show that $\mathrm{Ext}^r_U(F,\mathbb{G}_m)$ is torsion for $r \geq 2$. Clearly it suffices to show that $\mathrm{Ext}^r_U(F,g_*\mathbb{G}_m)$ and $\mathrm{Ext}^{r-1}_U(F,i_{v*}\mathbb{Z})$ are torsion for $r \geq 2$, but the first is isomorphic to $\mathrm{Ext}^r_\eta(g^*F,\mathbb{G}_m)$ and the second to $\mathrm{Ext}^{r-1}_v(i^*_vF,\mathbb{Z})$, and (I.0.10) shows that both groups are torsion for $r \geq 2$.

Note that $\mathrm{Ext}^r_U(\mathbb{Z},\mathbb{G}_m) = H^r(U,\mathbb{G}_m)$, and so for $F = \mathbb{Z}$ the lemma can be read off from (2.1). It follows that the lemma is true for all constant $\mathbb{Z}$-constructible sheaves F. Next suppose that F is locally constant and so becomes constant on some finite Galois covering $\pi\colon V \longrightarrow U$ with Galois group G, say. Then (0.2) gives a spectral sequence

$$H^r(G,\mathrm{Ext}^s_V(F,\mathbb{G}_m)) \Rightarrow \mathrm{Ext}^{r+s}_U(F,\mathbb{G}_m),$$

which shows that $\mathrm{Ext}^r_U(F,\mathbb{G}_m)$ differs from $H^0(G,\mathrm{Ext}^r_V(F,\mathbb{G}_m))$ by a finite group. Therefore the lemma holds for locally constant sheaves. Finally, let F be an arbitrary $\mathbb{Z}$-constructible sheaf, and let V be an open subset of U on which F is locally constant. Write j and i for the inclusions of V and its complement into U. The Ext sequence of

$$0 \longrightarrow j_!j^*F \longrightarrow F \longrightarrow i_*i^*F \longrightarrow 0$$

can be identified with

$$\dots \longrightarrow \mathrm{Ext}^{r-1}_{U-V}(i^*F,\mathbb{Z}) \longrightarrow \mathrm{Ext}^r_U(F,\mathbb{G}_m) \longrightarrow \mathrm{Ext}^r_V(F,\mathbb{G}_m) \longrightarrow \dots ,$$

from which the result follows.

Lemma 3.7. *Let*

$$0 \longrightarrow F' \longrightarrow F \longrightarrow F'' \longrightarrow 0$$

be an exact sequence of $\mathbb{Z}$*-constructible sheaves on* U. *If* (3.1) *holds for two out of the three sheaves* F', F, *and* F", *then it also holds for the third.*

Proof: Because $\mathrm{Ext}^1_U(F',\mathbb{G}_m)$ is finitely generated, its image in the torsion group $\mathrm{Ext}^2_U(F",\mathbb{G}_m)$ is finite. Therefore the sequence

$$\cdots \longrightarrow \mathrm{Ext}^r(F",\mathbb{G}_m) \longrightarrow \mathrm{Ext}^r(F,\mathbb{G}_m) \longrightarrow \mathrm{Ext}^r(F',\mathbb{G}_m) \longrightarrow \cdots$$

remains exact after the first six terms have been replaced by their completions. The lemma can now be proved by an easy five-lemma argument.

Lemma 3.8. *Let* V *be a nonempty open subscheme of* U, *and let* F *be a* $\mathbb{Z}$*-constructible sheaf on* U; *the theorem is true for* F *on* U *if and only if it is true for the restriction of* F *to* V.

Proof: Write j for the open immersion $V \hookrightarrow U$ and i for the complementary closed immersion $U - V \hookrightarrow U$. Then $\mathrm{Ext}^r_U(j_!F|V,\mathbb{G}_m) = \mathrm{Ext}^r_V(F|V,\mathbb{G}_m)$, and $H^r_c(U,j_!F|V) = H^r_c(V,F|V)$ by (2.3d), and so $\hat{\alpha}^r(U,j_!F|V)$ can be identified with $\hat{\alpha}^r(V,F|V)$. Therefore the theorem is true for $j_!(F|V)$ on U if and only if it is true for F|V on V. Now (3.7) and the exact sequence

$$0 \longrightarrow j_!j^*F \longrightarrow F \longrightarrow i_*i^*F \longrightarrow 0$$

show that the theorem is true for F on U if and only if it is true for $j_!(F|V)$ on U.

Proposition 3.9. *Let* $\pi\colon U' \longrightarrow U$ *be the normalization of* U *in a finite Galois extension* K' *of* K.

(a) *There is a canonical norm map* $\mathrm{Nm}\colon \pi_*\mathbb{G}_{m,U'} \longrightarrow \mathbb{G}_{m,U}$.

(b) *For every* $\mathbb{Z}$*-constructible sheaf* F *on* U', *the composite*

$$N\colon \mathrm{Ext}^r_{U'}(F,\mathbb{G}_m) \longrightarrow \mathrm{Ext}^r_U(\pi_*F,\pi_*\mathbb{G}_m) \xrightarrow{\mathrm{Nm}} \mathrm{Ext}^r_U(\pi_*F,\mathbb{G}_m)$$

is an isomorphism.

Proof: (a) Let $V \longrightarrow U$ be étale. Then V is an open subset of the normalization of U in some finite separable K-algebra L. By definition $\pi_*\mathbb{G}_m(V) = \Gamma(V', \mathcal{O}_{V'}^\times)$ where $V' \overset{\text{df}}{=} V\times_U U'$. As V' is étale over U', it is normal, and as it is finite over V, it must be the integral closure of V in the finite Galois L-algebra $K'\otimes_K L$. Consequently, the norm map $K'\otimes_K L \longrightarrow L$ induces a map $\Gamma(V,\pi_*\mathbb{G}_m) \longrightarrow \Gamma(V,\mathbb{G}_m)$, and for varying V these maps define a canonical map of sheaves Nm: $\pi_*\mathbb{G}_{m,U'} \longrightarrow \mathbb{G}_{m,U}$.

(b) When $U' \longrightarrow U$ is étale, the norm map agrees with that defined in [Milne (1980), V.1.12], and in this case the result is proved (ibid., V.1.13). For a sheaf of the form $i_{v*}F$, $v \in U$, the result again follows from [Milne (1980), V.1.13] because the sequence of maps can be identified with

$$\mathrm{Ext}^{r-1}_{\pi^{-1}(v)}(F,\mathbb{Z}) \longrightarrow \mathrm{Ext}^{r-1}_v(F,\pi_*\mathbb{Z}) \xrightarrow{\mathrm{Tr}} \mathrm{Ext}^{r-1}_v(\pi_*F,\mathbb{Z}).$$

For the general case, choose an open subscheme $j\colon V \hookrightarrow U$ such that $\pi^{-1}(V)$ is étale over V, and note that we have shown that the map N is an isomorphism for $j_!(F|V)$ (cf. 0.1a) and the quotient of F be $j_!(F|V)$. The result now follows by a five-lemma argument.

Corollary 3.10. *Let* $\pi\colon U' \longrightarrow U$ *be as in the proposition. For any* $\mathbb{Z}$*-constructible sheaf* F' *on* U', $\alpha^r(U',F)^\wedge$ *is an isomorphism if and only if* $\alpha^r(U,\pi_*F)^\wedge$ *is an isomorphism.*

Proof: It is not difficult to check from the definition of N that

$$\begin{array}{ccccc}
\mathrm{Ext}^r_{U'}(F,\mathbb{G}_m) & \times & H^{3-r}_c(U',F) & \longrightarrow & H^3_c(U',\mathbb{G}_m) \\
\downarrow N & & \uparrow \approx & & \downarrow \mathrm{Nm} \\
\mathrm{Ext}^r_U(\pi_*F,\mathbb{G}_m) & \times & H^{3-r}_c(U,\pi_*F) & \longrightarrow & H^3_c(U,\mathbb{G}_m)
\end{array}$$

commutes. If π is étale, then Nm: $H^3_c(U',\mathbb{G}_m) \longrightarrow H^3_c(U,\mathbb{G}_m)$ is easily seen to be an isomorphism. This implies that it is always an isomorphism because the groups are unchanged when U is replaced by an open subscheme. The corollary is now obvious.

Remark 3.11. In the proposition, the extension of K was assumed to be Galois (rather than separable) only to simplify the proof.

Lemma 3.12. (a) *The group* $H^r_c(U,F) = 0$ *for* $r > 4$ *when* F *is* $\mathbb{Z}$*-constructible, and for* $r > 3$ *when* F *is constructible.*

(b) *If* K *has no real primes, then* $\mathrm{Ext}^r_U(F,\mathbb{G}_m) = 0$ *for* $r > 4$ *and all constructible sheaves* F.

Proof: (a) We first prove the statement for constructible sheaves. After replacing U by an open subscheme (see 2.3d), we can assume that F is locally constant and, in the number field case, that $mF = 0$ for some integer m that is invertible on U. We have to show (see 2.3a) that $H^r(U,F) \longrightarrow \bigoplus_{v \text{ real}} H^r(K_v,F)$ is an isomorphism for $r \geq 3$. But (2.9) allows us to identify this with a map $H^r(G_S,M) \longrightarrow \bigoplus_{v \text{ real}} H^r(K_v,M)$, and (I.4.10) shows that the map is an isomorphism for $r \geq 3$ except possibly when K is a function field and the order of M is divisible by p. In the last case we can assume that F is killed by some power of p and have to show that $H^r(U,F) = 0$ for $r > 3$. From the cohomology sequence of

$$0 \longrightarrow \mathbb{Z}/p\mathbb{Z} \longrightarrow \mathcal{O}_U \xrightarrow{\wp} \mathcal{O}_U \longrightarrow 0$$

we see that $H^r(U,\mathbb{Z}/p\mathbb{Z}) = 0$ for $r \geq 2$. In general there will be a finite étale covering $\pi\colon U' \longrightarrow U$ of degree d prime to p such that $F|U'$ has a composition series whose quotients are isomorphic to $\mathbb{Z}/p\mathbb{Z}$. Then $H^r(U',F|U') = 0$ for $r \geq 2$, and as the composite

$$H^r(U,F) \longrightarrow H^r(U,\pi_*F|U') \xrightarrow{Nm} H^r(U,F)$$

is multiplication by d (see [Milne (1980), V.1.12]), this proves that $H^r(U,F) = 0$ for $r \geq 2$.

Now let F be $\mathbb{Z}$-constructible. Then F_{tors} is constructible, and so it suffices to prove the result for F/F_{tors}: we can assume that F is torsion-free. Clearly, it suffices to show that $H^r_c(U,F)$ is torsion for $r > 4$. The same arguments as above show that it suffices to prove that $H^r(U,F)$ is torsion when F is locally constant, but in this case the result is proved in (2.9).

(b) Because K has no real primes, $H^r(U,F) = H^r_c(U,F) = 0$ for $r > 3$. If F has support on a closed subscheme Z, the lemma is obvious from the isomorphism $\mathrm{Ext}^{r-1}_Z(F,\mathbb{G}_m) \xrightarrow{\approx} \mathrm{Ext}^r(i_*F,\mathbb{G}_m)$ of (3.5). As usual, this allows us to assume that F is locally constant. Then $\mathcal{E}xt^r_U(F,\mathbb{G}_m) = 0$ for $r > 1$, and for $r = 0,1$, it is torsion, and is therefore a direct limit of constructible sheaves (0.9). Hence $H^r(U,\mathcal{E}xt^s_U(F,\mathbb{G}_m)) = 0$ for $r > 3$, and so $\mathrm{Ext}^r(F,\mathbb{G}_m) = 0$ for $r > 4$.

Lemma 3.13. *Assume that* $\hat{\alpha}^r(X,\mathbb{Z})$ *is an isomorphism for all* r *whenever* K *has no real primes. Then Theorem* 3.1 *is true.*

Proof: The assumption and (3.8) imply that the theorem is true for constant sheaves on any open $U \subset X$ whenever K has no real primes. We shall use induction on r to prove the theorem for all pairs (U,F) where U is such that some prime is invertible on it and F is a locally constant sheaf on U. Assume that $\hat{\alpha}^r(U,F)$ is an isomorphism for all such pairs when $r < r_0$. Lemma 3.12 shows that the assumption is fulfilled if $r_0 = -1$. Then there will exist a finite étale covering $\pi: U' \longrightarrow U$ such that U' is the normalization of U in field K' with no real primes and F becomes constant on U'. Let $F_* = \pi_*\pi^*F$; then the trace map [Milne (1980), V.1.12] $F_* \longrightarrow F$ is surjective (on stalks it is just summation, $F_{\bar{v}}^d \xrightarrow{\Sigma} F_{\bar{v}}$), and we write F' for its kernel. From the commutative diagram

$$\begin{array}{ccccccc} \mathrm{Ext}_U^{r_0-1}(F_*,\mathbb{G}_m) & \longrightarrow & \mathrm{Ext}_U^{r_0-1}(F',\mathbb{G}_m) & \longrightarrow & \mathrm{Ext}_U^{r_0}(F,\mathbb{G}_m) & \longrightarrow & \mathrm{Ext}_U^{r_0}(F_*,\mathbb{G}_m) \\ \downarrow\approx & & \downarrow\approx & & \downarrow & & \downarrow\approx \\ H^{4-r_0}(U,F_*)^* & \longrightarrow & H^{4-r_0}(U,F')^* & \longrightarrow & H^{3-r_0}(U,F)^* & \longrightarrow & H^{3-r_0}(U,F_*)^* \end{array}$$

we see that $\hat{\alpha}^{r_0}(U,F)\colon \mathrm{Ext}^{r_0}(F,\mathbb{G}_m) \longrightarrow H^{3-r_0}(U,F)^*$ is injective. (For $r = 0,1$, it is necessary to replace the groups on the top row with their completions; see the proof of (3.7).) Since F' is also locally constant, $\hat{\alpha}^{r_0}(U,F')$ is also injective, and the five-lemma implies that $\hat{\alpha}^{r_0}(U,F)$ is an isomorphism.

Finally, (3.8) shows that if the theorem is true for all locally constant $\mathbb{Z}$-constructible sheaves, then it is true for all $\mathbb{Z}$-constructible sheaves.

For a constructible sheaf F, we define $\beta^r(U,F)\colon H_c^r(U,F) \longrightarrow \mathrm{Ext}_U^{3-r}(F,\mathbb{G}_m)^*$ to be the dual of $\alpha^r(U,F)$.

Lemma 3.14. *For any $\mathbb{Z}$-constructible sheaf F on U, there is a finite surjective map $\pi_1\colon U_1 \longrightarrow U$, a finite map $\pi_2\colon U_2 \longrightarrow U$ with finite image, constant $\mathbb{Z}$-constructible sheaves F_i on U_i, and an injective map $F \longrightarrow \oplus_i \pi_{i*}F_i$.*

Proof: Let V be an open subset of U such that F|V is locally constant. Then there is a finite extension K' of K such that the normalization $\pi\colon V' \longrightarrow V$ of V in K' is étale over V and F|V' is constant. Let $\pi_1\colon U_1 \longrightarrow U$ be the normalization of U in K', and let F_1 be the constant sheaf on U_1 corresponding to the group $\Gamma(V',F|V')$. Then the canonical map $F|V \longrightarrow \pi_*F|V'$ extends to a map $\alpha\colon F \longrightarrow \pi_{1*}F_1$ whose kernel has support on U - V. Now take U_2 to be an étale covering of U - V on which the inverse image of F on V - U becomes a constant sheaf, and take F_2 to be the direct image of this constant sheaf.

Note that Lemma 3.13 shows that it suffices to prove Theorem 3.1 under the assumption that K has no real primes. From now until the end of the proof of the theorem we shall make this assumption.

Lemma 3.15. (a) *Let* r_o *be an integer* ≥ 1. *If for all K, all constructible sheaves F on X, and all* $r < r_o$, $\beta^r(X,F)$ *is an isomorphism, then* $\beta^{r_o}(X,F)$ *is injective.*

(b) *Assume that for all K, all constructible sheaves F on X, and all* $r < r_o$, $\beta^r(X,F)$ *is an isomorphism; further assume that* $\beta^{r_o}(X,\mathbb{Z}/m\mathbb{Z})$ *is an isomorphism whenever* $\mu_m(K) = \mu_m(K_s)$. *Then* $\alpha^{r_o}(X,F)$ *is an isomorphism for all X and all constructible sheaves F.*

Proof: (a) Let F be a constructible sheaf on some X, and let $c \in H^{r_o}(X,F)$. There exists an embedding $F \hookrightarrow I$ of F into a torsion flabby sheaf I on X. According to (0.9), I is a direct limit of constructible sheaves. As $H^{r_o}(X,I) = 0$, and cohomology commutes with direct limits, this implies that there is a constructible sheaf F_* on X and an embedding $F \hookrightarrow F_*$ such that c maps to zero in $H^{r_o}(X,F_*)$. Let Q be the cokernel of $F \longrightarrow F_*$. Then Q is constructible, and a chase in the diagram

$$\begin{array}{ccccccc} H^{r_o-1}(X,F_*) & \longrightarrow & H^{r_o-1}(X,Q) & \longrightarrow & H^{r_o}(X,F) & \longrightarrow & H^{r_o}(X,F_*) \\ \downarrow\approx & & \downarrow\approx & & \downarrow & & \downarrow \\ \mathrm{Ext}_X^{4-r_o}(F_*,\mathbb{G}_m)^* & \longrightarrow & \mathrm{Ext}_X^{4-r_o}(Q,\mathbb{G}_m)^* & \longrightarrow & \mathrm{Ext}_X^{3-r_o}(F,\mathbb{G}_m)^* & \longrightarrow & \mathrm{Ext}_X^{3-r_o}(F_*,\mathbb{G}_m)^* \end{array}$$

shows that $\beta^{r_o}(c) \neq 0$. Since the argument works for all c, this shows that $\beta^{r_o}(X,F)$ is injective.

(b) Let F be a constructible sheaf on X. For a suitably small open subset U of X, there will exist a finite Galois extension K' of K such that the normalization U' of U in K' is étale over U, F|U' is constant, and $\mu_m(K) = \mu_m(K_s)$ for some m with mF = 0. In the construction of the preceding lemma, we can take U_1 to be the normalization of X in K'. Let $F_* = \pi_{1*}F_1 \oplus \pi_{2*}F_2$; then (3.10) and (3.5) show respectively that $\beta^{ro}(X,\pi_{1*}F)$ and $\beta^{ro}(X,\pi_{2*}F)$ are isomorphisms. The map $F \longrightarrow F_*$ is injective, and we can construct a diagram similar to the above, except that now we know that $\beta^{ro}(X,F_*)$ is an isomorphism. Therefore $\beta^{ro}(X,F)$ is injective, and as Q is constructible we have also that $\beta^{ro}(X,Q)$ is injective. The five-lemma now shows that $\beta^{ro}(U,F)$ is an isomorphism.

Lemma 3.16. *Theorem 3.1 is true for all constructible sheaves F on X.*

Proof: We prove $\beta^r(X,F)$ is an isomorphism by induction on r. For $r < 0$ it is an isomorphism by (3.12).

To compute the group $\mathrm{Ext}^r_X(\mathbb{Z}/m\mathbb{Z},\mathbb{G}_m)$, we use the exact sequence

$$\cdots \longrightarrow \mathrm{Ext}^r_X(\mathbb{Z}/m\mathbb{Z},\mathbb{G}_m) \longrightarrow H^r(X,\mathbb{G}_m) \xrightarrow{m} H^r(X,\mathbb{G}_m) \longrightarrow \cdots .$$

By definition $H^0(X,\mathbb{Z}/m\mathbb{Z}) = \mathbb{Z}/m\mathbb{Z}$, and it follows from (2.2b) that $\mathrm{Ext}^3(\mathbb{Z}/m\mathbb{Z},\mathbb{G}_m) = m^{-1}\mathbb{Z}/\mathbb{Z}$. The pairing is the obvious one, and so $\beta^0(X,\mathbb{Z}/m\mathbb{Z})$ is an isomorphism. Now (3.15b) implies that $\beta^0(X,F)$ is an isomorphism for all F.

Lemma 3.15a shows that $\beta^1(X,F)$ is always injective. The order of $H^1(X,\mathbb{Z}/m\mathbb{Z})$ is equal to the degree of the maximal unramified abelian extension of K of exponent m. By class field theory, this is also the order of $\mathrm{Pic}(X)^{(m)} = [\mathrm{Ext}^2_X(\mathbb{Z}/m\mathbb{Z},\mathbb{G}_m)]$, and so $\beta^1(X,\mathbb{Z}/m\mathbb{Z})$ is an isomorphism for all X. It follows that $\beta^1(X,F)$ is an isomorphism for all X and F.

Lemma 3.15a again shows that $\beta^2(X,F)$ is always injective, but to proceed further, we need (yet) another lemma.

Lemma 3.17. *For any sheaf F on X and element* $c \in \operatorname{Ext}^1_X(F,\mathbb{G}_m)$, *there exists a surjective map* $F' \longrightarrow F$ *such that* c *maps to zero in* $\operatorname{Ext}^1_X(F',\mathbb{G}_m)$. *The sheaf* F' *can be chosen to be constructible (and killed by a power of* ℓ*) if* F *is constructible (and killed by a power of* ℓ*).*

Proof: To c there corresponds an extension

$$0 \longrightarrow \mathbb{G}_m \longrightarrow F' \longrightarrow F \longrightarrow 0,$$

and c obviously maps to zero in $\operatorname{Ext}^1_X(F',\mathbb{G}_m)$. If F is constructible (and killed by a power of ℓ), then $\operatorname{Ext}^1_X(F,\mathbb{G}_m)$ is torsion, and so c arises from an element c' of $\operatorname{Ext}^1(F,\mu_n)$ for some n (which is a power of ℓ). In this case we can take $F' \longrightarrow F$ to be the map in the extension of F by μ_n corresponding to c'.

Now c be nonzero element of $\operatorname{Ext}^1_X(F,\mathbb{G}_m)$, choose a map $F' \longrightarrow F$ as in the lemma, and let Q be its kernel. There is a diagram

$$\begin{array}{ccccccc} \operatorname{Ext}^0_X(F',\mathbb{G}_m) & \longrightarrow & \operatorname{Ext}^0_X(Q,\mathbb{G}_m) & \longrightarrow & \operatorname{Ext}^1_X(F,\mathbb{G}_m) & \longrightarrow & \operatorname{Ext}^1_X(F',\mathbb{G}_m) \\ \downarrow\approx & & \downarrow\approx & & \downarrow & & \downarrow \\ H^3(X,F') & \longrightarrow & H^3(X,Q)^* & \longrightarrow & H^2(X,F)^* & \longrightarrow & H^2(X,F')^*. \end{array}$$

A diagram chase shows that c does not map to zero under $\alpha^1(X,F)\colon \operatorname{Ext}^1_X(F,\mathbb{G}_m) \longrightarrow H^2(X,F)^*$. Since this holds for all c, it follows that $\alpha^1(X,F)$ is injective, and therefore that its dual $\beta^2(X,F)$ is surjective. This proves that $\beta^2(X,F)$ is an isomorphism.

Next, for $r \geq 3$, $H^r(X,\mathbb{Z}/m\mathbb{Z}) \approx H^r(X,\mu_m)$ if $(m,\operatorname{char}(K)) = 1$ and m contains the mth roots of 1 because $\mathbb{Z}/m\mathbb{Z}$ then differs from μ_m by a sheaf with support on the closed subset of primes dividing m. Hence $H^3(X,\mathbb{Z}/m\mathbb{Z})$ has order m and $H^r(X,\mathbb{Z}/m\mathbb{Z}) = 0$ for $r > 3$ (by (2.2b) again). As $\mathrm{Ext}^0(\mathbb{Z}/m\mathbb{Z},\mathbb{G}_m) = \mu_m$, which also has order m, this completes the proof for sheaves killed by some integer m prime to the characteristic. For the remaining sheaves it follows from the fact that $\mathrm{Ext}^0(\mathbb{Z}/p^n,\mathbb{G}_m) = 0$ for all n.

We now complete the proof of Theorem 3.1 by proving that $\hat{\alpha}^r(X,\mathbb{Z})$ is an isomorphism for all r (recall that we are assuming K has no real primes). We are concerned with the maps

$$\alpha^r(X,\mathbb{Z})\colon H^r(X,\mathbb{G}_m) \longrightarrow H^{3-r}(X,\mathbb{Z})^*,\ r \neq 0,1,$$

$$\hat{\alpha}^r(X,\mathbb{Z})\colon H^r(X,\mathbb{G}_m)^\wedge \longrightarrow H^{3-r}(X,\mathbb{Z})^*,\ r = 0,1.$$

Consider the diagram

$$\begin{array}{ccccccccc}
0 \longrightarrow & H^r(X,\mathbb{G}_m)^\wedge & \longrightarrow & \varprojlim_n \mathrm{Ext}^{r+1}(\mathbb{Z}/n\mathbb{Z},\mathbb{G}_m) & \longrightarrow & \varprojlim_n H^{r+1}(X,\mathbb{G}_m)_n & \longrightarrow 0 \\
& \downarrow & & \downarrow \approx & & \downarrow & \\
& H^{3-r}(X,\mathbb{Z})^* & \longrightarrow & H^{2-r}(X,\mathbb{Q}/\mathbb{Z})^* & \longrightarrow & H^{2-r}(X,\mathbb{Q})^* &
\end{array}$$

For $2-r \geq 1$ (that is, for $r \leq 1$), $H^{2-r}(X,\mathbb{Q}) = 0 = H^{3-r}(X,\mathbb{Q})$ (see 2.10), and so $H^{2-r}(X,\mathbb{Q}/\mathbb{Z}) \longrightarrow H^{3-r}(X,\mathbb{Z})$ is an isomorphism. For $r \leq 1$, $H^{r+1}(X,\mathbb{G}_m)$ is finitely generated, and so $\varprojlim H^{r+1}(X,\mathbb{G}_m)_n = 0$ (see I.0.19). Therefore it is obvious from the diagram that $\hat{\alpha}^r(X,\mathbb{Z})$ is an isomorphism for $r \leq 1$. For $r = 3$, the map is the obvious isomorphism $\mathbb{Q}/\mathbb{Z} \longrightarrow \mathbb{Z}^*$, and for all other values of r, both groups are zero.

Remark 3.18. For a locally Noetherian scheme Y, let Y_{sm} denote the category of smooth schemes over Y endowed with the étale topology. Let $f\colon Y_{sm} \longrightarrow Y_{et}$ be the morphism of sites defined by the identity map. Then f_* is exact and preserves injectives ([Milne (1980),

III.3.1]), and it follows that $\mathrm{Ext}^r_{Y_{sm}}(f^*F, F') = \mathrm{Ext}^r_{Y_{et}}(F, f_*F)$ for all sheaves F on Y_{et}, all sheaves F' on Y_{sm}, and all r. Therefore U_{et} can be replaced by U_{sm} in the above results provided one defines a $\mathbb{Z}$-constructible sheaf on Y_{sm} to be the inverse image by f of a $\mathbb{Z}$-constructible sheaf on Y_{et}.

Notes: Corollary 3.2 in the number field case is the original theorem of Artin and Verdier (announcement in [Artin and Verdier (1964)]). As far as I know, no complete proof of the theorem has been published before, but [Mazur (1973)] contains most of the ingredients. It and the notes of a 1964 seminar by Mazur were sources for this section. The proof of the theorem for a constructible sheaf F over X in lemmas 3.14 through 3.17 follows the original proof. We note that Theorem 3.1 improves the original theorem in three respects: by taking into account the archimedean primes, it is able to handle the 2-torsion; it includes the function field case; and it allows the sheaves to be $\mathbb{Z}$-constructible instead of constructible. To my knowledge, several people have extended the original theorem to the function field case, but the only published account is in [Deninger (1984)]. The extension to $\mathbb{Z}$-constructible sheaves was carried out in [Deninger (1986)]. In [Zink (1978)] there is an alternative method of obviating the problem with 2-primary components in the original theorem.

§4 Global results: complements

This section is concerned with various improvements of Theorem 3.1. We also discuss its relation to the theorems in Chapter I. The notations are the same as in the preceding two sections.

Sheaves without sections with finite support

Let F be a sheaf on an open subscheme U of X. For any V étale over U, a section $s \in \Gamma(V,F)$ is said to have *finite support* if $s_{\bar{v}} = 0$ for all but finitely many $v \in V$.

Proposition 4.1. *Let F be a $\mathbb{Z}$-constructible sheaf on an open affine subscheme U of X. If F has no sections with finite support, then* $\mathrm{Ext}^1_U(F,\mathbb{G}_m)$ *and* $H^2_c(U,F)$ *are finite, and* $\alpha^1(U,F)$ *is an isomorphism.*

Proof: Note that $\mathrm{Hom}_U(\mathbb{Z},\mathbb{G}_m) = \mathcal{O}_U^\times$, which is finitely generated, and that $\mathrm{Ext}^1_U(\mathbb{Z},\mathbb{G}_m) = \mathrm{Pic}(U)$, which is finite (because, in the function field case, it is a quotient of $\mathrm{Pic}^0(X)$). It follows immediately that $\mathrm{Ext}^1_U(F,\mathbb{G}_m)$ is finite if F is constant. As we observed in (3.14), there is a finite surjective map $\pi\colon U' \longrightarrow U$, a constant $\mathbb{Z}$-constructible sheaf F' on U', and a morphism $F \longrightarrow \pi_*F'$ whose kernel has support on a proper closed subset of U. As F has no sections with finite support, we see that the map must be injective. Let F" be its cokernel. In the exact sequence

$$\mathrm{Ext}^1_U(\pi_*F',\mathbb{G}_m) \longrightarrow \mathrm{Ext}^1_U(F,\mathbb{G}_m) \longrightarrow \mathrm{Ext}^2(F'',\mathbb{G}_m),$$

$\mathrm{Ext}^1_U(\pi_*F',\mathbb{G}_m) = \mathrm{Ext}^1_{U'}(F',\mathbb{G}_m)$ (see 3.9) and so is finite, and $\mathrm{Ext}^2(F'',\mathbb{G}_m)$ is torsion, and so $\mathrm{Ext}^1_U(F,\mathbb{G}_m)$ (being finitely generated) has finite image in it. This proves that $\mathrm{Ext}^1_U(F,\mathbb{G}_m)$ is finite, and Theorem 3.1 implies that $\alpha^1(X,F)$ is an isomorphism. It follows that $H^2_c(U,F)$ is also finite.

Nonconstructible sheaves

We say that a sheaf F on $U \subset X$ is *countable* if F(V) is countable for all V étale over U. For example, any sheaf defined by a group scheme of finite type over U is countable. Fix a separable closure K_s of K. If F is countable, then there are only countably many pairs (s,V) with V an open subset of the normalization of U in a finite subextension of K_s and $s \in F(V)$. Therefore the construction in (0.9) expresses F as a countable union of $\mathbb{Z}$-constructible sheaves (of constructible sheaves if F is torsion).

Proposition 4.2. *Let F be a countable sheaf on an open subscheme U of X, and consider the map* $\alpha^r(U,F)\colon \mathrm{Ext}^r_U(F,\mathbb{G}_m) \longrightarrow H^{3-r}(U,F)^*$.

(a) *For* $r \leq 2$, *the kernel of* $\alpha^r(U,F)$ *is divisible, and it is uncountable when nonzero; for* $r = 0$ *or* $r > 4$, $\alpha^r(U,F)$ *is injective.*

(b) *For* $r \geq 2$, $\alpha^r(U,F)$ *is surjective.*

(c) *If F is torsion, then* $\alpha^r(U,F)$ *is an isomorphism for all r.*

(d) *If U is affine and F has no sections with finite support, then* $\alpha^2(U,F)$ *is an isomorphism and* $\alpha^1(U,F)$ *is surjective.*

Proof: Write F as a countable union of $\mathbb{Z}$-constructible subsheaves, $F = \bigcup F_i$. Then (see I.0.21 and I.0.22) there is an exact sequence

$$0 \longrightarrow \varprojlim{}^{(1)}\mathrm{Ext}^{r-1}_U(F_i,\mathbb{G}_m) \longrightarrow \mathrm{Ext}^r_U(F,\mathbb{G}_m) \longrightarrow \varprojlim \mathrm{Ext}^r_U(F_i,\mathbb{G}_m) \longrightarrow 0,$$

and $\varprojlim^{(1)}\mathrm{Ext}^{r-1}_U(F_i,\mathbb{G}_m)$ is divisible (and uncountable when nonzero) if each group $\mathrm{Ext}^{r-1}(F_i,\mathbb{G}_m)$ is finitely generated, and it is zero if each group $\mathrm{Ext}^{r-1}(F_i,\mathbb{G}_m)$ is finite. Theorem 3.1 provides us with a map $\mathrm{Ext}^r(F_i,\mathbb{G}_m) \longrightarrow H^r(U,F_i)^*$ which is injective for all r and is surjective for $r \geq 2$; it is an isomorphism for any r for which the groups $\mathrm{Ext}^r(F_i,\mathbb{G}_m)$ are finite. On passing to the inverse limit, we obtain a map $\varprojlim \mathrm{Ext}^r_U(F_i,\mathbb{G}_m) \longrightarrow H^{3-r}_c(U,F)^*$ with the similar properties. The proposition is now obvious from (3.1) and (4.1).

Corollary 4.3. *Let* G *be a separated group scheme of finite type over an open affine subscheme* U *of* X. *Then* $\alpha^2(U,G)$: $\mathrm{Ext}^2_U(G,\mathbb{G}_m) \longrightarrow H^1_c(U,G)^*$ *is an isomorphism. If* G *defines a torsion sheaf, then* $\alpha^r(U,G)$ *is an isomorphism for all* r.

Proof: If a section s of G over V agrees with the zero section on an open subset of V, then it agrees on the whole of V (because G is separated over V). Thus G (when regarded as a sheaf) has no sections with support on a finite subscheme, and the corollary results immediately from part (d) of the proposition.

Example 4.4. In particular,

$\mathrm{Ext}^2_U(A,\mathbb{G}_m) \xrightarrow{\approx} H^1_c(U,A)^*$, A a semi-abelian scheme over U,

$\mathrm{Ext}^2_U(\mathbb{G}_m,\mathbb{G}_m) \xrightarrow{\approx} H^1_c(U,\mathbb{G}_m)^*$, and

$\mathrm{Ext}^r_U(\mathbb{G}_a,\mathbb{G}_m) \xrightarrow{\approx} H^{3-r}(U,\mathbb{G}_a)^*$ for all r when K has characteristic $p \neq 0$. (In fact the groups $\mathrm{Ext}^r_U(G,\mathbb{G}_m)$, computed for the small étale site, seem to be rather pathological. For example, if k is a finite field, then $\mathrm{Hom}_k(\mathbb{G}_m,\mathbb{G}_m) \supset \mathrm{Gal}(k_s/k) = \hat{\mathbb{Z}}$.)

Exercise 4.5. (a) Show that there are only countably many $\mathbb{Z}$-constructible sheaves on U. (Hint: Use Hermite's theorem.)

(b) Show that there are uncountably many countable sheaves on Spec $\mathbb{Z}$. (Hint: Consider sheaves of the form $\bigoplus_{p \text{ prime}} i_{p*}F_p$.)

Tori

We investigate the duality theorem when F is replaced by a torus. By a torus over a scheme Y, we mean a group scheme that becomes isomorphic to a product of copies of $\mathbb{G}_m$ on a finite étale covering of Y. The sheaf of characters $X^*(T)$ of T is the sheaf $V \mapsto \mathrm{Hom}_V(T,\mathbb{G}_m)$ (homomorphisms as group schemes). It is a locally constant $\mathbb{Z}$-constructible sheaf. In the next theorem, ^ denotes completion relative to the topology of subgroups of finite index.

Theorem 4.6. *Let* T *be a torus on an open subscheme* U *of* X.

(a) *The cup-product pairing*

$$H^r(U,T) \times H_c^{3-r}(U,X^*(T)) \longrightarrow H_c^3(U,\mathbb{G}_m) \approx \mathbb{Q}/\mathbb{Z}$$

induces isomorphisms

$$H^r(U,T)\hat{} \longrightarrow H_c^{3-r}(U,X^*(T))^* \quad \textit{for } r = 0,1,$$
$$H^r(U,T) \longrightarrow H_c^{3-r}(U,X^*(T))^* \quad \textit{for } r \geq 2.$$

If U *is affine, then* $H^1(U,T)$ *is finite.*

(b) *Assume that* K *is a number field. The cup-product pairing*

$$H^r(U,X^*(T)) \times H_c^{3-r}(U,T) \longrightarrow H_c^3(U,\mathbb{G}_m) \approx \mathbb{Q}/\mathbb{Z}$$

induces isomorphisms

$$H^r(U,X^*(T))\hat{} \longrightarrow H_c^{3-r}(U,T)^* \quad \textit{for } r = 0,1,$$
$$H^r(U,X^*(T)) \longrightarrow H_c^{3-r}(U,T)^* \quad \textit{for } r \geq 2.$$

Proof: (a) The sheaf $X^*(T)$ is locally isomorphic to $\mathbb{Z}^{\dim(T)}$, and so $\mathcal{E}xt_U^r(X^*(T),\mathbb{G}_m)$ is locally isomorphic to the sheaf associated with the presheaf $V \mapsto H^r(V,\mathbb{G}_m)^{\dim(T)}$. It is therefore zero for $r > 0$. As $\mathcal{E}xt_U^0(X^*(T),\mathbb{G}_m) = \mathcal{H}om_U(X^*(T),\mathbb{G}_m) = T$, the spectral sequence

$$H^r(U,\mathcal{E}xt_U^s(X^*(T),\mathbb{G}_m) \Rightarrow \mathrm{Ext}_U^{r+s}(X^*(T),\mathbb{G}_m)$$

gives isomorphisms $H^r(U,T) \xrightarrow{\approx} \mathrm{Ext}_U^r(X^*(T),\mathbb{G}_m)$ for all r. Thus (a) follows from (3.1) and (4.1).

(b) Consider the diagram

$$\begin{array}{ccccccc}
\prod_{v\notin U} H^{r-1}(K_v,X^*(T)) & \longrightarrow & H_c^r(U,X^*(T)) & \longrightarrow & H^r(U,X^*(T)) & \longrightarrow & \prod_{v\notin U} H^r(K_v,X^*(T)) \\
\downarrow & & \downarrow & & \downarrow & & \downarrow \\
\prod_{v\notin U} H^{3-r}(K_v,T)^* & \longrightarrow & H^{3-r}(U,T)^* & \longrightarrow & H_c^{3-r}(U,T)^* & \longrightarrow & \prod_{v\notin U} H^{2-r}(K_v,T))^*.
\end{array}$$

Replace the groups H^0 and $H_c^1(U,X^*(T))$ with their completions. Then

(I.2.4) shows that the maps $\prod H^r(K_v,X^*(T)) \longrightarrow \prod H^{3-r}(K_v,T)^*$ are isomorphisms (complete $H^0(K_v,X^*(T))$), and (a) shows that the maps $H^r_c(U,X^*(T)) \longrightarrow H^{3-r}(U,T)^*$ are isomorphisms (complete $H^r_c(U,X^*(T))$ for $r = 0,1$). Now the five-lemma shows that $H^r(U,X^*(T)) \longrightarrow H^{3-r}_c(U,T)^*$ (complete $H^0(U,X^*(T))$) is an isomorphism for all r.

Corollary 4.7. *Assume K is a number field. There are canonical isomorphisms*

$$D^r(U,X^*(T)) \longrightarrow D^{3-r}(U,T)^*$$

where

$$D^r(U,X^*(T)) = \operatorname{Im}(H^r_c(U,X^*(T)) \longrightarrow H^r(U,X^*(T)),\ r \neq 0,$$
$$D^0(U,X^*(T)) = \operatorname{Im}(H^0_c(U,X^*(T))^\wedge \longrightarrow H^0(U,X^*(T))^\wedge),\ r = 0,$$
$$D^r(U,T) = \operatorname{Im}(H^r_c(U,T) \longrightarrow H^r(U,T)),\ r \neq 0,1,$$
$$D^r(U,T) = \operatorname{Im}(H^r_c(U,T)^\wedge \longrightarrow H^r(U,T)^\wedge),\ r = 0,1.$$

Proof: Part (a) of the theorem and (I.2.4) show that the dual of the sequence

$$0 \longrightarrow D^r(U,X^*(T)) \longrightarrow H^r(U,X^*(T)) \longrightarrow \oplus\, H^r(K_v,X^*(T))$$

(complete the groups for $r = 0$) is an exact sequence

$$\oplus\, H^{2-r}(K_v,T) \longrightarrow H^{3-r}_c(U,T) \longrightarrow D^r(U,X^*(T))^* \longrightarrow 0,$$

(complete the groups for $3-r = 0,1$) which identifies $D^r(U,X^*(T))^*$ with $D^{3-r}(U,T)$.

Duality for Exts of tori

We wish to interprete (4.6b) in terms of Exts, but for this we shall need to use the big étale site X_{Et} on X and the flat site X_{fl}. Recall that for any locally Noetherian scheme Y, Y_{Et} is the category of schemes locally of finite type over Y endowed with the étale topology, and Y_{fl} is the same category of schemes endowed with the flat topology. Also f denotes the morphism $Y_{fl} \longrightarrow Y_{Et}$ which is the identity map on the underlying categories. For the rest of this section, $\mathcal{E}xt^r_{Y_{fl}}(F,F')$ denotes the sheaf on Y_{Et} associated with the presheaf $V \mapsto \mathrm{Ext}^r_{V_{fl}}(F,F')$. Note that $V \mapsto \mathrm{Hom}_{V_{fl}}(F,F')$ is already a sheaf, and so $\Gamma(V,\mathcal{H}om_{V_{fl}}(F,F')) = \mathrm{Hom}_{V_{fl}}(F,F')$.

Proposition 4.8. *For any sheaf* F *on* Y_{Et} *and smooth group scheme* G *of finite type over* Y, *there is a spectral sequence*

$$H^r(Y_{Et},\mathcal{E}xt^s_{Y_{fl}}(f^*F,G)) \Longrightarrow \mathrm{Ext}^{r+s}_{Y_{Et}}(F,G).$$

Proof: If F' is an injective sheaf on Y_{fl}, then f_*F' is also injective [Milne (1980), III.1.20], and so $\mathcal{H}om_{Y_{Et}}(F,f_*F')$ is flabby (ibid. III.1.23). Hence $H^r(Y_{Et},\mathcal{H}om_{Y_{Et}}(F,f_*F')) = 0$ for $r > 0$. But for any V locally of finite type over Y,

$$\Gamma(V,\mathcal{H}om_{Y_{Et}}(F,f_*F')) = \mathrm{Hom}_{V_{Et}}(F,f_*F') = \mathrm{Hom}_{V_{fl}}(f^*F,F')$$
$$= \Gamma(V,\mathcal{H}om_{Y_{fl}}(f^*F,F')).$$

Therefore $H^r(Y_{Et},\mathcal{H}om_{Y_{fl}}(f^*F,F')) = 0$ for $r > 0$, which means that $\mathcal{H}om_{Y_{fl}}(f^*F,F')$ is acyclic for $\Gamma(Y_{Et},-)$. Next note that

$$\Gamma(Y_{Et},\mathcal{H}om_{Y_{fl}}(f^*F,F')) = \mathrm{Hom}_{Y_{fl}}(f^*F,F') = \mathrm{Hom}_{Y_{Et}}(F,f_*F'),$$

and so there is a spectral sequence

$$H^r(Y_{Et},\mathcal{E}xt^s_{Y_{fl}}(f^*F,F')) \Longrightarrow R^{r+s}\alpha(F')$$

where $\alpha = \mathrm{Hom}_{Y_{Et}}(F,-)\circ f_*$. There is an obvious spectral sequence $\mathrm{Ext}^r_{Y_{Et}}(F,R^sf_*F') \Longrightarrow R^{r+s}\alpha(F')$. On replacing F' with G in this spectral sequence and using that $R^sf_*G = 0$ for $s > 0$ (ibid. III.3.9), we find that $R^{r+s}\alpha(G) = \mathrm{Ext}^{r+s}_{Y_{Et}}(F,G)$. The result follows.

Corollary 4.9. *For any sheaf* F *on* Y_{et} *and smooth group scheme* G *on* Y, *there is a spectral sequence*

$$H^r(Y_{et}, \mathcal{E}xt^s_{Y_{fl}}(f^*F,G)) \Rightarrow \mathrm{Ext}^{r+s}_{Y_{et}}(F,G)$$

where f *now denotes the obvious morphism* $Y_{fl} \longrightarrow Y_{et}$ *and* $\mathcal{E}xt^s_{Y_{fl}}(f^*F,G)$ *denotes the sheaf on* Y_{et} *associated with* $V \mapsto \mathrm{Ext}^s_{V_{fl}}(f^*F,F')$.

Proof: Let $f'\colon Y_{Et} \longrightarrow Y_{et}$ be the obvious morphism. For any sheaves F on Y_{et} and F' on Y_{Et}, $\mathrm{Hom}_{Y_{Et}}(f'^*F,F') = \mathrm{Hom}_{Y_{et}}(F,f'_*F')$. As f'_* is exact and preserves injectives, this shows that $\mathrm{Ext}^r_{Y_{Et}}(f'^*F,F') = \mathrm{Ext}^r_{Y_{et}}(F,f_*F')$ for all r. Moreover the sheaf $\mathcal{E}xt^s_{Y_{fl}}(f^*F,\mathbb{G}_m)$ of the corollary is the restriction to Y_{et} of the corresponding sheaf in (4.8), and so the result follows from (4.8) because Y_{Et} and Y_{et} yield the same cohomology groups (see [Milne (1980), III.3.1]).

Proposition 4.10. *Let* Y *be a regular scheme, and let* p *be a prime such that* p! *is invertible on* Y. *Then* $\mathcal{E}xt^r_{Y_{fl}}(T,\mathbb{G}_m) = 0$ *for* $0 < r < 2p-1$.

Proof: Since this is a local question, we can assume that $T = \mathbb{G}_m$. From [Breen (1969), §7], we know that $\mathcal{E}xt^r_{U_{fl}}(\mathbb{G}_m,\mathbb{G}_m)$ is torsion for $r \geq 1$. Let ℓ be prime, and consider the sequence

$$\cdots \longrightarrow \mathcal{E}xt^{r-1}_{Y_{fl}}(\mu_\ell,\mathbb{G}_m) \longrightarrow \mathcal{E}xt^r_{Y_{fl}}(\mathbb{G}_m,\mathbb{G}_m) \xrightarrow{\ \ell\ } \mathcal{E}xt^r_{Y_{fl}}(\mathbb{G}_m,\mathbb{G}_m) \longrightarrow \cdots .$$

If Y is connected, then this sequence starts as

$$0 \longrightarrow \mathbb{Z} \xrightarrow{\ \ell\ } \mathbb{Z} \longrightarrow \mathbb{Z}/\ell\mathbb{Z} \xrightarrow{\ 0\ } \mathcal{E}xt^1_{Y_{fl}}(\mathbb{G}_m,\mathbb{G}_m) \xrightarrow{\ \ell\ } \cdots .$$

Therefore, $\mathcal{E}xt^1_{Y_{fl}}(\mathbb{G}_m,\mathbb{G}_m) = 0$. We shall complete the proof by showing that $\mathcal{E}xt^r_{Y_{fl}}(\mu_\ell,\mathbb{G}_m) = 0$ for all ℓ if $0 < r < 2p - 2$.

If ℓ is invertible on Y, then μ_ℓ is locally isomorphic to $\mathbb{Z}/\ell\mathbb{Z}$, and so $\mathcal{E}xt^r_{Y_{fl}}(\mu_\ell,\mathbb{G}_m)$ is locally isomorphic to $\mathcal{E}xt^r_{Y_{fl}}(\mathbb{Z}/\ell\mathbb{Z},\mathbb{G}_m)$. There is an exact sequence

$$\cdots \longrightarrow \mathcal{E}xt^r_{Y_{fl}}(\mathbb{Z}/\ell\mathbb{Z},\mathbb{G}_m) \longrightarrow \mathcal{E}xt^r_{Y_{fl}}(\mathbb{Z},\mathbb{G}_m) \longrightarrow \mathcal{E}xt^r_{Y_{fl}}(\mathbb{Z},\mathbb{G}_m) \longrightarrow \cdots .$$

But $\mathcal{E}xt^r_{Y_{fl}}(\mathbb{Z},\mathbb{G}_m)$ is the sheaf (for the étale topology) associated with the presheaf $V \mapsto H^r(V_{et},\mathbb{G}_m)$; it is therefore zero for $r > 0$ and equal to $\mathbb{G}_m$ for $r = 0$. The sequence therefore shows that $\mathcal{E}xt^r_{Y_{fl}}(\mathbb{Z}/\ell\mathbb{Z},\mathbb{G}_m) = 0$ for $r > 0$.

Next assume that ℓ is not invertible on Y. Our assumption implies that for all primes $q < p$, $q\mu_\ell = \mu_\ell$. Therefore the main theorem of [Breen (1975)] shows that $\mathcal{E}xt^r_{Y_{fl}}(\mu_\ell,\mathbb{G}_m) = 0$ for $1 < r < 2p - 2$, and for $r = 1$ the sheaf is well-known to be zero (see [Milne (1980), III.4.17]).

Theorem 4.11. *Assume that K is a number field. Let T be a torus on an open subscheme U of X, and assume that 6 is invertible on U.*

(a) *The group* $\mathrm{Ext}^r_{U_{Et}}(T,\mathbb{G}_m)$ *is finitely generated for* $r = 0$, *finite for* $r = 1$, *and torsion of cofinite type for* $r = 2, 3$.

(b) *The map* $\alpha^r(U,F)\colon \mathrm{Ext}^r_{U_{Et}}(T,\mathbb{G}_m) \longrightarrow H^{3-r}_c(U_{et},T)^*$ *is an isomorphism for* $0 < r \le 4$, *and* $\alpha^0(U,T)$ *defines an isomorphism* $\mathrm{Hom}_U(T,\mathbb{G}_m)^\wedge \longrightarrow H^3_c(U,T)^*$ *(as usual, ^ denotes completion for the topology of finite subgroups).*

Proof: The lemma shows that the spectral sequence in (4.9) gives isomorphisms $\mathrm{Ext}^r_{U_{Et}}(T,\mathbb{G}_m) \xrightarrow{\approx} H^r(U,X^*(T))$ for $r \le 4$. Therefore the theorem follows from (4.6).

Remark 4.12. (a) The only reason we did not allow K to be a function field in (4.6b) and (4.11) is that this case involves additional complications with the topologies.

(b) It is likely that (4.11) holds with X_{Et} replaced by the smooth site X_{sm}. If one knew that the direct image functor f_*, where f is the obvious morphism : $X_{Et} \longrightarrow X_{sm}$, preserved injectives, then this would be obvious.

(c) It is not clear to the author whether or not pathologies of the type noted in [Breen (1969b)] should prevent $\mathrm{Ext}^r_{U_{Et}}(T,\mathbb{G}_m)$ being dual to $H^{3-r}_c(U,T)$ for all $r \geq 0$ (and without restriction on the residue characteristics).

Relations to the theorems in Galois cohomology

In this subsection, U is an open affine subscheme of X and S is the set of primes of K not corresponding to a point of U. We also make use of the notations in I.4; for example, $G_S = \pi_1(U,\bar{\eta})$. For a sheaf F on U, we write $F^D = \mathcal{H}om(F,\mathbb{G}_m)$. When M is a G_S-module such that $mM = 0$ for some integer m that is invertible on U, $M^D = \mathrm{Hom}(M,K_s^{\times})$.

Proposition 4.13. *Let F be a locally constant constructible sheaf on U such that $mF = 0$ for some m that is invertible on U, and let $M = F_{\bar{\eta}}$ and $N = M^D$ be the G_S-modules corresponding to F and F^D.*

(a) *The group $D^r(U,F) = Ш^r_S(K,M)$ and $D^r(U,F^D) = Ш^r_S(U,M^d)$; consequently, the pairing*

$$D^r(U,F) \times D^{3-r}(U,F^D) \longrightarrow \mathbb{Q}/\mathbb{Z}$$

of (3.4) *can be identified with a pairing*

$$Ш_S^r(K,M) \times Ш_S^{3-r}(K,M^D) \longrightarrow \mathbb{Q}/\mathbb{Z}.$$

(b) *The group* $\mathrm{Ext}_U^r(F,\mathbb{G}_m) = H^r(G_S,N)$ *and* $H_c^r(U,F) = \mathrm{Ext}_{G_S}^{r-1}(N,C_S)$; *consequently the pairing*

$$\mathrm{Ext}_U^r(F,\mathbb{G}_m) \times H_c^{3-r}(U,F) \longrightarrow \mathbb{Q}/\mathbb{Z},$$

of (3.1) *can be identified with a pairing*

$$H^r(G_S,N) \times \mathrm{Ext}_{G_S}^{2-r}(N,C_S) \longrightarrow \mathbb{Q}/\mathbb{Z}.$$

(c) *The long exact sequence*

$$\cdots \longrightarrow H_c^r(U,F) \longrightarrow H^r(U,F) \longrightarrow \bigoplus_{v\in S} H^r(K_v,F_v) \longrightarrow \cdots$$

can be identified with a long exact sequence

$$\cdots \longrightarrow H^{3-r}(G_S,N)^* \longrightarrow H^r(G_S,M) \longrightarrow \bigoplus_{v\in S} H^r(K_v,M) \longrightarrow \cdots .$$

Proof: (*a*) Compare the sequences

$$\begin{array}{ccccccc} 0 \longrightarrow & Ш_S^r(K,M) & \longrightarrow & H^r(G_S,M) & \longrightarrow & \bigoplus_{v\in S} H^r(K_v,M), \\ & \downarrow & & \downarrow \approx & & \downarrow \approx \\ 0 \longrightarrow & D^r(U,F) & \longrightarrow & H^r(U,F) & \longrightarrow & \bigoplus_{v\in S} H^r(K_v,F) \end{array}$$

the second of which arises from the sequence in (2.3a) and the definition of $D^r(U,F)$.

(b) As $\mathcal{E}xt_U^r(F,\mathbb{G}_m) = 0$ for $r > 0$, $\mathrm{Ext}_U^r(F,\mathbb{G}_m) = H^r(U,F^D)$, and (2.9) shows that $H^r(U,F^D) = H^r(G_S,N)$. The second isomorphism can be read off from the long exact $\mathrm{Ext}^r(F,-)$-sequence corresponding to the sequence of sheaves defined by the exact sequence of G_S-modules

$$0 \longrightarrow R_S^\times \longrightarrow \oplus K_v^* \longrightarrow C_S \longrightarrow 0.$$

(c) It follows from (2.9) that $H^r(U,F) = H^r(G_S,M)$, and it is obvious that $H^r(K_v,F) = H^r(K_v,M)$. According to (3.3), $H_c^r(U,F) =$

$H^{3-r}(U,F^D)^*$, and (2.9) again shows that $H^{3-r}(U,F^D) = H^{3-r}(U,N)$.

For a G_S-module M, write $M^d = \mathrm{Hom}(M,R_S^\times)$.

Proposition 4.14. *Let T be a torus on U, and let $X^*(T)$ be its sheaf of characters. If $M = X^*(T)_{\bar\eta}$ is the G_S-module corresponding to $X^*(T)$, then $M^d = T_{\bar\eta}$. For all ℓ that are invertible on U and all $r \geq 1$, $D^r(U,T)(\ell) = Ш_S^r(K,M)(\ell)$ and $D^r(U,X^*(T))(\ell) = Ш_S^r(U,M^d)(\ell)$; consequently, the pairing*

$$D^r(U,T)(\ell) \times D^{3-r}(U,X^*(T))(\ell) \longrightarrow (\mathbb{Q}/\mathbb{Z})(\ell)$$

of (4.8) can be identified for $r \geq 1$ with a pairing

$$Ш_S^r(K,M^d)(\ell) \times Ш_S^{3-r}(K,M)(\ell) \longrightarrow (\mathbb{Q}/\mathbb{Z})(\ell).$$

Proof: In the course of proving (2.9), we showed that $H^r(\tilde{U},\mathbb{G}_m)(\ell) = 0$ for all $r > 0$. Therefore $H^r(U,T)(\ell) = H^r(G_S,T)(\ell)$.

Remark 4.15. Presumably, the maps are the same as those in Chapter I. Once this has been checked, some of the results of each chapter can be deduced from the other. It is not surprising that there is an overlap between the two chapters: to give a constructible sheaf on X is the same as to give a G_S-module M for some finite set of nonarchimedean primes S together with $\mathrm{Gal}(K_{v,s}/K_v)$-modules M_v for each $v \in S$ and equivariant maps $M \longrightarrow M_v$ (see [Milne (1980), II.3.16]).

Galois cohomology has the advantage of being more elementary than étale cohomology, and one is not led to impose unnecessary restrictions (for example, that S is finite) as is sometimes required for the étale topology. Etale cohomology has the advantage that more machinery is available and the results are closer to those that algebraic topology would suggest.

Notes: Propositions 4.1 and 4.2 are taken from [Deninger (1986)].

§5 Global results: abelian schemes

The notations are the same as those listed at the start of §2. In particular, U is always an open subscheme or X. As in (I.6), we fix an integer m that is invertible on U and write $M^\wedge \overset{df}{=} \varprojlim M/m^n M$ for the m-adic completion of M.

Let $\mathcal{A}$ be an abelian scheme over U, and let A be its generic fibre. As $\mathcal{A}$ is proper over U, the valuative criterion of properness [Hartshorne (1977), II.4.7] shows that every morphism $\mathrm{Spec}(K) \longrightarrow A$ extends to a morphism $U \longrightarrow \mathcal{A}$, that is, $A(K) = \mathcal{A}(U)$. A similar statement holds for any V étale over U, which shows that $\mathcal{A}$ represents g_*A on U_{et}. In fact (see [Artin (1986), 1.4]) $\mathcal{A}$ represents g_*A on U_{sm}.

Proposition 5.1. (a) *The group* $H^0(U,\mathcal{A})$ *is finitely generated; for* $r > 0$, $H^r(U,\mathcal{A})$ *is torsion and* $H^r(U,\mathcal{A})(m)$ *is of cofinite-type; the map* $H^r(U,\mathcal{A})(m) \longrightarrow \prod_{v\ \mathrm{arch}} H^r(K_v,A)(m)$ *is surjective for* $r = 2$ *and an isomorphism for* $r > 2$.

(b) *For* $r < 0$, $\prod_{v\ \mathrm{arch}} H^{r-1}(K_v,A) \longrightarrow H^r_c(U,\mathcal{A})$ *is an isomorphism;* $H^0_c(U,A)$ *is finitely generated;* $H^1_c(U,\mathcal{A})$ *is an extension of a torsion group by a subgroup which has a natural compactification;* $H^2_c(U,\mathcal{A})$ *is torsion, and* $H^2_c(U,\mathcal{A})(m)$ *is of cofinite-type; for* $r \geq 3$, $H^r_c(U,\mathcal{A})(m) = 0$.

Proof: (a) The group $H^0(U,\mathcal{A}) = \mathcal{A}(U) = A(K)$, which the Mordell-Weil theorem states is finitely generated. As Galois cohomology groups are torsion in degree > 1, the Leray spectral sequence $H^r(U,R^sg_*A) \Rightarrow H^{r+s}(K,A)$ shows that the groups $H^r(U,g_*A)$ are torsion for $r > 0$ because $R^s g_* A$ is torsion for $s > 0$ and $H^r(K,A)$ is torsion for $r > 0$. Finally, the finiteness of $H^r(U,\mathcal{A})_m$ follows from the cohomology sequence of

$$0 \longrightarrow \mathcal{A}_m \longrightarrow \mathcal{A} \xrightarrow{m} \mathcal{A} \longrightarrow 0$$

because $H^r(U,\mathcal{A}_m)$ is finite for all r (by 2.13). On replacing m with m^n in this cohomology sequence and passing to the direct limit over n, we obtain an exact sequence

$$0 \longrightarrow H^{r-1}(U,\mathcal{A})\otimes\mathbb{Q}_m/\mathbb{Z}_m \longrightarrow H^r(U,\mathcal{A}(m)) \longrightarrow H^r(U,\mathcal{A})(m) \longrightarrow 0 \qquad \textbf{(5.1.1)}.$$

The first term in this sequence is zero for $r > 1$ because then $H^{r-1}(U,\mathcal{A})$ is torsion. Hence $H^r(U,\mathcal{A}(m)) \xrightarrow{\approx} H^r(U,\mathcal{A})(m)$ for $r \geq 2$. As $\mathcal{A}_{m^n}$ is locally constant, $H^r(U,\mathcal{A}_{m^n}) = H^r(G_S,\mathcal{A}_{m^n}(K_S))$ (by 2.9). Therefore $H^r(U,\mathcal{A}_{m^n}) \longrightarrow \prod_{v \text{ arch}} H^r(K_v,A_{m^n})$ is surjective for $r = 2$ (by I.4.16) and an isomorphism for $r \geq 3$ (by I.4.10c), and it follows that $H^r(U,\mathcal{A})(m) \longrightarrow \bigoplus_{v \text{ arch}} H^r(K_v,A)(m)$ has the same properties.

(b) All statements follow immediately from (a) and the exact sequence

$$\cdots \longrightarrow H^r_c(U,\mathcal{A}) \longrightarrow H^r(U,\mathcal{A}) \longrightarrow \bigoplus_{v \notin U} H^r(K_v,A) \longrightarrow \cdots .$$

The dual abelian scheme $\mathcal{A}^t$ to $\mathcal{A}$ is characterised by the fact that it represents the functor $V \mapsto \mathrm{Ext}^1_V(\mathcal{A},\mathbb{G}_m)$ on U_{sm} (generalized Barsotti-Weil formula [Oort (1966), III.18]). As $\mathcal{H}om(\mathcal{A},\mathbb{G}_m) = 0$, the local-global spectral sequence for Exts gives rise to a map $H^r(U,\mathcal{A}^t) \longrightarrow \mathrm{Ext}^{r+1}_U(\mathcal{A},\mathbb{G}_m)$ all r. On combining this with the pairing

$$\mathrm{Ext}^r_U(\mathcal{A},\mathbb{G}_m) \times H^{3-r}_c(U,\mathcal{A}) \longrightarrow H^3_c(U,\mathbb{G}_m) \approx \mathbb{Q}/\mathbb{Z}$$

we get a pairing

$$H^r(U,\mathcal{A}^t) \times H^{2-r}_c(U,\mathcal{A}) \longrightarrow H^3_c(U,\mathbb{G}_m) \approx \mathbb{Q}/\mathbb{Z}.$$

(For a symmetric definition of this pairing in terms of biextensions, see Chapter III.) We define

$$D^1(U,\mathcal{A}) = \mathrm{Im}(H^1_c(U,\mathcal{A}) \longrightarrow H^1(U,\mathcal{A})) = \mathrm{Ker}(H^1(U,\mathcal{A}) \longrightarrow \prod_{v \notin U} H^1(K_v,A)).$$

It is a torsion group, and $D^1(U,\mathcal{A})(m)$ is of cofinite-type.

Theorem 5.2. (a) *The group* $H^0(U,\mathcal{A}^t)(m)$ *is finite; the pairing*

$$H^0(U,\mathcal{A}^t)(m) \times H^2_c(U,\mathcal{A}) \longrightarrow \mathbb{Q}/\mathbb{Z}$$

is nondegenerate on the left and its right kernel is the m*-divisible subgroup of* $H^2_c(U,\mathcal{A})$.

(b) *The groups* $H^1(U,\mathcal{A}^t)(m)$ *and* $H^1_c(U,\mathcal{A})(m)$ *are of cofinite-type, and the pairing*

$$H^1(U,\mathcal{A}^t)(m) \times H^1_c(U,\mathcal{A})(m) \longrightarrow \mathbb{Q}/\mathbb{Z}$$

annihilates exactly the divisible subgroups.

(c) *If* $D^1(U,\mathcal{A}^t)(m)$ *is finite, then the compact group* $H^0(U,\mathcal{A}^t)^\wedge$ *is dual to the discrete torsion group* $H^2_c(U,\mathcal{A})$.

Proof: Because $H^r_c(U,\mathcal{A}_{m^n})$ is finite for all n and r, passage to the inverse limit in the sequences

$$0 \longrightarrow H^{r-1}_c(U,\mathcal{A})^{(m^n)} \longrightarrow H^r_c(U,\mathcal{A}_{m^n}) \longrightarrow H^r_c(U,\mathcal{A})_{m^n} \longrightarrow 0$$

yields an exact sequence

$$0 \longrightarrow H^{r-1}_c(U,\mathcal{A})^\wedge \longrightarrow H^r_c(U,T_m\mathcal{A}) \longrightarrow T_m H^r_c(U,\mathcal{A}) \longrightarrow 0, \qquad \textbf{(5.2.1)}$$

where we have written $H^r_c(U,T_m\mathcal{A})$ for $\varprojlim H^r_c(U,\mathcal{A}_{m^n})$. Note that $T_m H^r_c(U,\mathcal{A})$ is torsion-free and is nonzero only if the divisible subgroup of $H^r_c(U,\mathcal{A})(m)$ is nonzero. Corollary 3.3 provides us with nondegenerate pairings of finite groups

$$H^r(U,\mathcal{A}^t_{m^n}) \times H^{3-r}_c(U,\mathcal{A}_{m^n}) \longrightarrow \mathbb{Q}/\mathbb{Z},$$

and hence a nondegenerate pairing (of a discrete torsion group with a compact group)

$$H^r(U,\mathcal{A}^t(m)) \times H^{3-r}_c(U,T_m\mathcal{A}) \longrightarrow \mathbb{Q}/\mathbb{Z}.$$

For r = 0, this shows that the finite group $H^0(U,\mathcal{A}^t)(m) = H^0(U,\mathcal{A}^t(m)) = A(K)(m)$ is dual to $H^3_c(U,T_m\mathcal{A})$, and (5.2.1) shows that this last group equals $H^2_c(U,\mathcal{A})\hat{} = H^2_c(U,\mathcal{A})/H^2_c(U,\mathcal{A})_{m\text{-div}}$ because $H^3_c(U,\mathcal{A}) = 0$. This completes the proof of (a).

From (5.1.1) we obtain an isomorphism

$$H^1(U,\mathcal{A}^t(m))/H^1(U,\mathcal{A}^t(m))_{div} \xrightarrow{\approx} H^1(U,\mathcal{A}^t)(m)/H^1(U,\mathcal{A}^t)(m)_{div},$$

and the left hand group is dual to $H^2_c(U,T_m\mathcal{A})_{tor}$ (see I.0.20e). The sequence (5.2.1) gives an isomorphism $H^1_c(U,\mathcal{A})\hat{}_{tor} \xrightarrow{\approx} H^2_c(U,T_m\mathcal{A})_{tor}$, and $H^1_c(U,\mathcal{A})\hat{}_{tor} = H^1_c(U,\mathcal{A})(m)/H^1_c(U,\mathcal{A})(m)_{div}$. This completes the proof of (b).

For (c), consider the diagram

$$\begin{array}{ccccccccc}
0 \longrightarrow & H^0(U,\mathcal{A}^t)\hat{} & \longrightarrow & H^1(U,T_m\mathcal{A}^t) & \longrightarrow & T_mH^1(U,\mathcal{A}^t) & \longrightarrow 0 \\
 & \downarrow & & \downarrow\approx & & \downarrow & \\
0 \longrightarrow & H^2_c(U,\mathcal{A})(m)^* & \longrightarrow & H^2_c(U,\mathcal{A}(m))^* & \longrightarrow & (H^1_c(U,\mathcal{A})\otimes\mathbb{Q}_m/\mathbb{Z}_m)^* & \longrightarrow 0.
\end{array}$$

It shows that the map $H^0(U,\mathcal{A}^t)\hat{} \longrightarrow H^2_c(U,A)(m)^*$ is injective, and that it is an isomorphism if and only if $T_mH^1(U,\mathcal{A}^t) \longrightarrow (H^1_c(U,\mathcal{A})\otimes\mathbb{Q}_m/\mathbb{Z}_m)^*$ is injective. On applying $\mathrm{Hom}(\mathbb{Q}_m/\mathbb{Z}_m,-)$ to the exact sequence

$$0 \longrightarrow D^1(U,\mathcal{A}^t) \longrightarrow H^1(U,\mathcal{A}^t) \longrightarrow \prod_{v\notin U} H^1(K_v,A^t),$$

we obtain the top row of the following diagram:

$$\begin{array}{ccccc}
0 \longrightarrow T_mD^1(U,\mathcal{A}^t) \longrightarrow & T_mH^1(U,\mathcal{A}^t) & \longrightarrow & \prod T_mH^1(K_v,A^t) \\
 & \downarrow & & \downarrow\approx \\
 & (H^1_c(U,\mathcal{A})\otimes\mathbb{Q}_m/\mathbb{Z}_m))^* & \longrightarrow & \prod (H^0(K_v,A)\otimes\mathbb{Q}_m/\mathbb{Z}_m)^*.
\end{array}$$

Our assumption on $D^1(U,\mathcal{A}^t)$ implies that $T_mD^1(U,\mathcal{A}^t) = 0$, and so the diagram shows that $T_mH^1(U,\mathcal{A}^t) \longrightarrow (H^1_c(U,\mathcal{A})\otimes\mathbb{Q}_m/\mathbb{Z}_m)^*$ is injective. This completes the proof of (c).

Corollary 5.3. *The group* $D^1(U,\mathcal{A})$ *is torsion, and* $D^1(U,\mathcal{A}^t)(m)$ *is of cofinite-type; there is a canonical pairing*

$$D^1(U,\mathcal{A}^t) \times D^1(U,\mathcal{A}) \longrightarrow \mathbb{Q}/\mathbb{Z}$$

whose left and right kernels are the divisible subgroups of the two groups.

Proof: The first statement follows directly from the definition of $D^1(U,\mathcal{A})$. For the second statement, we use the commutative diagram

$$\begin{array}{ccccccc} 0 \longrightarrow & D^1(U,\mathcal{A}^t)(m) & \longrightarrow & H^1(U,\mathcal{A}^t)(m) & \longrightarrow & \prod_{v\notin U} H^1(K_v,A^t) \\ & & & \downarrow & & \downarrow \approx \\ 0 \longrightarrow & D^1(U,\mathcal{A})(m)^* & \longrightarrow & H^1_c(U,\mathcal{A})(m)^* & \longrightarrow & \prod_{v\notin U} H^0(K_v,A)^*. \end{array}$$

It demonstrates that there is a map $D^1(U,\mathcal{A}^t)(m) \longrightarrow D^1(U,\mathcal{A})(m)^*$ whose kernel obviously contains the divisible subgroup of $D^1(U,\mathcal{A}^t)(m)$. The kernel of the second vertical map is zero, and that of the first is divisible. A diagram chase now shows that the kernel $D^1(U,\mathcal{A}^t)(m) \longrightarrow D^1(U,\mathcal{A})^*$ is divisible. Because of the symmetry of the situation, this implies that the right kernel of the pairing is also divisible.

Exercise 5.4. Let $\varphi_D\colon \mathcal{A} \longrightarrow \mathcal{A}^t$ be the map defined by a divisor on A. Show that for all $a \in D^1(U,\mathcal{A})$, $\langle\varphi_D(a),a\rangle = 0$.

We now show how the above results can be applied to the Tate-Shafarevich group. Write S for the set of primes of K not corresponding to a point of U.

Lemma 5.5. *The map* $H^1(U,\mathcal{A}) \longrightarrow H^1(K,A)$ *induces isomorphisms*

$H^1(U,\mathscr{A}) \xrightarrow{\approx} H^1(G_S,A)$ *and* $D^1(U,\mathscr{A}) \xrightarrow{\approx} Ш^1(K,A)$.

Proof: Because $\mathscr{A} = g_*A$, and the Leray spectral sequence for g gives an exact sequence

$$0 \longrightarrow H^1(U,\mathscr{A}) \longrightarrow H^1(K,A) \longrightarrow \Gamma(U,R^1g_*A).$$

This sequence identifies $H^1(U,\mathscr{A})$ with the set of principal homogeneous spaces for A over K that are split by the inverse image of some étale cover of U, or equivalently, that have a point in $K_{v,un}$ for each v (recall that K_v is the field of fractions of $\mathcal{O}_v^h$). From the Hochschild-Serre spectral sequence for $K_{v,un}$ over K_v and (I.3.8), we see that the restriction map $H^1(K_v,A) \longrightarrow H^1(K_{v,un},A)$ is injective. Therefore we have an exact sequence

$$0 \longrightarrow H^1(U,\mathscr{A}) \longrightarrow H^1(K,A) \longrightarrow \bigoplus_{v\in U} H^1(K_v,A) \qquad (5.5.1).$$

We have seen (I.3.10) that the maps $H^1(K_v,A) \longrightarrow H^1(\hat{K}_v,A)$ are injective, and so on comparing this sequence with that in (I.6.5), we see immediately that $H^1(U,\mathscr{A})(m) = H^1(G_S,A)(m)$. On combining (5.5.1) with the exact sequence

$$0 \longrightarrow D^1(U,\mathscr{A}) \longrightarrow H^1(U,A) \longrightarrow \bigoplus_{v\notin U} H^1(K_v,A),$$

we obtain an exact sequence

$$0 \longrightarrow D^1(U,\mathscr{A}) \longrightarrow H^1(K,A) \longrightarrow \bigoplus_{\text{all } v} H^1(K_v,A),$$

and this shows that $D^1(U,A) = Ш^1(K,A)$.

Theorem 5.6. *Let A be an abelian variety over K.*

(a) *There is a canonical pairing*

$$Ш^1(K,A^t) \times Ш^1(K,A) \longrightarrow \mathbb{Q}/\mathbb{Z}$$

whose kernels are the divisible subgroups of each group.

(b) *Assume* $Ш^1(K,A)(m)$ *is finite. Then the dual of the exact se-*

quence

$$0 \longrightarrow Ш^1(K,A)(m) \longrightarrow H^1(K,A)(m) \longrightarrow \bigoplus_{\text{all } v} H^1(K_v,A)(m) \longrightarrow Б \longrightarrow 0$$

is an exact sequence

$$0 \longleftarrow Ш^1(K,A^t)(m) \longleftarrow H^1(K,A)(m)^* \longleftarrow \prod_{\text{all } v} A^t(K_v) \longleftarrow A^t(K)^\wedge \longleftarrow 0.$$

Proof: (a) Fix a prime ℓ and choose U so that ℓ is invertible on U and A has good reduction at all primes of U. Then A and A^t extend to abelian schemes on U, and the lemma shows that the pairing

$$D^1(U,\mathcal{A}^t)(\ell) \times D^1(U,\mathcal{A})(\ell) \longrightarrow \mathbb{Q}/\mathbb{Z}$$

of (5.2) can be identified with a pairing

$$Ш^1(K,\mathcal{A}^t)(\ell) \times Ш^1(K,\mathcal{A})(\ell) \longrightarrow \mathbb{Q}/\mathbb{Z}.$$

(b) Choose U so small that m is invertible on it and A has good reduction at all primes of U. The assumption implies that $Ш^1(K,A^t)$ is also finite. Therefore, (5.2) shows that the dual of the sequence

$$0 \longrightarrow Ш^1(K,A)(m) \longrightarrow H^1(U,\mathcal{A})(m) \longrightarrow \bigoplus_{v \notin U} H^1(K_v,A)(m) \longrightarrow H^2_c(U,\mathcal{A})(m) \longrightarrow \ldots$$

is an exact sequence

$$0 \longleftarrow Ш^1(K,\mathcal{A}^t)(m) \longleftarrow H^1(U,\mathcal{A})(m)^* \longleftarrow \prod_{v \notin U} A^t(K_v)^\wedge \longleftarrow A^t(K)^\wedge \longleftarrow \ldots .$$

Now pass to the direct limit in the first sequence and to the inverse limit in the second over smaller and smaller open sets U. According to (I.6.25), the map $A^t(K)^\wedge \longrightarrow \prod_{\text{all } v} A^t(K_v)^\wedge$ is injective, and so the result if obvious.

Remark 5.7. (a) We now have defined three pairings

$$Ш^1(K,A^t) \times Ш^1(K,A) \longrightarrow \mathbb{Q}/\mathbb{Z}.$$

For the sake of definiteness, we shall refer to the pairing in (5.6) as the *Cassels-Tate pairing*.

(b) For any abelian scheme $\mathcal{A}$ on a regular scheme Y, Breen's theorems [Breen (1969a), (1975)] imply that $\mathcal{E}xt^r_Y(\mathcal{A},\mathbb{G}_m) = 0$ for $r = 0$ or $1 < r < 2p-1$, where p is a prime such that p! is invertible on Y. Since $\mathcal{E}xt^1_{Y_{fl}}(\mathcal{A},\mathbb{G}_m) = \mathcal{A}^t$, we see that $\mathrm{Ext}^r_{U_{Et}}(\mathcal{A},\mathbb{G}_m) = H^{r-1}(U,\mathcal{A}^t)$ for $r \leq 4$, provided 6 is invertible on U. In particular $\mathrm{Ext}^2_{U_{Et}}(\mathcal{A},\mathbb{G}_m) = H^1(U,\mathcal{A}^t)$, which is countable. By way of contrast, (4.4) implies that $\mathrm{Ext}^2_{U_{et}}(\mathcal{A},\mathbb{G}_m)$ is countable if and only if the divisible subgroup of $H^1_c(U,\mathcal{A})$ is zero. Thus if $\mathrm{Ext}^2_{U_{Et}}(\mathcal{A},\mathbb{G}_m) = \mathrm{Ext}^2_{U_{et}}(\mathcal{A},\mathbb{G}_m)$ then the divisible subgroup of $Ш^1(K,A)$ is zero (and there is an effective procedure for finding the rank of A(K)!).

Finally we show that, by using étale cohomology, it is possible to simplify the last part of the proof of the compatibility of the conjecture of Birch and Swinnerton-Dyer with isogenies.

Let $f: A \longrightarrow B$ be an isogeny of degree prime to char K. The initial easy calculations in (I.7) showed that to prove the equivalence of the conjecture for A and B, one must show that

$$z(f(K)) = \prod_{v \notin U} z(f(K_v)).z(f^t(K)).z(Ш^1(f)).$$

In this formula, for a nonarchimedean prime v, K_v denotes the completion of K rather than the Henselization, but it is easy to see that this does not change the value of $z(f(K_v))$.

Let m be the degree of f, and choose an open scheme $U \subsetneq X$ on which m is invertible and which is such that f extends to an isogeny $f: \mathcal{A} \longrightarrow \mathcal{B}$ of abelian schemes over U. The exact commutative diagram

$$\begin{array}{ccccccccc} 0 \to \mathcal{A}(U) & \to & \prod_{v \notin U} H^0(K_v, A) & \to & H^1_c(U, \mathcal{A}) & \to & Ш^1(K, A) & \to & 0 \\ \downarrow & & \downarrow & & \downarrow & & \downarrow & & \\ 0 \to \mathcal{B}(U) & \to & \prod_{v \notin U} H^0(K_v, B) & \to & H^1_c(U, \mathcal{B}) & \to & Ш^1(K, B) & \to & 0. \end{array}$$

shows that $z(f(K)).z(H^1_c(f)) = \prod_{v \notin U} z(H^0(K_v, f)).z(Ш^1(f))$. To prove the compatibility, it therefore remains to show that

$$z(H^1_c(f)).z(f^t(K)) = \prod_{v \text{ arch}} z(H^0(K_v, f)).z(f(K_v))^{-1}.$$

Let F be the kernel of $f\colon \mathcal{A} \to \mathcal{B}$. The exact sequence

$$0 \to \operatorname{Coker}(H^0_c(f)) \to H^1_c(U, F) \to H^1_c(U, \mathcal{A}) \to H^1_c(U, \mathcal{B}) \to H^2_c(U, F)$$
$$\to H^2_c(U, \mathcal{A}) \to H^2_c(U, \mathcal{B}) \to H^3_c(U, F) \to 0$$

shows that

$$\chi_c(U, F)\ [H^0_c(U, F)]^{-1} = [\operatorname{Coker}(H^0_c(f))]^{-1} z(H^1_c(f))^{-1}.z(H^2_c(f)).$$

But

$$\chi_c(U, F)\ [H^0_c(U, F)]^{-1} = \prod_{v \text{ arch}} [F(K_v)]\ [H^0(K_v, F)]^{-1}$$

by 2.14b,

$$z(H^2_c(f)) = z(f^t(K))^{-1}$$

by duality, and

$$\operatorname{Coker} H^0_c(f) = \prod_{v \text{ arch}} \operatorname{Coker} H^{-1}(K_v, f)$$

because $\prod H^{-1}(K_v, A) \xrightarrow{\approx} H^0_c(U, \mathcal{A})$ and $\prod H^{-1}(K_v, B) \xrightarrow{\approx} H^0_c(U, \mathcal{B})$. Therefore

$$z(H^1_c(f)).z(f^t(K)) = \prod_{v \text{ arch}} [F(K_v)]^{-1}[H^0(K_v, F)][\operatorname{Coker}(H^{-1}(K_v, f))],$$

and it remains to show that for all archimedean primes v

$$[F(K_v)]^{-1}.z(f(K_v)) = [H^0(K_v, F)]^{-1}.z(H^0(K_v, f)).[\operatorname{Coker}(H^{-1}(K_v, f))].$$

From the exact sequence

$$0 \longrightarrow \text{Coker } H^{-1}(K_v, f) \longrightarrow H^0(K_v, F) \longrightarrow \text{Ker } H^0(K_v, f) \longrightarrow 0,$$

we see that this comes down to the obvious fact that Coker $f(K_v) \longrightarrow$ Coker $H^0(K_v, f)$ is an isomorphism (cf. I.3.7).

Notes: This section interpretes Tate's theorems on the Galois cohomology of abelian varietes over number fields [Tate (1962)] in terms of étale cohomology.

§6 Global results: singular schemes

We now let X be an integral scheme whose normalization is the spectrum of the ring of integers in K (number field case) or the unique complete smooth curve with K as its function field (function field case). The definition in §2 of cohomology groups with compact support also applies to singular X: for any sheaf F on an open subscheme U of K, there is an exact sequence

$$\cdots \longrightarrow H^r_c(U,F) \longrightarrow H^r(U,F) \longrightarrow \bigoplus_{v \in S} H^r(K_v,F) \longrightarrow \cdots$$

where S is the set of primes of K not corresponding to point of the normalization of U. Using (1.12) - (1.15), it is possible to prove an analogue of (2.3).

Let $u \in U$, and let $h \in K^\times$. Then h can be written $h = f/g$ with $f,g \in \mathcal{O}_u$, and we define

$$\text{ord}_u(h) = \text{length}(\mathcal{O}_u/(f)) - \text{length}(\mathcal{O}_u/(g)).$$

This determines a homomorphism $K^\times \longrightarrow \mathbb{Z}$ (see [Fulton (1984), 1.2]). Alternatively, we could define $\text{ord}_u(h) = \sum [k(v){:}k(u)]\text{ord}_v(h)$ where the sum is over the points of the normalization of U lying over u (ibid. 1.2.3). One can define similar maps for each closed point lying over u on scheme V étale over U and so obtain a homomorphism

$\mathrm{ord}_u\colon g_*\mathbb{G}_m \longrightarrow i_{u*}\mathbb{Z}$. Define $\mathbb{G}$ to be the complex of sheaves

$$g_*\mathbb{G}_m \xrightarrow{\sum \mathrm{ord}_u} \bigoplus_{u\in U^0} i_{u*}\mathbb{Z}$$

on U_{et}. Note that if U is smooth, then we can identify $\mathbb{G}$ with $\mathbb{G}_m$. We shall frequently make use of the fact that $R^s g_*\mathbb{G}_m = 0$ for $s > 0$. This follows from the similar statement for the normalization $\tilde{U}$ of U, and the fact that $\tilde{U} \longrightarrow U$ is finite.

Lemma 6.1. *For all open subschemes* U *of* X, *there is a canonical trace map* $\mathrm{Tr}\colon H^3_c(U,\mathbb{G}) \xrightarrow{\approx} \mathbb{Q}/\mathbb{Z}$ *such that*

(a) *whenever* U *is smooth,* Tr *is the map defined at the start of* §3;

(b) *whenever* $V \subset U$, *the diagram*

$$\begin{array}{ccc} H^3_c(V,\mathbb{G}) & \xrightarrow{\mathrm{Tr}} & \mathbb{Q}/\mathbb{Z} \\ \downarrow & & \| \\ H^3_c(U,\mathbb{G}) & \xrightarrow{\mathrm{Tr}} & \mathbb{Q}/\mathbb{Z} \end{array}$$

commutes.

Proof: The proof is similar to the smooth case. Let S be the set of primes of K not corresponding to a point in the normalization of U, and assume first that $U \neq X$. From the definition of $\mathbb{G}$, we obtain a cohomology sequence

$$0 \longrightarrow H^2(U,\mathbb{G}) \longrightarrow H^2(K,\mathbb{G}_m) \longrightarrow \bigoplus_{u\in U^0} \mathbb{Q}/\mathbb{Z} \longrightarrow H^3(U,\mathbb{G}_m) \longrightarrow 0.$$

The middle map sends an element a of Br(K) to $\sum \mathrm{inv}_u(a)$ where $\mathrm{inv}_u(a) = \sum_{v\mapsto u} [k(v):k(u)]\mathrm{inv}_v(a)$. The kernel-cokernel exact sequence of the pair of maps

$$\mathrm{Br}(K) \longrightarrow \bigoplus_{\text{all } v} \mathrm{Br}(K_v) \longrightarrow \bigoplus_{u\in U^0} \mathbb{Q}/\mathbb{Z}$$

provides us with the top row of the following diagram:

$$\begin{array}{ccccccc} & H^2(U,\mathbb{G}) & \longrightarrow & B & \longrightarrow & \mathbb{Q}/\mathbb{Z} & \longrightarrow 0 \\ & \| & & \downarrow & & \downarrow & \\ 0 \longrightarrow & H^2(U,\mathbb{G}) & \longrightarrow & \bigoplus_{v\in S} Br(K_v) & \longrightarrow & H^3_c(U,\mathbb{G}) & \longrightarrow 0. \end{array}$$

Here $B = \mathrm{Ker}(\bigoplus_{\text{all } v} Br(K_v) \longrightarrow \bigoplus_{v\in U^0} \mathbb{Q}/\mathbb{Z})$. Write $B = B'\oplus(\bigoplus_{v\in S} Br(K_v))$. Then B' maps to zero in $\mathbb{Q}/\mathbb{Z}$, and as it is the kernel of the middle vertical map, this shows that the map $\mathbb{Q}/\mathbb{Z} \longrightarrow H^3_c(U,\mathbb{G})$ is an isomorphism.

For U = X, one can remove a smooth point and prove as in (2.6) that $H^3_c(X,\mathbb{G}) = H^3_c(X-\{x\},\mathbb{G}_m)$.

As in §3, the trace map allows us to define maps

$$\alpha^r(U,F)\colon \mathrm{Ext}^r_U(F,\mathbb{G}_m) \longrightarrow H^{3-r}_c(U,F)^*$$

for any sheaf F on U.

Theorem 6.2. *Let F be a $\mathbb{Z}$-constructible sheaf on an open subset U of X. For $r \geq 2$, the groups $\mathrm{Ext}^r_U(F,\mathbb{G})$ are torsion of cofinite-type, and $\alpha^r(U,F)$ is an isomorphism. For r = 0,1, the groups $\mathrm{Ext}^r_U(F,\mathbb{G})$ are of finite-type, and $\alpha^r(U,F)$ defines isomorphisms*

$$\mathrm{Ext}^r_U(F,\mathbb{G})\hat{} \longrightarrow H^{3-r}_c(U,F)^*$$

where ^ denotes completion relative to the topology of subgroups of finite index. If F is constructible, then $\alpha^r(U,F)$ is an isomorphism of finite groups for all $r \in \mathbb{Z}$.

Proof: We begin by proving the theorem when F has support on a finite subset.

Lemma 6.3. *Theorem 6.2 is true if F has support on a proper closed subset of U.*

Proof: We can assume that our sheaf is i_*F where i is the inclusion of a single closed point v into U. From the analogue of (2.3) for singular schemes, we see that $H^r_c(U, i_*F) = H^r(v, F)$. As in (1.14), $\mathrm{Ext}^r_U(i_*F, g_*\mathbb{G}) = 0$, and so $\mathrm{Ext}^{r-1}_X(F, \mathbb{Z}) \longrightarrow \mathrm{Ext}^r_U(F, \mathbb{G})$ is an isomorphism. We have a commutative diagram

$$\begin{array}{ccccc} \mathrm{Ext}^r_U(F, \mathbb{G}) & \times & H^{3-r}_c(U, F) & \longrightarrow & H^3(U, \mathbb{G}) \\ \uparrow\approx & & \downarrow\approx & & \downarrow\approx \\ \mathrm{Ext}^{r-1}_{g_x}(M, \mathbb{Z}) & \times & H^{3-r}(g_x, M) & \longrightarrow & H^2(g_x, \mathbb{Z}) \end{array},$$

and so the theorem follows in this case from (I.1.10).

Now let F be a sheaf on U, and let $j: V \hookrightarrow U$ be a smooth open subscheme of U. For $F|V$, the theorem becomes (3.1). Since $\mathrm{Ext}^r_U(j_!F|V, \mathbb{G}) = \mathrm{Ext}^r_V(F, \mathbb{G})$ and $H^r_c(U, j_!F|V) = H^r_c(V, F)$, this implies that theorem is true for $j_!F|V$. The lemma shows that the theorem is true for i_*i^*F, and the two cases can be combined as in (3.7) to prove the general case.

Notes: Theorem 6.2 is proved in [Deninger (1986)] in the case that U = X and F is constructible.

§7 Global results: higher dimensions

The notations are the same as those in §2. Throughout, $\pi: Y \longrightarrow U$ will be a morphism of finite-type, and we define $H^r_c(Y, F) = H^r_c(U, R\pi_!F)$.

Proposition 7.1. *If F is constructible and mF = 0 for some integer m that is invertible on U, then the groups $H^r(Y, F)$ are finite.*

Proof: For any constructible sheaf F on Y, the sheaves $R^r\pi_*F$ are constructible (see [Deligne (1977), 1.1]). Therefore it suffices to prove the proposition for U itself, but we have already noted that (3.1) implies the proposition in this case.

Remark 7.2. In particular, the proposition shows that $H^1(Y,\mathbb{Z}/m\mathbb{Z})$ is finite for all m that are invertible on Y, and this implies that $\pi_1^{ab}(Y)^{(m)}$ is finite. Under some additional hypotheses, most notably that π is smooth, one knows [Katz and Lang (1981)] that the full group $\pi_1(Y)^{ab}$ is finite except for the part provided by constant field extensions in the function field case.

Let v be an archimedean prime of K. In the next proposition, we write Y_v for $Y\times_U \mathrm{Spec}\, K_v$ and $Y_{\bar{v}}$ for $Y\times_U \mathrm{Spec}\, K_{v,s}$.

Proposition 7.3. *Let* $\pi\colon Y \longrightarrow U$ *be proper and smooth, and let* F *be a locally constant constructible sheaf on* Y *such that* $mF = 0$ *for some* m *that is invertible on* U. *Then*

$$\chi(Y,F) = \prod_{v\ \mathrm{arch}} \frac{\chi(Y_{\bar{v}},F|Y_{\bar{v}})^{G_v}}{|\chi(Y_{\bar{v}},F|Y_{\bar{v}})|_v},$$

where $\chi(Y_{\bar{v}},F|Y_{\bar{v}})^{G_v} \overset{\mathrm{df}}{=} \prod_r [H^r(Y_{\bar{v}},F|Y_{\bar{v}})^{G_v}]^{(-1)^r}$.

Proof: The proper-smooth base change theorem [Milne (1980), VI.4.2] shows that the sheaves $R^s\pi_*F$ are locally constant and constructible for all r, and moreover that $(R^s\pi_*F)_{\bar{v}} = H^s(Y_{\bar{v}},F|Y_{\bar{v}})$ for all archimedean primes v. Therefore (2.13) shows that

$$\chi(G_S,R^s\pi_*F) = \prod_{v\ \mathrm{arch}} \frac{[H^s(Y_{\bar{v}},F|Y_{\bar{v}})^{G_v}]}{|[H^s(Y_{\bar{v}},F|Y_{\bar{v}})]|_v}.$$

On taking the alternating product of these equalities, we obtain the result.

Remark 7.4. (a) A similar result is true for $\chi_c(Y,F)$.

(b) The last result can be regarded as a formula expressing the trace of the identity map on Y in terms of the schemes $Y_{\bar{v}}$, v archimedean. For a similar result for other maps, see [Deninger (1986b)].

Before stating a duality theorem for sheaves on such a Y, we note a slight improvement of (3.1). Just as in the case of a single sheaf, there is a canonical pairing of (hyper-) Ext and (hyper-) cohomology groups

$$\mathrm{Ext}^r_U(F^\cdot,\mathbb{G}_m) \times H^{3-r}_c(U,F^\cdot) \longrightarrow H^3_c(U,\mathbb{G}_m),$$

for any complex of sheaves $F^\cdot$ on U. Consequently, there are also maps $\alpha^r(U,F^\cdot)\colon \mathrm{Ext}^r_U(F^\cdot,\mathbb{G}_m) \longrightarrow H^{3-r}_c(U,F^\cdot)^*$.

Lemma 7.5. *Let $F^\cdot$ be a complex of sheaves on U that is bounded below and such that $H^r(F^\cdot)$ is constructible for all r and zero for r >> 0. Then $\alpha^r(U,F^\cdot)\colon \mathrm{Ext}^r_U(F^\cdot,\mathbb{G}_m) \longrightarrow H^{3-r}_c(U,F^\cdot)^*$ is an isomorphism of finite groups.*

Proof: If F consists of a single sheaf, this is (3.1b). The general case follows from this case by a standard argument (see for example [Milne (1980), p280]).

We write $\mathrm{Ext}^r_{Y,m}(F,F')$ for the Ext group computed in the category of sheaves of $\mathbb{Z}/m\mathbb{Z}$-modules on Y_{et}. Let F be a sheaf killed by m; if F' is an m-divisible and $F' \longrightarrow I^\cdot$ is an injective resolution of F', then $F'_m \longrightarrow I^\cdot_m$ is an injective resolution of F'_m, and so

$$\mathrm{Ext}^r_{Y,m}(F,F'_m) \overset{\mathrm{df}}{=} H^r(\mathrm{Hom}_Y(F,I^{\cdot}_m)) = H^r(\mathrm{Hom}_Y(F,I^{\cdot})) \overset{\mathrm{df}}{=} \mathrm{Ext}^r_Y(F,F').$$

In particular, if F is killed by m and m is invertible on Y, then $\mathrm{Ext}^r_{Y,m}(F,\mu_m) = \mathrm{Ext}^r_Y(F,\mathbb{G}_m)$.

Let $\pi\colon Y \longrightarrow U$ be smooth and separated with fibres pure of dimension d, and let m be an integer that is invertible on U. Then there is a canonical trace map $R^{2d}\pi_!\mu_m^{\otimes d} \xrightarrow{\approx} \mathbb{Z}/m\mathbb{Z}$ [Milne (1980), p285]. On U, μ_m is locally isomorphic to $\mathbb{Z}/m\mathbb{Z}$, and so when the trace map is tensored with μ_m, it becomes an isomorphism $R^{2d}\pi_!\mu_n^{\otimes d+1} \xrightarrow{\approx} \mu_m$. As $R^r\pi_*\mu_n^{\otimes d+1} = 0$ for $r > 2d$, there is a canonical trace map $H^{2d+3}_c(Y,\mu_m^{\otimes d+1}) \xrightarrow{\approx} H^3_c(Y,\mu_m) \xrightarrow{\approx} \mathbb{Z}/m\mathbb{Z}$, and hence a pairing

$$\mathrm{Ext}^r_{Y,m}(F,\mu_m^{\otimes d+1}) \times H^{2d+3-r}_c(Y,F) \longrightarrow H^{2d+3}_c(Y,\mu_m^{\otimes d+1}) \approx \mathbb{Z}/m\mathbb{Z}.$$

Theorem 7.6. *Let* $Y \longrightarrow U$ *be a smooth separated morphism with fibres pure of dimension* d, *and let* F *be a constructible sheaf on* Y *such that* $mF = 0$ *for some* m *that is invertible on* U. *Then*

$$\mathrm{Ext}^r_{Y,m}(F,\mu_m^{\otimes d+1}) \times H^{2d+3-r}_c(Y,F) \longrightarrow H^{2d+3}_c(Y,\mu_m^{\otimes d+1}) \approx \mathbb{Z}/m\mathbb{Z}.$$

is a nondegenerate pairing of finite groups.

Proof: The duality theorem in [Artin, Grothendieck, and Verdier (1972/73), XVIII] shows that there is a canonical isomorphism

$$R\pi_!(R\mathcal{H}om_{Y,m}(F,\mu_m^{\otimes d+1}[2d])) \xrightarrow{\approx} R\mathcal{H}om_{U,m}(R\pi_!F,\mu_m).$$

(See also [Milne (1980), p285].) On applying $R\Gamma(U,-)$ to the left hand side, we get a complex of abelian groups whose r^{th} cohomology group is $\mathrm{Ext}^{r+2d}_{Y,m}(F,\mu_m^{\otimes d+1})$. On applying the same functor to the right hand side, we get a complex of abelian groups whose r^{th} cohomology group is $\mathrm{Ext}^r_{U,m}(R\pi_!F,\mu_m)$. This is equal to $\mathrm{Ext}^r_U(R\pi_!F,\mathbb{G}_m)$, and Lemma 7.5 shows that

$$\mathrm{Ext}^r_U(R\pi_!F,\mathbb{G}_m) \xrightarrow{\approx} \mathrm{Hom}(H^{3-r}_c(U,R\pi_!F),\mathbb{Z}/m\mathbb{Z}).$$

By definition, $H^{3-r}_c(U,R\pi_!F) = H^{3-r}_c(Y,F)$, and so this proves the theorem.

For any sheaf F on Y such that $mF = 0$, write $F(i) = F\otimes\mu_m^{\otimes i}$ and $F^D(i) = \mathcal{H}om(F,\mu_m^{\otimes i})$. Note that F^D in the old terminology is equal to $F^D(1)$ in the new.

Corollary 7.7. *Let* $\pi\colon Y \longrightarrow U$ *be a smooth separated morphism with fibres pure of dimension* d, *and let* F *be a locally constant constructible sheaf on* Y *such that* $mF = 0$ *for some* m *that is invertible on* U. *Then cup-product defines a nondegenerate pairing of finite groups*

$$H^r(Y,F^D(d+1)) \times H^{2d+3-r}_c(Y,F) \longrightarrow H^{2d+3}_c(Y,\mu_m^{\otimes d+1}) = \mathbb{Z}/m\mathbb{Z},$$

for all r.

Proof: In this case $\mathrm{Ext}^r_{Y,m}(F,\mu_m^{\otimes d+1}) = H^r(Y,F^D(d+1))$.

As usual, we let $\mathcal{K}_i$ be the sheaf on Y defined by the ith Quillen K-functor.

Corollary 7.8. *Let* F *be a constructible sheaf on* Y *such that* $\ell^nF = 0$ *for some prime* ℓ *invertible on* U. *Assume that* $H^{2d+2}_c(Y,\mathcal{K}_{2d+1})$ *is torsion. Then there is a trace map* $H^{2d+3}_c(Y,\mathcal{K}_{2d+1})(\ell) \xrightarrow{\approx} \mathbb{Q}_\ell/\mathbb{Z}_\ell$, *and the canonical pairing*

$$\mathrm{Ext}^r_Y(F,\mathcal{K}_{2d+1}) \times H^{2d+3-r}_c(Y,F) \longrightarrow H^{2d+3}_c(Y,\mathcal{K}_{2d+1})(\ell) \approx \mathbb{Q}_\ell/\mathbb{Z}_\ell$$

is a duality of finite groups.

Proof: Recall (1.19), that for any m that is invertible on U, there is an exact sequence of sheaves

$$0 \longrightarrow \mu_m^{\otimes i+1} \longrightarrow \mathcal{K}_{2i+1} \xrightarrow{m} \mathcal{K}_{2i+1} \longrightarrow 0.$$

Therefore $\mathrm{Ext}^r_{Y,\ell^n}(F,\mu_{\ell^n}^{\otimes d+1}) = \mathrm{Ext}^r_Y(F,\mathcal{K}_{2d+1})$. Also, there is an exact sequence

$$0 \to H_c^{2d+2}(Y,\mathcal{K}_{2d+1})\otimes\mathbb{Q}_\ell/\mathbb{Z}_\ell \longrightarrow H_c^{2d+3}(Y,\mu_{\ell^\infty}^{\otimes d+1}) \longrightarrow H_c^{2d+3}(Y,\mathcal{K}_{2d+1})(\ell) \to 0.$$

Since we have assumed $H_c^{2d+2}(Y,\mathcal{K}_{2d+1})$ to be torsion, the first term of this sequence is zero, and so we obtain isomorphisms $H_c^{2d+3}(Y,\mathcal{K}_{2d+1})(\ell) \approx H_c^{2d+3}(Y,\mu_{\ell^\infty}^{\otimes d+1}) \approx \mathbb{Q}_\ell/\mathbb{Z}_\ell$. The corollary now follows directly from (7.6).

Remark 7.9. For any regular scheme Y of finite type over a field, it is known [Milne (1986), 7.1] that $H^r(Y,\mathcal{K}_i)$ is torsion for $r > i$. (The condition that Y be of finite type over a field is only required so that Gersten's conjecture can be assumed.)

Aside 7.10. One would like to weaken the condition that Y is smooth over U in the above results to the condition that Y is regular. The purity conjecture in étale cohomology will be relevant for this. It states the following: Let $i\colon Z \longrightarrow Y$ be a closed immersion of regular local Noetherian schemes such that for each z in Z, the codimension of Z in Y at z is c, and let n be prime to the residue characteristics; then $(R^r i^!)\mathbb{Z}/m\mathbb{Z} = 0$ for $r \neq 2c$, and $(R^{2c} i^!)\mathbb{Z}/m\mathbb{Z} = \mu_n^{\otimes -c}$ [Artin, Grothendieck, and Verdier (1972/73), XIX].

The author is uncertain as to the exact conditions under which the proof of the conjecture is complete. See (ibid., XIX.2.1) and [Thomason (1984)].

The strategy for passing from the smooth case to the regular case is as follows: replace U by its normalization in Y, and note that the theorem will hold on an open subset V of Y; now examine the map $Y - V \longrightarrow U$. (Compare the proof of the Poincaré duality theorem VI.11.1 in [Milne (1980)], especially Step 3.)

In the case that $Y = U$, Corollary 7.8 is much weaker than Theorem 3.1 because it requires that F be killed by an integer that is invertible on U. We investigate some conjectures that lead to results that are true generalizations of (3.1). We first consider the problem of duality for p-torsion sheaves in characteristic p.

For a smooth variety Y over a field of characteristic $p \neq 0$, we let $W_n\Omega^i_{Y/k}$ be the sheaf of Witt differential i-forms of length n on Y [Illusie (1979)]. Define $\nu_n(i)$ to be the subsheaf of $W_n\Omega^i_{Y/k}$ of locally logarithmic differentials (see [Milne (1986a), §1]). The pairing

$$(\omega,\omega') \mapsto \omega\wedge\omega' : W_n\Omega^i \times W_n\Omega^j \longrightarrow W_n\Omega^{i+j}$$

induces a pairing $\nu_n(i) \times \nu_n(j) \longrightarrow \nu_n(i+j)$.

Theorem 7.11. *Let Y be a smooth complete variety of dimension d over a finite field k. Then there is a canonical trace map* $H^{d+1}(Y,\nu_n(d)) \xrightarrow{\approx} \mathbb{Z}/p^n\mathbb{Z}$, *and the cup-product pairing*

$$H^r(Y,\nu_n(i)) \times H^{d+1-r}(Y,\nu_n(d-i)) \longrightarrow H^{d+1}(Y,\nu_n(d)) \approx \mathbb{Z}/p^n\mathbb{Z}$$

is a duality of finite groups.

Proof: When $\dim Y \leq 2$ or $n = 1$, this is proved in [Milne (1976)]. The extension to the general case can be found in [Milne (1986a)].

Corollary 7.12. *The canonical pairing*

$$\mathrm{Ext}^r_{Y,p^n}(\mathbb{Z}/p^n\mathbb{Z},\nu_n(d)) \times H^{d+1-r}(Y,\mathbb{Z}/p^n\mathbb{Z}) \longrightarrow H^{d+1}(Y,\nu_n(d)) \approx \mathbb{Z}/p^n\mathbb{Z}$$

is a duality of finite groups.

Proof: One sees easily that $\mathrm{Ext}^r_{Y,p^n}(\mathbb{Z}/p^n\mathbb{Z}, \nu_n(d)) = H^r(Y, \nu_n(d))$, and so this follows immediately from the proposition.

Corollary 7.13. *Let Y be a smooth complete variety of dimension* $d \lesssim 2$ *over a finite field k. Then there is a canonical trace map* $H^{d+2}(Y, \mathcal{K}_d)(p) \xrightarrow{\approx} \mathbb{Q}_p/\mathbb{Z}_p$, *and for any n there are a nondegenerate pairings of finite groups*

$$\mathrm{Ext}^r_Y(\mathbb{Z}/p^n\mathbb{Z}, \mathcal{K}_d)) \times H^{d+2-r}(Y, \mathbb{Z}/p^n\mathbb{Z}) \longrightarrow H^{d+2}(Y, \mathcal{K}_d)(p) \approx \mathbb{Q}_p/\mathbb{Z}_p.$$

Proof: The key point is that there is an exact sequence

$$0 \longrightarrow \mathcal{K}_i \xrightarrow{p^n} \mathcal{K}_i \longrightarrow \nu_n(i) \longrightarrow 0.$$

for $i \lesssim 2$. When $i = 0, 1$, the exactness of the sequence is obvious; when $i = 2$, its exactness at the first and second terms follows from theorems of Suslin [Suslin (1983a)] and Bloch [Bloch and Kato (1986)] respectively. Now the corollary can be derived in the same manner as (7.8); in particular the trace map is obtained from the maps $H^{d+1}(Y, \nu_\infty(d)) \xrightarrow{\approx} H^{d+2}(Y, \mathcal{K}_d)(p)$ and $H^{d+1}(Y, \nu_\infty(d)) \xrightarrow{\approx} (\mathbb{Q}/\mathbb{Z})(p)$, where $\nu_\infty(i) = \varinjlim \nu_n(i)$.

Remark 7.14. It should be possible to extend the last three results to noncomplete varieties Y by using (in the proof of 7.11) cohomology with compact support for quasicoherent sheaves (see [Deligne (1966)] or [Hartshorne (1972)]). However, one problem in extending them to all constructible sheaves is that the purity theorem for the sheaves $\nu_n(i)$ is weaker than its analogue for the sheaves $\mu_n(i)$ (see [Milne (1986), §2]). Nevertheless, I conjecture that for any constructible sheaf F of $\mathbb{Z}/p^n\mathbb{Z}$-modules on a smooth variety Y of dimension d over a finite field,

$$\mathrm{Ext}^r_{Y,p^n}(F,\nu_n(d)) \times H^{d+1-r}_c(Y,F) \longrightarrow H^{d+1}_c(Y,\nu_n(d)) \approx \mathbb{Z}/p^n\mathbb{Z}$$

is a nondegenerate pairing of finite groups.

I do not conjecture that the (7.13) holds for varieties of all dimensions.

Let Y be a smooth complete surface over a finite field. Then we have dualities:

$$\mathrm{Ext}^r_Y(\mathbb{Z}/p^n\mathbb{Z},\mathcal{K}_2) \times H^{4-r}(Y,\mathbb{Z}/p^n\mathbb{Z}) \longrightarrow H^4(Y,\mathcal{K}_2)(p) \approx \mathbb{Q}_p/\mathbb{Z}_p,\ p = \text{char } k,$$

$$\mathrm{Ext}^r_Y(\mathbb{Z}/\ell^n\mathbb{Z},\mathcal{K}_3) \times H^{5-r}(Y,\mathbb{Z}/\ell^n\mathbb{Z}) \longrightarrow H^5(Y,\mathcal{K}_3)(\ell) \approx \mathbb{Q}_\ell/\mathbb{Z}_\ell,\ \ell \neq \text{char } k.$$

These are similar, but the numbers do not agree! It appears that in order to obtain a uniform statement, the sheaves $\mathcal{K}_i$ will have to be replaced by the objects $\mathbb{Z}(i)$ conjectured in [Lichtenbaum (1984)] to exist in the derived category of $\mathbf{S}(Y_{et})$ for any regular scheme Y. (Beilinson has independently conjectured the existence of similar objects in $\mathbf{S}(Y_{Zar})$.) These are to have the following properties:

(a) $\mathbb{Z}(0) = \mathbb{Z}$, $\mathbb{Z}(1) = \mathbb{G}_m[-1]$.

(b_ℓ) For $\ell \neq p$, there is a distinguished triangle

$$\mu^{\otimes i}_{\ell^n}[-1] \longrightarrow \mathbb{Z}(i) \xrightarrow{\ell^n} \mathbb{Z}(i) \longrightarrow \mu^{\otimes i}_{\ell^n}.$$

(This implies that there is an exact sequence

$$\dots\longrightarrow H^r(Y,\mathbb{Z}(i)) \xrightarrow{\ell^n} H^r(Y,\mathbb{Z}(i)) \longrightarrow H^r(Y,\mu^{\otimes i}_{\ell^n}) \longrightarrow \dots .)$$

(c) There are canonical pairings $\mathbb{Z}(i) \times \mathbb{Z}(j) \longrightarrow \mathbb{Z}(i+j)$.

(d) $H^{2r-j}(\mathbb{Z}(i)) = \mathrm{Gr}^r_\gamma \mathcal{K}_j$ up to small torsion, and $H^r(\mathbb{Z}(i)) = 0$ for $r > i$ or $r < 0$ (also $H^0(\mathbb{Z}(i)) = 0$ except when $i = 0$).

(e) If Y is a smooth complete variety over a finite field, then $H^r(Y,\mathbb{Z}(i))$ is torsion for all $r \neq 2i$, and $H^{2r}(Y,\mathbb{Z}(r))$ is finitely generated.

In the present context, it is natural to ask that the complex have the following additional properties when Y is a variety over a field of characteristic p:

(b_p) There is a distinguished triangle

$$\nu_n(i)[-i-1] \longrightarrow \mathbb{Z}(i) \xrightarrow{p^n} \mathbb{Z}(i) \longrightarrow \nu_n(i)[-i].$$

(This implies that there is an exact sequence

$$\ldots \longrightarrow H^r(Y,\mathbb{Z}(i)) \xrightarrow{p^n} H^r(Y,\mathbb{Z}(i)) \longrightarrow H^{r-i}(Y,\nu_n(i)) \longrightarrow \ldots\ .)$$

(f) (Purity) If $i\colon Z \hookrightarrow Y$ is the inclusion of a smooth closed subscheme of codimension c into a smooth scheme and $j > c$, then $Ri^!\mathbb{Z}(j) = \mathbb{Z}(j-c)[-2c]$.

Theorem 7.15. *Let $\pi\colon Y \longrightarrow U$ be smooth and proper with fibres pure of dimension d. Let ℓ be a prime, and assume that either ℓ is invertible on U or ℓ = char K and Y is complete. Assume that there exist complexes $\mathbb{Z}(i)$ satisfying (b_ℓ) and that $H_c^{2d+3}(Y,\mathbb{Z}(d+1))$ is torsion. Then there is a canonical isomorphism*

$$H_c^{2d+4}(Y,\mathbb{Z}(d+1))(\ell) \xrightarrow{\approx} (\mathbb{Q}/\mathbb{Z})(\ell),$$

and the cup-product pairing

$$H^r(Y,\mathbb{Z}(i))(\ell) \times H_c^{2d+4-r}(Y,\mathbb{Z}(d+1-i))(\ell) \longrightarrow H_c^{2d+4}(Y,\mathbb{Z}(d+1))(\ell) \approx \mathbb{Q}/\mathbb{Z}(\ell)$$

annihilates only the divisible subgroups.

Proof: Assume first that $\ell \neq$ char(K). The same argument as in the proof of (7.8) shows the existence of an isomorphism

$$H_c^{2d+3}(Y,\mu_{\ell^\infty}^{\otimes d+1}) \xrightarrow{\approx} H_c^{2d+4}(Y,\mathbb{Z}(d+1)).$$

This proves that a trace map exists. Now the exact sequence

$$0 \longrightarrow H^{r-1}(Y,\mathbb{Z}(i))\otimes(\mathbb{Q}/\mathbb{Z})(\ell) \longrightarrow H^{r-1}(Y,\mu_{\ell^\infty}^{\otimes i}) \longrightarrow H^r(Y,\mathbb{Z}(i))(\ell) \longrightarrow 0$$

shows that $H^{r-1}(Y,\mu_{\ell^\infty}^{\otimes i})$ modulo its divisible subgroup is isomorphic to $H^r(Y,\mathbb{Z}(i))(\ell)$ modulo its divisible subgroup. Similarly

$$0 \longrightarrow H_c^r(Y,\mathbb{Z}(i))^\wedge \longrightarrow \varprojlim H_c^r(Y,\mu_{\ell^m}^{\otimes i}) \longrightarrow T_\ell H_c^{r+1}(Y,\mathbb{Z}(i)) \longrightarrow 0$$

shows that $H_c^r(Y,\mathbb{Z}(i))(\ell)$ modulo its divisible subgroup is isomorphic to the torsion subgroup of $\varprojlim H_c^r(Y,\mu_{\ell^m}^{\otimes i})$. Now the theorem follows from (7.7) using (I.0.20e).

The proof when $\ell = p$ is similar.

Consider the following statement:

(*) for any smooth variety Y of dimension d over a finite field and any constructible sheaf F on Y, there is a duality of finite groups

$$\mathrm{Ext}_Y^r(F,\mathbb{Z}(d)) \times H_c^{2d+2}(Y,F) \longrightarrow H_c^{2d+2}(Y,\mathbb{Z}(d)) \approx \mathbb{Q}/\mathbb{Z}.$$

When F is killed by some m prime to the characteristic of k and we assume (b) and that $H_c^{2d+1}(Y,\mathbb{Z}(d))$ is torsion, this can be derived from (7.6) in the same way as (7.8). When F is a p-primary sheaf, it is necessary to assume the conjectured statement in (7.14).

Theorem 7.16. *Let* $\pi: Y \longrightarrow U$ *be a smooth proper morphism with fibres of dimension* d. *Assume there exist complexes* $\mathbb{Z}(i)$ *satisfying the conditions* (a), (b), *and* (f); *also assume* (*) *above, and that* $H_c^{2d+3}(Y,\mathbb{Z}(d+1))$ *is torsion, so that there exists a canonical trace map* $H_c^{2d+4}(Y,\mathbb{Z}(d+1)) \xrightarrow{\approx} \mathbb{Q}/\mathbb{Z}$. *Then for any locally constant constructible sheaf* F *on* Y, *there is a nondegenerate pairing of finite groups*

$$\mathrm{Ext}_Y^r(F,\mathbb{Z}(d)) \times H_c^{2d+4-i}(Y,F) \longrightarrow H_c^{2d+4}(Y,\mathbb{Z}(d)) \approx \mathbb{Q}/\mathbb{Z}.$$

Proof: First assume F has support on Y_Z for some closed subscheme Z of U, and write i for the closed immersion $Y_Z \hookrightarrow Y$. The spectral sequence $\mathrm{RHom}_{Y_Z}(F, Ri^!\mathbb{Z}(d+1)) = \mathrm{RHom}_Y(i_*F, \mathbb{Z}(d+1))$ shows that $\mathrm{Ext}^{r-2}_{Y_Z}(F, \mathbb{Z}(d-1)) = \mathrm{Ext}^r_Y(i_*F, \mathbb{Z}(i))$. Thus this case of the theorem follows from the induction assumption.

Next suppose that $mF = 0$ for some m that is invertible on U. In this case then the theorem can be deduced from (7.6) in the same way as (7.8).

Next suppose that $p^nF = 0$, where $p = \mathrm{char}\, K$. In this case the statement reduces to (*).

The last two paragraphs show that the theorem holds for the restriction of F to Y_V for some open subscheme V of U, and this can be combined with the statement proved in the first paragraph to obtain the full theorem.

In (4.11), we have shown that there is a nondegenerate pairing

$$\mathrm{Ext}^r_U(F\otimes\mathbb{Z}(1), \mathbb{Z}(1)) \times H^{4-r}_c(U, F\otimes\mathbb{Z}(1)) \longrightarrow H^4_c(\mathbb{Z}(1)) \approx \mathbb{Q}/\mathbb{Z}.$$

for any torsion-free $\mathbb{Z}$-constructible sheaf F on U. For a finite field k, we also know that there is a nondegenerate pairing

$$\mathrm{Ext}^r_k(F\otimes\mathbb{Z}(0), \mathbb{Z}(0)) \times H^2(k, F\otimes\mathbb{Z}(0)) \longrightarrow H^2(k, \mathbb{Z}(0)) \approx \mathbb{Q}/\mathbb{Z}.$$

This suggests the following conjecture.

Conjecture 7.16. *For any regular scheme* Y *of finite-type over* Spec $\mathbb{Z}$ *and of (absolute) dimension* d, *there is a canonical trace map* $H^{2d+2}_c(Y, \mathbb{Z}(d)) \xrightarrow{\approx} \mathbb{Q}/\mathbb{Z}$. *For any locally constant $\mathbb{Z}$-constructible sheaf* F *on* Y, *the canonical pairing*

$$\mathrm{Ext}^r_Y(F\otimes^L\mathbb{Z}(i), \mathbb{Z}(d)) \times H^{2d+2-r}_c(Y, F\otimes^L\mathbb{Z}(i)) \longrightarrow H^{2d+2}_c(Y, \mathbb{Z}(d)) \approx \mathbb{Q}/\mathbb{Z}$$

induces isomorphisms

$$\mathrm{Ext}^r_Y(F\otimes^L\mathbb{Z}(i),\mathbb{Z}(d)) \xrightarrow{\approx} H^{2d+2-r}_c(Y,F\otimes^L\mathbb{Z}(i))^*,\quad r \neq 2(d-i)$$

$$\mathrm{Ext}^{2d-2i}_Y(F\otimes^L\mathbb{Z}(i),\mathbb{Z}(d))^\wedge \xrightarrow{\approx} H^{2+2i}_c(Y,F\otimes^L\mathbb{Z}(i))^*.$$

The conjecture has obvious implications for higher class field theory.

Finally, we mention that [Lichtenbaum (1986)] has suggested a candidate for $\mathbb{Z}(2)$ and [Bloch (1986)] has suggested candidates for $\mathbb{Z}(r)$, all r. Also [Kato (1985/6)] has generalized (5.11) to a relative theorem, and in the case of a surface [Etesse (1986a,b)] has generalized it to other sheaves.

Notes: This section owes much to conversations with Lichtenbaum and to his criticisms of an earlier version.

CHAPTER III

FLAT COHOMOLOGY

This chapter is concerned with duality theorems for the flat cohomology groups of finite flat group schemes or Néron models of abelian varieties. In §1 — §4, the base scheme is the spectrum of the ring of integers in a number field or a local field of characteristic zero (with perfect residue field of nonzero characteristic). In the remaining sections, the base scheme is the spectrum of the rings of integers in a local field of nonzero characteristic or a curve over a finite field (or, more generally, a perfect field of nonzero characteristic). The appendices discuss various aspects of the theory of finite group schemes and Néron models.

The prerequisites for this chapter are the same as for the last: a basic knowledge of the theory of sites, as may be obtained from reading Chapters II and III of [Milne (1980)]. All schemes are endowed with the flat topology.

The results of the chapter are more tentative than those in the first two chapters. One problem is that we do not yet know what is the correct analogue for the flat site of the notion of a constructible sheaf. The examples of [Shatz (1966)] show that for any non-perfect field K, there exist torsion sheaves F over K such that $H^r(K_{fl},F)$ is nonzero for arbitrarily high values of r. In particular, no duality theorem can hold for all finite sheaves over such a field. We are thus forced to restrict our attention to sheaves that are represented by finite flat group schemes or are slight generalizations of such sheaves. Another problem is that for a finite flat group scheme N over an algebraically closed field k, the groups $\mathrm{Ext}_k^r(N,\mathbb{G}_m)$ computed in the category of flat sheaves over k need not vanish for $r > 0$ (see [Breen (1969b)]); they therefore do not agree with the same groups computed in the category of commutative algebraic groups over k, which vanish for $r > 0$.

§0 Preliminaries

We begin by showing that some of the familiar constructions for the étale site can also be made for the flat site.

Cohomology with support on a closed subscheme

Consider the diagram

$$Z \xrightarrow{i} X \xleftarrow{j} U$$

in which i and j are closed and open immersions respectively, and X is the disjoint union of i(Z) and j(U).

Lemma 0.1. *The functor* $j^*: \mathrm{S}(X_{fl}) \longrightarrow \mathrm{S}(U_{fl})$ *has an exact left adjoint* $j_!$.

Proof: For any presheaf P on U, we can define a presheaf $j_!P$ on X as follows: for any morphism $V \longrightarrow X$ of finite type, set

$$\Gamma(V,j_!P) = \oplus P(V_f)$$

where the sum is over all maps $f \in \mathrm{Hom}_X(V,U)$ and V_f denotes V regarded as a scheme over U by means of f. One checks easily that $j_!$ is left adjoint to the restriction functor $j^p: \mathrm{P}(X_{fl}) \longrightarrow \mathrm{P}(U_{fl})$ and that it is exact. Let a be the functor sending a presheaf on X_{fl} to its associated sheaf. Then the functor

$$S(U_{fl}) \hookrightarrow P(U_{fl}) \xrightarrow{j_!} P(X_{fl}) \xrightarrow{a} S(X_{fl}),$$

is easily seen to be left adjoint to j^*. It is therefore right exact. But it is also a composite of left exact functors, which shows that it is exact.

Lemma 0.2. *There is a canonical exact sequence*

$$0 \longrightarrow j_!j^*\mathbb{Z} \longrightarrow \mathbb{Z} \longrightarrow i_*i^*\mathbb{Z} \longrightarrow 0.$$

Proof: The maps are the adjunction maps. We explicitly compute the two end terms. Let $\varphi: V \longrightarrow X$ be a scheme of finite type over U. When V is connected, φ factors through U in at most one way. Therefore, in this case, the presheaf $j_!\mathbb{Z}$ takes the value $\mathbb{Z}$ on V if $\varphi(V) \subset j(U)$ and takes the value 0 otherwise. It follows that $j_!\mathbb{Z}$ is the sheaf

$$V \mapsto \Gamma(V, j_!\mathbb{Z}) = \mathrm{Hom}(\pi_0'(V), \mathbb{Z})$$

where $\pi_0'(V)$ is the subset of $\pi_0(V)$ of connected components of V whose structure morphisms factor through U. Since $\Gamma(V,\mathbb{Z}) = \mathrm{Hom}(\pi_0(V),\mathbb{Z})$, it is obvious that $j_!j^*\mathbb{Z} \longrightarrow \mathbb{Z}$ is injective.

For any V, $\Gamma(V, i_*\mathbb{Z}) = \mathrm{Hom}(\pi_0(\varphi^{-1}Z), \mathbb{Z})$, which is zero if and only if $\varphi(V) \cap Z = \emptyset$. It clear from this that the sequence is exact at its middle term, and that an element of $\Gamma(V, i_*\mathbb{Z})$ lifts to $\Gamma(V_i, \mathbb{Z})$ for each V_i in an appropriate Zariski open covering of V. This completes the proof.

The map $F \mapsto \mathrm{Ker}(\Gamma(X,F) \longrightarrow \Gamma(U,F))$ defines a left exact functor $S(X_{fl}) \longrightarrow$ **Ab**, and we write $H^r_Z(X,-)$ for its rth right derived functor.

Proposition 0.3. *Let* F *be a sheaf on* X_{fl}.

(a) *For all* r, $H^r_Z(X,F) = Ext^r_X(i_*\mathbb{Z},F)$.

(b) *For all* r, $Ext^r_X(j_!\mathbb{Z},F) = Ext^r_U(\mathbb{Z},j^*F)$.

(c) *There is a long exact sequence*

$$\dots \longrightarrow H^r_Z(X,F) \longrightarrow H^r(X,F) \longrightarrow H^r(U,F) \longrightarrow \dots .$$

Proof: (a) On applying Hom(-,F) to the exact sequence in (0.2), we get an exact sequence

$$0 \longrightarrow Hom_X(i_*\mathbb{Z},F) \longrightarrow Hom_X(\mathbb{Z},F) \longrightarrow Hom_U(\mathbb{Z},j^*F)$$

or,

$$0 \longrightarrow Hom_X(i_*\mathbb{Z},F) \longrightarrow \Gamma(X,F) \longrightarrow \Gamma(U,F).$$

Therefore $Hom_X(i_*\mathbb{Z},F) \xrightarrow{\approx} H^0_Z(X,F)$, and on taking the right derived functors we obtain the result.

(b) Note that, because it has an exact left adjoint, j^* preserves injectives. It is also exact [Milne (1980), p68]. Therefore we may derive the equality $Hom_X(j_!\mathbb{Z},F) = Hom_U(\mathbb{Z},j^*F)$ and obtain an isomorphism $Ext^r_X(j_!\mathbb{Z},F) \approx Ext^r_U(\mathbb{Z},j^*F)$.

(c) The $Ext_X(-,F)$-sequence arising from the exact sequence in (0.2) is the required sequence.

Cohomology with compact support

Let X be the spectrum of the ring of integers in a number field or else a complete smooth curve over a perfect field, and let K be the field of rational functions on X. For any open subscheme U of X and sheaf F on U_{fl}, we shall define *cohomology groups with compact support* $H^r_c(X_{fl},F)$ having properties similar to their namesakes for the étale topology. In particular, they will be related to the usual cohomology groups by an exact sequence

$$\dots \longrightarrow H^r_c(U,F) \longrightarrow H^r(U,F) \longrightarrow \bigoplus_{v\in X-U} H^r(K_v,F) \longrightarrow \dots$$

where K_v is the field of fractions of the Henselization $\mathcal{O}_v^h$ of $\mathcal{O}_v$.

Let Z be the complement of U in X, and let $Z' = \bigcup_{v\in X-U} \operatorname{Spec} K_v$ (disjoint union). Then $Z' = \varprojlim V\times_X U$, where the limit is over the étale neighbourhoods V of Z in X:

$$\begin{array}{ccccc} & & V & \longleftarrow & V\times_X U \\ & \nearrow & \downarrow \text{étale} & & \downarrow \\ Z & \longrightarrow & X & \longleftarrow & U\ . \end{array}$$

Let i' be the canonical map $i'\colon Z' \longrightarrow U$, and let $F \longrightarrow I^{\cdot}(F)$ be an injective resolution of F on U_{fl}. Then i'^{*} is exact and preserves injectives, and so $F|Z' \longrightarrow I^{\cdot}(F)|Z'$ is an injective resolution of $F|Z'$. There is an obvious restriction map

$$u\colon \Gamma(U,I^{\cdot}(F)) \longrightarrow \Gamma(Z',I^{\cdot}(F)|Z'),$$

and we define $H_c(U,F)$ to be the translate $C^{\cdot}(u)[-1]$ of its mapping cone. Finally, we set $H^r_c(U,F) = H^r(H_c(U,F))$.

Proposition 0.4. (a) *For any sheaf* F *on an open subscheme* $U \subset X$, *there is an exact sequence*

$$\dots \longrightarrow H^r_c(U,F) \longrightarrow H^r(U,F) \longrightarrow \bigoplus_{v\in X-U} H^r(K_v,F_v) \longrightarrow \dots\ .$$

(b) *For any short exact sequence*

$$0 \longrightarrow F' \longrightarrow F \longrightarrow F'' \longrightarrow 0$$

of sheaves on U, *there is a long exact sequence of cohomology groups*

$$\dots \longrightarrow H^r_c(U,F') \longrightarrow H^r_c(U,F) \longrightarrow H^r_c(U,F'') \longrightarrow \dots\ .$$

(c) *For any sheaf* F *on* U *and open subscheme* V *of* U, *there is an exact sequence*

$$\ldots \longrightarrow H_c^r(V,F|V) \longrightarrow H_c^r(U,F) \longrightarrow \bigoplus_{v\in U-V} H^r(\mathcal{O}_v^h,F) \longrightarrow \ldots .$$

(d) *If F is the inverse image of a sheaf* F_0 *on* U_{et}, *or if F is represented by a smooth algebraic space, then* $H_c^r(U_{fl},F) = H^r(X_{et},j_!F)$.

(e) *For any sheaves F and G on U, there are canonical pairings*

$$\mathrm{Ext}_U^r(F,G) \times H_c^s(U,F) \longrightarrow H_c^{r+s}(U,G).$$

Proof: (a) Directly from the definition of $H_c(U,F)$, we see that there is a distinguished triangle

$$H_c(U,F) \longrightarrow \Gamma(U,I^{\cdot}(F)) \longrightarrow \Gamma(Z',I^{\cdot}(F)|Z') \longrightarrow H_c(U,F)[1].$$

As $H^r(\Gamma(U,I^{\cdot}(F))) = H^r(U,F)$ and $H^r(\Gamma(Z',I^{\cdot}(F)|Z')) = H^r(Z',F|Z') = \bigoplus_{v\in X-U} H^r(K_v,F)$, we see that the required sequence is simply the cohomology sequence of this triangle.

(b) From the morphism

$$\begin{array}{ccccccccc} 0 \longrightarrow & \Gamma(U,I^{\cdot}(F')) & \longrightarrow & \Gamma(U,I^{\cdot}(F)) & \longrightarrow & \Gamma(U,I^{\cdot}(F'')) & \longrightarrow 0 \\ & \downarrow u' & & \downarrow u & & \downarrow u'' & \\ 0 \longrightarrow & \Gamma(Z',I^{\cdot}(F')|Z') & \longrightarrow & \Gamma(Z',I^{\cdot}(F)|Z') & \longrightarrow & \Gamma(Z',I^{\cdot}(F'')|Z') & \longrightarrow 0 \end{array}$$

of short exact sequences of complexes, we may deduce (II.0.10a) the existence of a distinguished triangle

$$H_c(U,F'')[-1] \longrightarrow H_c(U,F') \longrightarrow H_c(U,F) \longrightarrow H_c(U,F'').$$

This yields the required exact sequence.

(c) Let $F \longrightarrow I^{\cdot}(F)$ be an injective resolution of F on U, and consider the maps

$$\Gamma(V,I^{\cdot}(F)) \xrightarrow{a} \bigoplus_{v\notin V} \Gamma(K_v,I^{\cdot}(F)) \xrightarrow{b} \bigoplus_{v\notin U} \Gamma(K_v,I^{\cdot}(F)) \bigoplus_{v\in U-V} \Gamma_v(\mathcal{O}_v^h,I^{\cdot}(F))[1].$$

The map b is such that $H^r(b)$ is the sum of the identity maps

$$H^r(K_v,F) \longrightarrow H^r(K_v,F) \qquad (v \in X - U)$$

and the maps in the complex

$$H^r(\mathcal{O}_v^h, F) \longrightarrow H^r(K_v, F) \longrightarrow H_v^{r+1}(\mathcal{O}_v^h, F) \qquad (v \in U - V).$$

From these maps, we get a distinguished triangle (II.0.10c)

$$C^{\cdot}(b)[-1] \longrightarrow C^{\cdot}(a) \longrightarrow C^{\cdot}(b\circ a) \longrightarrow C^{\cdot}(b).$$

Clearly $C^{\cdot}(a) = H_c(V,F)[1]$. We shall show that there exist isomorphisms $C^{\cdot}(b\circ a) \approx H_c(U,F)[1]$ and $C^{\cdot}(b) \approx \bigoplus_{v\in U-V} \Gamma(\mathcal{O}_v^h, I^{\cdot}(F))[1]$ (in the derived category). Thus the cohomology sequence of this triangle is the required sequence.

From

$$\Gamma(V, I^{\cdot}(F)) \xrightarrow{b\circ a} \bigoplus_{v\notin U} \Gamma(K_v, I^{\cdot}(F)) \bigoplus_{v\in U-V} \Gamma_v(\mathcal{O}_v^h, I^{\cdot}(F))[1]$$
$$\xrightarrow{c} \bigoplus_{v\in U-V} \Gamma_v(\mathcal{O}_v^h, I^{\cdot}(F))[1]$$

we get a distinguished triangle

$$C^{\cdot}(c)[-1] \longrightarrow C^{\cdot}(b\circ a) \longrightarrow C^{\cdot}(c\circ b\circ a) \longrightarrow C^{\cdot}(c).$$

But $C^{\cdot}(c\circ b\circ a) \approx \Gamma(U, I^{\cdot}(F))[1]$ and $C^{\cdot}(c) \approx \bigoplus_{v\notin U} \Gamma(K_v, I^{\cdot}(F))[1]$, which shows that $C^{\cdot}(b\circ a) \approx H_c(U,F)[1]$.

Finally, the statement about $C^{\cdot}(b)$ is obvious from the distinguished triangles (for $v \in U - V$)

$$\Gamma(\mathcal{O}_v^h, I^{\cdot}(F)) \longrightarrow \Gamma(K_v, I^{\cdot}(F)) \longrightarrow \Gamma_v(\mathcal{O}_v^h, I^{\cdot}(F))[1] \longrightarrow \Gamma(\mathcal{O}_v^h, I^{\cdot}(F))[1].$$

(d) Since $H^r(U_{et}, F) \xrightarrow{\approx} H^r(U_{fl}, F)$ and

$$H^r((\mathrm{Spec}\ K_v)_{et}, F) \xrightarrow{\approx} H^r((\mathrm{Spec}\ K_v)_{fl}, F)$$

[Milne (1980), III.3], this follows from a comparison of the sequence in (0.4a) and with the corresponding sequence for the étale topology.

(e) Let $c' \in H_c^r(U,F)$, and regard it as a homotopy class of maps of degree r

$$c'\colon \mathbb{Z} \longrightarrow H_c(U,F)[r].$$

Let $c \in \mathrm{Ext}^r_U(F,F')$, and regard it as a homotopy class of maps of degree s,

$$c\colon I^{\cdot}(F) \longrightarrow I^{\cdot}(F')[s].$$

On restricting, we get a similar class of maps

$$c|Z'\colon I^{\cdot}(F)|Z' \longrightarrow (I^{\cdot}(F')|Z')[s].$$

The last two maps combine to give a morphism $H_c(U,F) \longrightarrow H_c(U,F')[s]$, and we define $\langle c,c'\rangle$ to be the composite of this map with c.

Remark 0.5. (a) Let $i\colon Z \hookrightarrow Y$ be a closed immersion, and let F be a sheaf on Z. The proof in [Milne (1980), II.3.6] of the exactness of i_* for the étale topology (hence the equality $H^r(U_{et},i_*F) = H^r(Z_{et},F)$), fails for the flat topology.

(b) Note that the sequence in (c) has the same form as (II.2.3d) except that in the latter sequence it has been possible to replace $H^r(\mathcal{O}^h_v,F)$ with $H^r(v,i^*F)$. In the case of the flat topology, this is also possible if F is represented by a smooth algebraic space [Milne (1980), III.3.11].

Remark 0.6. (a) In the case that X is the spectrum of the ring of integers in a number field, it is natural to replace $H_c(U,F)$ with the mapping cone of $\bigoplus_{v \text{ arch}} S^{\cdot}(K_v,F_v) \longrightarrow H_c(U,F)$. Then the cohomology groups with compact support fit into an exact sequence

$$\cdots \longrightarrow H^r_c(U,F) \longrightarrow H^r(U,F) \longrightarrow \oplus\, H^r(K_v,F) \longrightarrow \cdots$$

where the sum is now over all primes of K, including the archimedean primes, not in U, and for archimedean v,

$$H^r(K_v,F) = H^r_T(\mathrm{Gal}(K_{v,s}/K_v),F(K_{v,s})).$$

(b) In the definition of $H^r_c(U,F)$ it is possible to replace K_v with its completion. Then the sequence in (0.4a) will be exact with K_v the completion of K at v, and the sequence in (0.4c) will be exact with $\mathcal{O}_v^h$ replaced with $\hat{\mathcal{O}}_v$. This approach has the disadvantage that the groups no longer agree with the étale groups (that is, (0.4d) will no longer hold in general).

The definition given here of cohomology groups with compact support is simple and leads quickly to the results we want. I do not know whether there is a more natural definition nor in what generality it is possible to define such groups.

Topological duality for vector spaces

Let k be a finite field with k elements, and let V be a locally compact topological vector space over k. Write $V^\vee$ for the topological linear dual of V, $V^\vee = \mathrm{Hom}_{k,\mathrm{cts}}(V,k)$.

Proposition 0.7. *The pairing*

$$V^\vee \times V \longrightarrow \mathbb{C}^\times, \quad (f,v) \mapsto \exp\left(\frac{2\pi i}{p}\,\mathrm{Tr}_{k/\mathbb{F}_p} f(v)\right)$$

identifies $V^\vee$ *with the Pontryagin dual of* V.

Proof: Let V^* be the Pontryagin dual of V. The pairing identifies $V^\vee$ with a subspace of V^*. Clearly the elements of $V^\vee$ separate points in V, and so $V^\vee$ is dense in V^*. But $V^\vee$ is locally compact, and so it is an open subset of its closure in V^*; hence it is open in V^*. As it is a subgroup, this implies that it is also closed in V^* and so equals V^*.

Let R be a complete discrete valuation ring of characteristic p $\neq 0$ having a finite residue field k, and let K be the field of fractions of R. The choice of a uniformizing parameter t for R determines an isomorphism $K \xrightarrow{\approx} k((t))$ carrying R onto $k[[t]]$. Define a residue map res: $K \longrightarrow k$ by setting $\mathrm{res}(\sum a_i t^i) = a_{-1}$.

Corollary 0.8. *Let V be a free R-module of finite rank, and let $V^\vee$ be its R-linear dual. Then the pairing*

$$V^\vee \times (V\otimes K)/V \longrightarrow \mathbb{C}^\times, \quad (f,v) \mapsto \exp\left(\frac{2\pi i}{p}\mathrm{Tr}_{k/\mathbb{F}_p}(\mathrm{res}(f(v)))\right)$$

identifies $V^\vee$ with the Pontryagin dual of $(V\otimes K)/V$.

Proof: Consider first the case that V = R. Then R is isomorphic (as a topological k-vector space) to the direct product of countably many copies of k, and K/R is isomorphic to the direct sum of countably many copies of k. The pairing $(a,b) \mapsto \mathrm{res}(ab)$: $R \times K/R \longrightarrow k$ identifies R with the k-linear topological dual of K/R, and so (0.7) shows that $R \xrightarrow{\approx} (K/R)^*$. In the general case, the pairing

$$V^\vee \times (V\otimes K)/V \longrightarrow k, \quad (f,v) \mapsto \mathrm{res}(f(v))$$

similarly identifies $V^\vee$ with k-linear topological dual of $(V\otimes K)/V$, and so again the result follows from the proposition.

The Frobenius morphism

For any scheme S of characteristic $p \neq 0$, the *absolute Frobenius map* F_{abs}: $S \longrightarrow S$ is defined to be the identity map on the underlying topological space and $a \mapsto a^p$ on $\mathcal{O}_S$. It is functorial in the sense that for any morphism π: $X \longrightarrow S$, the diagram

$$\begin{array}{ccc} X & \xleftarrow{F_{abs}} & X \\ \downarrow \pi & & \downarrow \pi \\ S & \xleftarrow{F_{abs}} & S \end{array}$$

commutes, but it does not commute with base change. The *relative Frobenius map* $F_{X/S}$ is defined by the diagram

$$\begin{array}{ccccc} & & \overset{F_{abs}}{\longleftarrow\!\!\!-\!\!\!-\!\!\!-\!\!\!-\!\!\!-\!\!\!-} & & \\ X & \longleftarrow & X^{(p)} & \overset{F_{X/S}}{\longleftarrow} & X, \\ \downarrow\pi & & \downarrow\pi^{(p)} & \swarrow & \\ S & \overset{F_{abs}}{\longleftarrow} & S & & \end{array} \qquad X^{(p)} \overset{\mathrm{df}}{=} X\times_{S,F_{abs}} S.$$

For any morphism $T \longrightarrow S$, $F_{X/S}\times_T id = F_{X/T}$.

A scheme S is said to be *perfect* if $F_{abs}: S \longrightarrow S$ is an isomorphism. For example, an affine scheme Spec R is perfect if the pth power map $a \mapsto a^p: R \longrightarrow R$ is an isomorphism. In the case that S is perfect, it is possible to identify $\pi^{(p)}: X^{(p)} \longrightarrow S$ with $F_{abs}^{-1}\circ\pi: X \longrightarrow S$ and $F_{X/S}$ with F_{abs}.

A finite group scheme N over a scheme S is said to have *height* h if $F_{N/S}^h: N \longrightarrow N^{(p^h)}$ is zero but $F_{N/S}^{h-1}$ is not zero. For any flat group scheme N, there is a canonical morphism $V = V_N: N^{(p)} \longrightarrow N$, called the *Verschiebung* (see [Demazure and Gabriel (1970), IV, 3, 4]).

The Oort-Tate classification of group schemes of order p

Let $\Lambda = \mathbb{Z}[\zeta,(p(p-1))^{-1}]\cap\mathbb{Z}_p$, where ζ is a primitive p^{th} root of 1, and the intersection is taken inside $\mathbb{Q}_p$. We consider only schemes X such that the unique morphism $X \longrightarrow \mathrm{Spec}\ \mathbb{Z}$ factors through Spec Λ. For example, X can be any scheme of characteristic p because Λ has $\mathbb{F}_p$ as a residue field. The following statements classify the finite flat group schemes of order p over X (see [Oort and Tate, (1970)], or [Shatz (1986), §4]).

(0.9a) It is possible to associate a finite flat group scheme $N_{a,b}^{\mathcal{L}}$ of order p over X with each triple $(\mathcal{L},a,b)$ comprising an invertible sheaf $\mathcal{L}$ over X, an element $a \in \Gamma(X,\mathcal{L}^{\otimes p-1})$, and an element $b \in \Gamma(X,\mathcal{L}^{\otimes 1-p})$ such that $a\otimes b = w_p$ for a certain universal element w_p.

(0.9b) Every finite flat group scheme of order p over X is of the form $N^{\mathcal{L}}_{a,b}$ for some triple $(\mathcal{L},a,b)$.

(0.9c) There is an isomorphism $N^{\mathcal{L}}_{a,b} \xrightarrow{\approx} N^{\mathcal{L}'}_{a',b'}$ if and only if there is an isomorphism $\mathcal{L} \xrightarrow{\approx} \mathcal{L}'$ carrying a to a' and b to b'.

(0.9d) For all X-schemes Y,

$$N^{\mathcal{L}}_{a,b}(Y) = \{ y \in \Gamma(Y,\mathcal{L}\otimes\mathcal{O}_Y) \mid y^{\otimes p} = a\otimes y \}.$$

(0.9e) The Cartier dual of $N^{\mathcal{L}}_{a,b}$ is $N^{\mathcal{L}^\vee}_{b,a}$.

(0.9f) When X has characteristic p, $w_p = 0$. If $a = 0$ in this case, then N has height one, and the p-Lie algebra of $N^{\mathcal{L}}_{a,b}$ is $\mathcal{L}$ with the p-power map $f \mapsto f^{(p)} = b\otimes f$.

Duality for unipotent perfect group schemes

When the ground field is finite, our duality theorems will be for the cohomology groups endowed with the structure of a topological group. When the ground field is not finite, it will be necessary to endow the cohomology groups with a stronger structure, namely the structure of a perfect pro-algebraic group, and replace Pontryagin duality with Breen-Serre duality. We now describe this last duality.

Let S be a perfect scheme of characteristic $p \neq 0$. The *perfection* X^{pf} of an S-scheme X is the projective limit of the system

$$X_{red} \xleftarrow{F} X_{red}^{(p^{-1})} \xleftarrow{F} \cdots \xleftarrow{F} X_{red}^{(p^{-n})} \xleftarrow{F} \cdots .$$

It is a perfect scheme, and has the universal property that $\mathrm{Hom}_S(X,Y) = \mathrm{Hom}_S(X^{pf},Y)$ for any perfect S-scheme Y.

Let S be the spectrum a perfect field k. A perfect S-scheme X is said to be *algebraic* if it is the perfection of a scheme of finite type over S. From the corresponding fact for the algebraic group schemes over S, one sees easily that the perfect algebraic group schemes over X form an abelian category. Define the *perfect site* S_{pf} to be that whose underlying category consists of all algebraic perfect S-schemes and whose covering families are the surjective families of étale morphisms. For any algebraic group scheme G ove S, the sheaf on S_{pf} defined by G is represented by G^{pf}. One checks easily, that any sheaf on S_{pf} that is an extension of perfect algebraic group schemes is itself represented by a perfect algebraic group scheme.

We write $\mathbf{S}(p^n)$ for the category of sheaves on S_{pf} killed by p^n.

Theorem 0.10. *For all* $r > 0$, $\mathrm{Ext}^r_{\mathbf{S}(p)}(\mathbb{G}_a^{pf},\mathbb{G}_a^{pf}) = 0$.

Proof: See [Breen (1981)], where the result is proved with the base scheme the spectrum of any perfect ring of characteristic $p \neq 0$.

Lemma 0.11. *Let* $f\colon S_{fl} \longrightarrow S_{pf}$ *be the morphism of sites defined by the identity map. For any affine commutative algebraic group scheme* G *on* S, f_*G *is represented by* G^{pf} *and* $R^r f_* G = 0$ *for* $r > 0$.

Proof: We have already observed that f_*G is represented by G^{pf}. If G is smooth, then $R^r f_* G = 0$ for $r > 0$ because of the coincidence of flat and étale cohomology groups of smooth group schemes [Milne (1980), III.3.9]. We calculate $R^r f_* G$ for $G = \alpha_p$ and μ_p by using the exact sequences

$$0 \longrightarrow \mu_p \longrightarrow \mathbb{G}_m \xrightarrow{F} \mathbb{G}_m \longrightarrow 0$$

$$0 \longrightarrow \alpha_p \longrightarrow \mathbb{G}_a \xrightarrow{F} \mathbb{G}_a \longrightarrow 0.$$

Since F is an automorphism of $\mathbb{G}_m^{pf}$ and $\mathbb{G}_a^{pf}$, we have $R^r f_* \mu_p = 0 = R^r f_* \alpha_p$ for all r. The general case follows from these case because, locally for the étale topology, any G has a composition series whose quotients are $\mathbb{G}_m$, $\mathbb{G}_a$, μ_p, α_p, or an étale group scheme.

By a *p-primary* group scheme, we mean a group scheme killed by a power of p.

Lemma 0.12. *Let G be a perfect p-primary affine algebraic group scheme on S; let U be its identity component, and let D = G/U. Then D is étale and U has a composition series whose quotients are all isomorphic to* $\mathbb{G}_a^{pf}$.

Proof: This is an immediate consequence of the exactness of f_* and the structure theorem for affine commutative algebraic group schemes on S.

Let $G(p^n)$ be the category of perfect affine algebraic group schemes on S killed by p^n, and let $G(p^\infty) = \cup G(p^n)$. Note that (0.12) shows that $G(p^\infty)$ can also be described as the category of perfect unipotent group schemes on S.

Lemma 0.13. *Let G be a perfect unipotent group scheme on S; let U be its identity component, and let D = G/U.*

(a) *If G is killed by* p^n, *then there exists a canonical isomorphism*

$$R\mathcal{H}om_{S(p^n)}(G, \mathbb{Z}/p^n\mathbb{Z}) \xrightarrow{\approx} R\mathcal{H}om_S(G, \mathbb{Q}_p/\mathbb{Z}_p).$$

(b) $\mathcal{H}om_S(G, \mathbb{Q}_p/\mathbb{Z}_p) \xrightarrow{\approx} \mathcal{H}om_S(D, \mathbb{Q}_p/\mathbb{Z}_p)$, *which equals* D^*, *the Pontryagin dual of* D.

(c) $\mathcal{E}xt^1_S(G, \mathbb{Q}_p/\mathbb{Z}_p) \xrightarrow{\approx} \mathcal{E}xt^1_S(U, \mathbb{Q}_p/\mathbb{Z}_p)$, *which is represented by a connected unipotent perfect group scheme; if U has the structure of a* $W_n(k)$*-module, then there is a canonical isomorphism of* $W_n(k)$*-modules*

$$\mathcal{E}xt^1_S(G, \mathbb{Q}_p/\mathbb{Z}_p) \xrightarrow{\approx} \mathcal{H}om_{W_n(k)}(U, W_n(\mathcal{O}_S)).$$

(d) $\mathcal{E}xt^r_S(G, \mathbb{Q}_p/\mathbb{Z}_p) = 0$ *for* $r > 1$.

Proof: (a) Choose an injective resolution $I^\cdot$ of $\mathbb{Q}_p/\mathbb{Z}_p$. The usual argument in the case of abelian groups shows that an injective sheaf is divisible. Therefore the kernel $I^\cdot_{p^n}$ of $p^n\colon I^\cdot \to I^\cdot$ is a resolution of $\mathbb{Z}/p^n\mathbb{Z}$, and it is obvious that it is an injective resolution in $S(p^n)$. Consequently

$$R\mathcal{H}om_{S(p^n)}(G,\mathbb{Z}/p^n\mathbb{Z}) = \mathcal{H}om^{\cdot}_{S(p^n)}(G,I^{\cdot}_{p^n}) = \mathcal{H}om^{\cdot}_{S}(G,I^{\cdot}) = R\mathcal{H}om_{S}(G,\mathbb{Q}_p/\mathbb{Z}_p).$$

(b,c,d) When $D = \mathbb{Z}/p\mathbb{Z}$, the sequence

$$0 \to \mathcal{H}om_S(\mathbb{Z}/p\mathbb{Z},\mathbb{Q}/\mathbb{Z}) \to \mathcal{H}om_S(\mathbb{Z},\mathbb{Q}/\mathbb{Z}) \xrightarrow{p} \mathcal{H}om_S(\mathbb{Z},\mathbb{Q}/\mathbb{Z}) \to \mathcal{E}xt^1_S(\mathbb{Z}/p\mathbb{Z},\mathbb{Q}/\mathbb{Z}) \to 0$$

shows that

$$R\mathcal{H}om_S(D,\mathbb{Q}/\mathbb{Z}) = \mathcal{H}om_S(D,\mathbb{Q}/\mathbb{Z}) = D^*.$$

A general étale group D is locally (for the étale topology) an extension of copies of $\mathbb{Z}/p\mathbb{Z}$, and so the same equalities holds for it.

If $U = \mathbb{G}_a^{pf}$, then (0.10) shows that the exact sequence

$$0 \to \mathcal{H}om_{S(p)}(\mathbb{G}_a^{pf},\mathbb{G}_a^{pf}) \xrightarrow{1-F} \mathcal{H}om_{S(p)}(\mathbb{G}_a^{pf},\mathbb{G}_a^{pf}) \to \mathcal{E}xt^1_{S(p)}(\mathbb{G}_a^{pf},\mathbb{Z}/p\mathbb{Z}) \to \dots$$

yields an isomorphism

$$R\mathcal{H}om_{S(p)}(\mathbb{G}_a^{pf},\mathbb{Z}/p\mathbb{Z}) \approx G[-1]$$

where G is the cokernel of $1 - F$. It is well-known (see for example [Serre (1960)]) that $G = \mathbb{G}_a^{pf}$. Therefore $\mathcal{E}xt^1_S(\mathbb{G}_a^{pf},\mathbb{Z}/p\mathbb{Z}) = \mathcal{E}xt^1_{S(p)}(\mathbb{G}_a^{pf},\mathbb{Q}/\mathbb{Z})$ is connected, and $\mathcal{E}xt^r_S(\mathbb{G}_a^{pf},\mathbb{Q}/\mathbb{Z}) = 0$ for $r \neq 1$. A general connected unipotent group U is an extension of copies of $\mathbb{G}_a^{pf}$, and so $\mathcal{E}xt^r_S(U,\mathbb{Q}/\mathbb{Z})$ is connected for $r = 1$ and zero for $r \neq 1$.

Suppose that U has the structure of a $W_n(\mathcal{O}_S)$-module. The Artin-Schreier sequence

$$0 \longrightarrow \mathbb{Z}/p^n\mathbb{Z} \longrightarrow W_n(\mathcal{O}_S) \xrightarrow{F-1} W_n(\mathcal{O}_S) \longrightarrow 0$$

gives a morphism $W_n(\mathcal{O}_S) \longrightarrow (\mathbb{Z}/p^n\mathbb{Z})[1]$ in the derived category of $S(p^n)$, and hence a canonical homomorphism

$$\mathcal{H}om_{W_n(\mathcal{O}_S)}(U,W_n(\mathcal{O}_S)) \longrightarrow R\mathcal{H}om_{S(p^n)}(U,\mathbb{Z}/p^n\mathbb{Z}),$$

which is $W_n(\mathcal{O}_S)$-linear for the given structure on U. One checks easily that $\mathcal{E}xt^1_{W_n(\mathcal{O}_S)}(U, W_n(\mathcal{O}_S)) = 0$: as U has a filtration by sub-$W_n(\mathcal{O}_S)$-modules such that the quotients are linearly isomorphic to $\mathbb{G}_a^{pf}$, it suffices to show that this homomorphism is an isomorphism for $U = \mathbb{G}_a^{pf}$, which is assured by (0.10). It follows that the homomorphism is an isomorphism.

The assertions (b), (c), and (d) now result in the general case from making use of the exact sequence

$$0 \longrightarrow U \longrightarrow G \longrightarrow D \longrightarrow 0.$$

For any perfect connected unipotent group U, we write $U^\vee$ for $\mathcal{E}xt^1_S(U, \mathbb{Q}_p/\mathbb{Z}_p)$. Let $D^b(G(p^\infty))$ be the full subcategory of the derived category of $S(S_{pf})$ consisting of those bounded complexes whose cohomology lies in $G(p^\infty)$. For any $G^\cdot$ in $D^b(G(p^\infty))$, define $G^{\cdot t} = R\mathcal{H}om_S(G^\cdot, \mathbb{Q}_p/\mathbb{Z}_p)$.

Theorem 0.14. *For any $G^\cdot$ in $D^b(G(p^\infty))$, $G^{\cdot t}$ also lies $D^b(G(p^\infty))$, and there is a canonical isomorphism $G^\cdot \xrightarrow{\approx} G^{\cdot tt}$. There exist canonical exact sequences*

$$0 \longrightarrow U^{r+1}(G^\cdot)^\vee \longrightarrow H^{-r}(G^{\cdot t}) \longrightarrow D^r(G^\cdot)^* \longrightarrow 0,$$

where $U^r(G^\cdot)$ is the identity component of $H^r(G^\cdot)$ and $D^r(G^\cdot) = H^r(G^\cdot)/U^r(G^\cdot)$.

Proof: The cohomlogy sheaves of $R\mathcal{H}om_S(G^\cdot, \mathbb{Q}_p/\mathbb{Z}_p)$ are the abutment of the spectral sequence

$$E_2^{r,s} = \mathcal{E}xt^r_S(H^{-s}(G^\cdot), \mathbb{Q}_p/\mathbb{Z}_p) \Rightarrow \mathcal{E}xt^{r+s}_S(G^\cdot, \mathbb{Q}_p/\mathbb{Z}_p).$$

After (0.13), $E_2^{r,s} = 0$ for $r \neq 0,1$, so that the spectral sequence reduces to a family of short exact sequences

$$0 \to \mathcal{E}xt^1_S(H^{r+1}(G^\cdot),\mathbb{Q}_p/\mathbb{Z}_p) \to \mathcal{E}xt^{-r}_S(G^\cdot,\mathbb{Q}_p/\mathbb{Z}_p) \to \mathcal{H}om_S(H^r(G^\cdot),\mathbb{Q}_p/\mathbb{Z}_p) \to 0,$$

which are the required exact sequences. They imply moreover that $G^{\cdot t}$ is in $D^b(G(p^\infty))$ because $G(p^\infty)$ is stable under extension. Finally one shows that the homomorphism of biduality $G^\cdot \longrightarrow G^{\cdot tt}$ is an isomorphism by reducing the question to the cases of $\mathbb{Z}/p\mathbb{Z}$ and $\mathbb{G}_a$, which both follow directly from (0.13).

Remark 0.15. Denote by $\mathrm{Ext}^r_k(G,H)$ the Ext group computed in the category of affine perfect group schemes over k. Each such Ext group can be given a canonical structure as a perfect group scheme, and (0.10) implies that, when G and H are unipotent, $\mathrm{Ext}^r_k(G,H)$ agrees with $\mathcal{E}xt^r_S(G,H)$.

Pairings in the derived category

We review some of the basic definitions concerning pairings in the derived category. For more details, see [Gamst and Hoechsmann (1970)] or [Hartshorne (1966)].

Fix a scheme X, endow it with a Grothendieck topology, and write S(X) for the resulting category of sheaves. Write C(X) for the category of complexes in S(X), and K(X) for category with the same objects but whose morphisms are homotopy classes of maps in C(X). The derived category D(X) is obtained from K(X) by formally inverting quasi-isomorphisms. Thus, for example, a map $A^\cdot \longrightarrow B^\cdot$ and a quasi-isomorphism $B^\cdot \xleftarrow{\sim} C^\cdot$ define a morphism $A^\cdot \longrightarrow C^\cdot$ in D(X). As usual, $C^+(X)$, $C^-(X)$, and $C^b(X)$ denote respectively the categories of complexes bounded below, bounded above, and bounded in both directions. We use similar notations for the homotopy and derived categories. Since S(X) has enough injectives, for every $A^\cdot$ in $C^+(X)$, there is a quasi-isomorphism $A^{\cdot} \xrightarrow{\sim} I(A^{\cdot})$ with $I(A^{\cdot})$ a complex of injectives, and there is a canonical equivalence of categories $\mathbf{I}^+(X) \longrightarrow \mathbf{D}^+(X)$, where $\mathbf{I}^+(X)$ is the full subcategory of $\mathbf{K}^+(X)$ whose objects are complexes of injective objects.

Recall that a sheaf P is flat if $-\otimes P\colon \mathbf{S}(X) \longrightarrow \mathbf{S}(X)$ is exact. For any bounded-above complex $A^{\cdot}$, there is a quasi-isomorphism $P(A^{\cdot}) \xrightarrow{\sim} A^{\cdot}$ with $P(A^{\cdot})$ a complex of flat sheaves. If $B^{\cdot}$ is a second bounded-above complex, then $P(A^{\cdot})\otimes B^{\cdot}$ is a well-defined object of $D(X)$, which is denoted by $A^{\cdot}\otimes^{\mathbf{L}}B^{\cdot}$. Despite appearances, there is a canonical isomorphism $A^{\cdot}\otimes^{\mathbf{L}}B^{\cdot} \approx B^{\cdot}\otimes^{\mathbf{L}}A^{\cdot}$.

Let M and N be flat sheaves on X, and let $A^{\cdot}$ and $B^{\cdot}$ be objects of $\mathbf{C}^b(X)$. There is a canonical pairing

$$\mathrm{Ext}_X^r(M,A^{\cdot}) \times \mathrm{Ext}_X^s(N,B^{\cdot}) \longrightarrow \mathrm{Ext}_X^{r+s}(M\otimes N, A^{\cdot}\otimes^{\mathbf{L}}B^{\cdot})$$

that can be defined as follows: represent elements $f \in \mathrm{Ext}_X^r(M,A^{\cdot})$ and $g \in \mathrm{Ext}_X^s(N,B^{\cdot})$ as homotopy classes of maps $f\colon M \longrightarrow I(A^{\cdot})[r]$ and $g\colon N \longrightarrow I(B^{\cdot})[s]$; then $f\otimes g$ is represented by

$$M\otimes N \longrightarrow I(A^{\cdot})[r]\otimes I(B^{\cdot})[s] \xleftarrow{\sim} P(I(A^{\cdot}))\otimes I(B^{\cdot})[r+s].$$

The pairing is natural, bi-additive, associative, and symmetric (up to the usual signs). It also behaves well with respect to boundary maps [Gamst and Hoechsmann, ibid.]. A similar discussion applies when N is not flat — replace it with a flat resolution.

Consider a map $A^{\cdot}\otimes^{\mathbf{L}}B^{\cdot} \longrightarrow G^{\cdot}$. The above discussion gives a pairing

$$H^r(X,A^{\cdot}) \times \mathrm{Ext}_X^s(M,B^{\cdot}) \longrightarrow \mathrm{Ext}_X^{r+s}(M,G^{\cdot}),$$

There is also the usual (obvious) pairing

$$\mathrm{Ext}_X^r(B^{\cdot},G^{\cdot}) \times \mathrm{Ext}_X^s(M,B^{\cdot}) \longrightarrow \mathrm{Ext}_X^{r+s}(M,G^{\cdot}).$$

The map $A^{\cdot}\otimes^{\mathbf{L}}B^{\cdot} \longrightarrow G$ define a map $A^{\cdot} \longrightarrow \mathcal{H}om(B^{\cdot},C^{\cdot})$ and hence edge morphisms $H^r(X,A^\cdot) \longrightarrow \mathrm{Ext}^r_X(B^\cdot,G^\cdot)$.

Theorem 0.16. *The following diagram commutes:*

$$\begin{array}{ccccc} H^r(X,A^\cdot) & \times & \mathrm{Ext}^s_X(M,B^\cdot) & \longrightarrow & \mathrm{Ext}^{r+s}_X(M,G^\cdot) \\ \downarrow & & \| & & \downarrow \\ \mathrm{Ext}^r_X(B^\cdot,G^\cdot) & \times & \mathrm{Ext}^s_X(M,B^\cdot) & \longrightarrow & \mathrm{Ext}^{r+s}_X(M,G^\cdot). \end{array}$$

Proof: See [Gamst and Hoechsmann (1970)].

Notes: The definition of cohomology groups with compact support for the flat topology is new, and will play an important role in this chapter.

The duality for unipotent perfect group schemes has its origins in a remark of Serre [Serre (1960), p55] that Ext's in the category of unipotent perfect group schemes over an algebraically closed field can be used to define an autoduality of the category. For a detailed exposition in this context, see [Bégueri (1980), §1]; Serre in fact worked with the equivalent category of quasi-algebraic groups. The replacement of Ext's in the category of perfect group schemes with Ext's in the category of sheaves, which is essential for the applications we have in mind, is easy once one has Breen's vanishing theorem (0.10). Our exposition of the autoduality is based on [Berthelot (1981), II] (which, in turn, is based on [Milne (1976)]).

Most of the rest of the material is standard.

§1 Local results: mixed characteristic, finite group schemes

Throughout this section, R will be a Henselian discrete valuation ring with finite residue field k and field of fractions K of characteristic zero. In particular, R is excellent. We use the same notations as in (II.1): for example, X = Spec R and i and j are the inclusions of the closed point x and the open point u of X into X. The characteristic of k will be denoted by p and the maximal ideal of R by m.

Lemma 1.1. *Let* N *be a finite flat group scheme over* R.

(a) *The map* $N(R) \longrightarrow N(K)$ *is a bijective, and* $H^1(X,N) \longrightarrow H^1(K,N_K)$ *is injective; for* $r \geq 2$, $H^r(X,N) = 0$.

(c) *The boundary map* $H^r(K,N) \longrightarrow H^{r+1}_x(X,N)$ *defines isomorphisms* $H^1(K,N)/H^1(R,N) \xrightarrow{\approx} H^2_x(X,N)$ *and* $H^2(K,N) \xrightarrow{\approx} H^3_x(X,N)$; *for* $r \neq 2,3$, $H^r_x(X,N) = 0$.

Proof: (a) As N is finite, it is the spectrum of a finite R-algebra A. The image of any R-homomorphism $A \longrightarrow K$ is finite over R and is therefore contained in R. This shows that N(R) = N(K). An element c of $H^1(X,N)$ is represented by a principal homogeneous space P over X [Milne (1980), III.4.3], and c = 0 if and only if P(R) is nonempty. Again P is the spectrum of a finite R-algebra, and so if P has a point in K then it already has a point in R.

From Appendix A, we know that there is an exact sequence

$$0 \longrightarrow N \longrightarrow G \longrightarrow G' \longrightarrow 0$$

in which G and G' are smooth group schemes of finite type over X. According to [Milne (1980), III.3.11], $H^r(X,G) = H^r(k,G_0)$ and $H^r(X,G') = H^r(k,G_0')$ for $r > 0$, where G_0 and G_0' are the closed fibres of G and G' over X. The five lemma therefore shows that $H^r(X,N) \xrightarrow{\approx} H^r(k,N_0)$ for $r > 1$. We now use that there is an exact sequence

$$0 \longrightarrow N_0 \longrightarrow G_1 \longrightarrow G_1' \longrightarrow 0$$

with G_1 and G_1' smooth connected group schemes over k (for example, abelian varieties). By Lang's lemma, $H^r(k,G_1) = 0 = H^r(k,G_1')$ for $r > 0$, and it follows that $H^r(k,N_o) = 0$ for $r > 1$.

(b) This follows from the first statement because of the exact sequence

$$\cdots \longrightarrow H_x^r(X,N) \longrightarrow H^r(X,N) \longrightarrow H^r(K,N) \longrightarrow \cdots$$

and the fact that $H^r(K,N) = 0$ for $r > 2$ (K has cohomological dimension 2).

Remark 1.2. The proof of the lemma does not use that K has characteristic zero. The same argument as in the proof of (a) shows that $H^r(X,N) = 0$ for $r \geq 2$ if N is a finite flat group scheme over any Noetherian Henselian local ring with finite residue field.

Let F be a sheaf on X. The pairing

$$\mathrm{Ext}_X^r(F,\mathbb{G}_m) \times \mathrm{Ext}_X^s(i_*\mathbb{Z},F) \longrightarrow \mathrm{Ext}_X^{r+s}(i_*\mathbb{Z},\mathbb{G}_m)$$

can be identified with a pairing

$$\mathrm{Ext}_X^r(F,\mathbb{G}_m) \times H_x^s(X,F) \longrightarrow H_x^{r+s}(X,\mathbb{G}_m);$$

see (0.3a). Since $\mathbb{G}_m$ is a smooth group scheme, the natural map $H_x^r(X_{et},\mathbb{G}_m) \longrightarrow H_x^r(X_{fl},\mathbb{G}_m)$ is an isomorphism for all r, and so (see II.1) there is a canonical trace map $H_x^3(X_{fl},\mathbb{G}_m) \xrightarrow{\approx} \mathbb{Q}/\mathbb{Z}$. Let N be a finite group scheme over X. The sheaf defined by the Cartier dual N^D of N can be identified with $\mathcal{H}om(N,\mathbb{G}_m)$, and the pairing $N^D \times N \longrightarrow \mathbb{G}_m$ defines a pairing $H^r(X,N^D) \times H_x^{3-r}(X,N) \longrightarrow H_x^3(X,\mathbb{G}_m) = \mathbb{Q}/\mathbb{Z}$.

This can also be defined using the edge morphisms $H^s(X,N^D) \longrightarrow \mathrm{Ext}_X^s(N,\mathbb{G}_m)$ and the Ext-pairing (0.16).

Theorem 1.3. *For any finite flat group scheme* N *on* X,

$$H^r(X,N^D) \times H^{3-r}_x(X,N) \longrightarrow H^3_x(X,\mathbb{G}_m) = \mathbb{Q}/\mathbb{Z}$$

is a nondegenerate pairing of finite groups, all r.

Proof: We shall give two proofs, but first we list some corollaries.

Corollary 1.4. *For any finite flat group scheme* N *over* X, $H^1(X,N^D)$ *is the exact annihilator of* $H^1(X,N)$ *in the pairing*

$$H^1(K,N^D) \times H^1(K,N) \longrightarrow H^2(K,\mathbb{G}_m) = \mathbb{Q}/\mathbb{Z}$$

of (I.2.3).

Proof: The diagram

$$\begin{array}{ccccc} H^1(X,N^D) & \times & H^1(X,N) & \longrightarrow & H^2(X,\mathbb{G}_m) = 0 \\ \downarrow & & \downarrow & & \downarrow \\ H^1(K,N^D) & \times & H^1(K,N) & \longrightarrow & H^2(K,\mathbb{G}_m) = \mathbb{Q}/\mathbb{Z} \end{array}$$

shows that $H^1(X,N^D)$ and $H^1(X,N)$ annihilate each other in the pairing. For r = 1, (1.1) allows us to identify the pairing in the theorem with

$$H^1(X,N^D) \times H^1(K,N)/H^1(X,N) \longrightarrow H^2(K,\mathbb{G}_m).$$

Thus we see that the nondegeneracy of the pairing in this case is equivalent to the statement of the corollary.

Corollary 1.5. *Let* N *be a finite flat group scheme on* X. *For all* $r < 2p-2$,

$$\mathrm{Ext}^r_X(N,\mathbb{G}_m) \times H^{3-r}_x(X,N) \longrightarrow H^3_x(X,\mathbb{G}_m) = \mathbb{Q}/\mathbb{Z}$$

is a nondegenerate pairing of finite groups.

Proof: Let N(p) be the p-primary component of N. According to [Breen (1975)], $\mathcal{E}xt^r_X(N(p),\mathbb{G}_m) = 0$ for $1 \leq r < 2p-2$, and as we explained in the proof of (II.4.10), this implies that $\mathcal{E}xt^r_X(N,\mathbb{G}_m) = 0$ for $1 < r < 2p-2$ ($\mathcal{E}xt^1_X(N,\mathbb{G}_m) = 0$ by [Milne (1980), III.4.17], and $\mathcal{E}xt^r_X(N(\ell),\mathbb{G}_m) = 0$ for $r > 1$ and $\ell \neq p$ because $N(\ell)$ is locally constant for the étale topology). Hence $H^r(X,N^D) = \mathrm{Ext}^r(N,\mathbb{G}_m)$ for $r < 2p-2$.

Write $f\colon X_{fl} \longrightarrow X_{et}$ for the morphism defined by the identity map.

Corollary 1.6. *Let N be a quasi-finite flat group scheme over X whose p-primary component N(p) is finite over X. Let N^D be the complex of sheaves such that*

$$N^D(\ell) = \begin{cases} \mathcal{H}om_{X_{fl}}(N(\ell),\mathbb{G}_m) & \ell = p \\ f^*R\mathcal{H}om_{X_{et}}(N(\ell),\mathbb{G}_m) & \ell \neq p \end{cases}.$$

Then

$$H^r(X,N^D) \times H^{3-r}_x(X,N) \longrightarrow H^3_x(X,\mathbb{G}_m) = \mathbb{Q}/\mathbb{Z}$$

is a nondegenerate pairing of finite groups.

Proof: For each $\ell \neq p$, $H^s_x(X,N(\ell)) = H^s_x(X_{et},N(\ell))$ and $H^r(X,N^D(\ell)) = H^r(X_{et},R\mathcal{H}om_{X_{et}}(N(\ell),\mathbb{G}_m)) = \mathrm{Ext}^r_{X_{et}}(N(\ell),\mathbb{G}_m)$. Therefore, for the prime-to-p components of the groups, the corollary follows from (II.1.8). For the p component it follows immediately from the theorem.

Question 1.7. Does there exist a single statement that fully generalizes both (1.3) and (II.1.8b)?

The first proof of Theorem 1.3

The first proof is very short, but makes use of (A.6). We begin by proving a duality result for abelian schemes.

Proposition 1.8. *Let $\mathcal{A}$ be an abelian scheme over X, and let $\mathcal{A}^t$ be its dual. Then the pairing*

$$H^r(X,\mathcal{A}^t) \times H_x^{2-r}(X,\mathcal{A}) \longrightarrow H_x^3(X,\mathbb{G}_m) = \mathbb{Q}/\mathbb{Z}$$

defined by the canonical biextension $\mathcal{A}^t \otimes^L \mathcal{A} \longrightarrow \mathbb{G}_m$ (see Appendix C) induces an isomorphism $H^0(X,\mathcal{A}^t)^\wedge \longrightarrow H_x^2(X,\mathcal{A})^*$ (^ denotes the completion for the profinite topology); for $r \neq 0$, both groups are zero.

Proof: Let A and $\mathcal{A}_0$ be the open and closed fibres respectively of $\mathcal{A}/X$. Then $H^r(X,\mathcal{A}) = H^r(x,\mathcal{A}_0)$ for $r > 0$ (see [Milne (1980), I.3.11]), and $H^r(X,\mathcal{A}_0) = 0$ for $r > 0$ by Lang's lemma. Moreover, $\mathcal{A}(X) = A(K)$ because $\mathcal{A}$ is proper over X. Therefore $H_x^r(X,\mathcal{A})$ is zero for $r \leq 1$ and equals $H^{r-1}(K,A)$ for $r > 1$. Hence $H_x^2(X,\mathcal{A}) = H^1(K,A)$, and $H_x^r(X,\mathcal{A}) = 0$ for all other values of r. Consequently, when $r = 0$, the pairing becomes

$$H^0(K,A^t) \times H^2(K,A) \longrightarrow H^2(K,\mathbb{G}_m) = \mathbb{Q}/\mathbb{Z},$$

and both groups are zero for all other values of r. The proposition now follows from (I.3.4).

We now prove (1.3). Note that the Lemma 1.1 implies that $H^0(X,N)$ and $H^1(X,N)$ are finite (N_K is a finite étale group scheme). According to (A.6) and (A.7), N can be embedded into an exact sequence

$$0 \longrightarrow N \longrightarrow \mathcal{A} \longrightarrow \mathcal{B} \longrightarrow 0$$

with $\mathcal{A}$ and $\mathcal{B}$ abelian schemes over X. This leads to an exact cohomology sequence

$$0 \longrightarrow H_x^2(X,N) \longrightarrow H_x^2(X,\mathcal{A}) \longrightarrow H_x^2(X,\mathcal{B}) \longrightarrow H_x^3(X,N) \longrightarrow 0,$$

which we regard as a sequence of discrete groups.

There is a dual exact sequence

$$0 \longrightarrow N^D \longrightarrow \mathcal{B}^t \longrightarrow \mathcal{A}^t \longrightarrow 0$$

(see Appendix C), which leads to a cohomology sequence

$$0 \longrightarrow H^0(X,N^D) \longrightarrow H^0(X,\mathcal{B}^t) \longrightarrow H^0(X,\mathcal{A}^t) \longrightarrow H^1(X,N^D) \longrightarrow 0.$$

The two middle terms of the sequence have natural topologies, and the two end terms inherit the discrete topology. Therefore the sequence remains exact after the middle two terms have been completed. The theorem now follows from the diagram

$$\begin{array}{ccccccccc} 0 \longrightarrow & H^0(X,N^D) & \longrightarrow & H^0(X,\mathcal{B}^t)^\wedge & \longrightarrow & H^0(X,\mathcal{A}^t)^\wedge & \longrightarrow & H^1(X,N^D) & \longrightarrow 0 \\ & \downarrow & & \downarrow \approx & & \downarrow \approx & & \downarrow & \\ 0 \longrightarrow & H^3_x(X,N)^* & \longrightarrow & H^2_x(X,\mathcal{A})^* & \longrightarrow & H^2_x(X,\mathcal{A})^* & \longrightarrow & H^2_x(X,N^D)^* & \longrightarrow 0. \end{array}$$

The second proof of Theorem 1.3

The second proof will use p-divisible groups, for whose basic theory we refer the reader to [Tate (1967b)] or [Shatz (1986)]. In order to simplify the argument, we shall assume throughout that R is complete.

Let $H = (H_\nu, i_\nu)_{\nu\geq 1}$ be a p-divisible group over X. Let L be a finite extension of K, and let R_L be the integral closure of R in L. Then the group of points $H(R_L)$ of H with values in R_L is defined to be $\varprojlim_i H(R_L/m_L^i)$, where $H(R_L/m_L^i) \overset{df}{=} \varinjlim_\nu H_\nu(R_L/m_L^i)$. Let $M_H = \bigcup H(R_L)$ where L runs over the finite extensions of K contained in K_s. Then M_H becomes a discrete module under the obvious action of $\mathrm{Gal}(K_s/K)$.

Lemma 1.9. (a) *The group* H(R) *is compact if and only if its torsion subgroup is finite.*

(b) *The group of elements of* M_H *fixed by* G_K *is* $H(R)$.

(c) *The sequence of* $Gal(K_s/K)$-*modules*

$$0 \longrightarrow H_\nu(K_s) \longrightarrow M_H \xrightarrow{p^\nu} M_H \longrightarrow 0$$

is exact.

Proof: (a) Let H^o be the identity component of H. Then $H^o(R)$ is an open subgroup of $H(R)$, and $H(R)/H^o(R)$ is torsion. Since $H^o(R)$ is compact and its torsion subgroup is finite ($H(R)$ is isomorphic to $R^{\dim(R)}$), the assertion is obvious.

(b) It suffices to show that $H(R_L)^G = H(R)$ for L a finite Galois extension of K with Galois group G. When we write $H = \mathrm{Spf}\, A$, $H(R_L)$ is the set of continuous homomorphisms $A \longrightarrow R_L$, and so the assertion is obvious.

(c) The sequences

$$0 \longrightarrow H_\nu(R_L/m^iR_L) \longrightarrow H(R_L/m^iR_L) \xrightarrow{p^\nu} H(R_L/m^iR_L)$$

are exact, and so on passing to the inverse limit, we obtain an exact sequence

$$0 \longrightarrow H_\nu(R_L) \longrightarrow H(R_L) \xrightarrow{p^\nu} H(R_L).$$

The term $H_\nu(R_L)$ has its usual meaning, and we have observed in (1.1) that $H_\nu(R_L) = H_\nu(L)$. Therefore on passing to the direct limit we obtain an exact sequence

$$0 \longrightarrow H_\nu(K_s) \longrightarrow M_H \xrightarrow{p^\nu} M_H.$$

It remains to show that $p\colon M_H \longrightarrow M_H$ is surjective. If H is étale, then $M_H = H(k_s)$, which is obviously divisible by p. If H is connected, say $H = \mathrm{Spf}\, A$, then the map $p\colon H \longrightarrow H$ turns A into a free A-module of finite rank, and so the divisibity is again obvious. The general case now follows from the fact that

$$0 \longrightarrow H^{o}(R_L) \longrightarrow H(R_L) \longrightarrow H^{et}(R_L) \longrightarrow 0$$

is exact for all L (see [Tate (1967b), p168]).

Let H^t be the p-divisible group dual to H (ibid. 2.3), so that $(H^t)_\nu = H^D_\nu$.

Proposition 1.10. *Assume that the torsion subgroups of* H(R) *and* $H^t(R)$ *are both finite. Then there is a canonical pairing*

$$H^1(K, M_{H^t}) \times H(R) \longrightarrow \mathbb{Q}/\mathbb{Z}$$

which identifies the discrete group $H^1(K, M_{H^t})$ *with the dual of the compact group* H(R).

Proof: From the cohomology sequence of the sequence in (1.9c), we get an exact sequence

$$0 \longrightarrow H(R)^{(p^\nu)} \longrightarrow H^1(K, H_\nu) \longrightarrow H^1(K, M_H)_{p^\nu} \longrightarrow 0.$$

I claim that the first map in the sequence factors through $H^1(X, H_\nu) \hookrightarrow H^1(K, H_\nu)$. Note first that it is possible to define H(R') for any finite flat R-algebra R'. Let P ∈ H(R); then the inverse image P under p^ν: H ⟶ H (regarded as map of functors of finite flat R-algebras) is a principal homomgeneous space for H_ν over X whose generic fibre represents the image of P in $H^1(K, H_\nu)$. This proves the claim.

As we observed in the proof of (1.4), the images of $H^1(X, H^D_\nu)$ and $H^1(X, H_\nu)$ annihilate each other in the nondegenerate pairing

$$H^1(K, H^D_\nu) \times H^1(K, H_\nu) \longrightarrow H^2(K, \mathbb{G}_m) = \mathbb{Q}/\mathbb{Z}.$$

Therefore, the images of $H^t(R)^{(p^\nu)}$ and $H(R)^{(p^\nu)}$ annihilate each other under the same pairing, and so the diagram

$$\begin{array}{ccccccc} 0 \longrightarrow & H(R)^{(p^\nu)} & \longrightarrow & H^1(K,H_\nu) & & \\ & & & \downarrow \approx & & \\ 0 \longrightarrow & (H^1(K,M_{H^t})_{p^\nu})^* & \longrightarrow & H^1(K,H^t_\nu)^* & \longrightarrow & H^t(R)^{(p^\nu)*} \end{array}$$

shows that the pairing induces an injection $H(R)^{(p^\nu)} \hookrightarrow (H^1(K,M_{H^t})_{p^\nu})^*$.
In the limit this becomes an injection

$$\varprojlim H(R)^{(p^\nu)} \hookrightarrow (\varinjlim (H^1(K,M_{H^t})_{p^\nu})^* = H^1(K,M_{H^t})^*,$$

and because of our assumption on $H(R)$, $\varprojlim H(R)^{(p^\nu)} = H(R)$.

We therefore have a injection $H(R) \longrightarrow H^1(K,M_{H^t})^*$, and to prove that it is surjective, it suffices to show that $[H(R)^{(p)}] = [H^1(K,M_{H^t})_p]$. This we do using an argument similar to that in the proof of (I.3.2). From (I.2.8), we know that

$$\chi(K,H_1) = (R:pR)^{-h} = \chi(K,H^t_1)$$

where h is the common height of H and H^t. The logarithm map [Tate (1967b), 2.3] and our assumptions on $H(R)$ and $H^t(R)$ show that $H(R)$ and $H^t(R)$ contain subgroups of finite index isomorphic to R^d and $R^{d'}$ repectively where d and d' are the dimensions of H and H^t. Therefore

$$[H(R)^{(p)}]/[H(R)_p] = (R:pR)^d, \qquad [H^t(R)^{(p)}]/[H^t(R)_p] = (R:pR)^{d'}.$$

From the cohomology sequence of

$$0 \longrightarrow H^t_1(K_s) \longrightarrow M_{H^t} \xrightarrow{p} M_{H^t} \longrightarrow 0$$

we see that

$$\chi(K,H^t_1) = \frac{[H^t(R)_p] \quad [H^2(K,H^t_1)]}{[H^t(R)^{(p)}] \; [H^1(K,M_{H^t})_p]},$$

or
$$\frac{1}{(R:pR)^h} = \frac{1}{(R:pR)^{d'}} \frac{[H^0(K,H_1)]}{[H^1(K,M_{H^t})_p]}.$$

But d + d' = h [Tate (1967b), Pptn 3], and so this shows that

$$[H^1(K,M_{H^t})_p] = (R:pR)^d\ [H(R)_p] = [H(R)^{(p)}],$$

as required.

Remark 1.11. The proposition is false without the condition that the torsion subgroups of H(R) and $H^t(R)$ are finite. For example, if $H = (\mathbb{Z}/p^\nu\mathbb{Z})_{\nu\geq 1}$, then H(R) ($= \mathbb{Q}_p/\mathbb{Z}_p$) and $H^1(K,M_{H^t})$ are both infinite and discrete, and so can not be dual. If $H = (\mu_{p^\nu})_{\nu\geq 1}$, then

$$H(R) = \{a \in R \mid a \equiv 1 \bmod m\}$$

and $H^1(K,M_{H^t}) = \mathrm{Hom}(\mathrm{Gal}(K_s/K),\mathbb{Q}_p/\mathbb{Z}_p)$, which are not (quite) dual.

We now complete the second proof of (1.3). For r = 0, the pairing can be identified with the pairing

$$H^0(K,N^D) \times H^2(K,N) \longrightarrow H^2(K,\mathbb{G}_m) = \mathbb{Q}/\mathbb{Z}$$

of (I.2.3), and for r ≠ 0,1 both groups are zero. This leaves the case r = 1, and we saw in the proof of (1.4) that this case is equivalent to the statement that $H^1(X,N^D)$ and $H^1(X,N)$ are exact annihilators in the duality between $H^1(K,N^D)$ and $H^1(K,N)$. We know that the groups in question do annihilate each other, and so

$$[H^1(K,N)] \geq [H^1(X,N)][H^1(X,N^D)],$$

and to show that they are exact annihilators it suffices to prove that equality holds.

According to (A.4) and (A.7), there is an exact sequence

$$0 \longrightarrow N \longrightarrow H \xrightarrow{\varphi} H' \longrightarrow 0$$

with H and H' p-divisible groups. Moreover from the construction of the sequence, it is clear that H, H', and their duals satisfy the hypotheses of (1.10). Write $H^r(X,H)$ for $\varinjlim H^r(X,H_\nu)$, and let

$H'(R)^{(\varphi)} = \mathrm{Coker}(\varphi\colon H(R) \longrightarrow H'(R))$ and

$H^1(X,H)_\varphi = \mathrm{Ker}(\varphi\colon H^1(X,H) \longrightarrow H^1(X,H'))$.

The cohomology sequence of the above sequence and its dual

$$0 \longrightarrow N^D \longrightarrow H'^t \xrightarrow{\varphi^t} H^t \longrightarrow 0$$

show that

$$[H^1(X,N)] = [H'(R)^{(\varphi)}][H^1(X,H)_\varphi] \geq [H'(R)^{(\varphi)}] \qquad \text{and}$$

$$[H^1(X,N^D)] = [H^t(R)^{(\varphi^t)}][H^1(X,H'^t)_{\varphi^t}] \geq [H^t(R)^{(\varphi^t)}].$$

On combining the three inequalities, we find that

$$[H^1(K,N)] \geq [H^1(X,N)][H^1(X,N^D)] \geq [H'(R)^{(\varphi)}]\,[H^t(R)^{(\varphi^t)}].$$

It follows from (1.10) that $H'^t(R) \xrightarrow{\varphi^t} H^t(R)$ is dual to $H^1(K,M_H) \xrightarrow{\varphi} H^1(K,M_{H'})$, and so $[H^t(R)^{(\varphi^t)}] = [H^1(K,M_H)_\varphi]$. But $[H^1(K,N)] = [H'(R)^{(\varphi)}][H^1(K,M_H)_\varphi]$, from which it follows that all of the above inequalities are equalities. This completes the second proof of (1.3).

Remark 1.12. The above argument shows that for any isogeny $\varphi\colon H \longrightarrow H'$ of p-divisible groups over X, $H'(R)^{(\varphi)} \xrightarrow{\approx} H^1(X,\mathrm{Ker}(\varphi))$ and $H^1(X,H)_\varphi = 0$, provided the torsion subgroups of H(R) and $H^t(R)$ are finite.

A duality theorem for p-divisible groups

If A is an abelian variety over K, then (I.3.4) shows that there is an exact sequence

$$0 \longrightarrow A(K)\otimes\mathbb{Q}_p/\mathbb{Z}_p \longrightarrow H^1(K,A(p)) \longrightarrow A^t(K)^*(p) \longrightarrow 0.$$

Our next result is the analogue of this for p-divisible groups. Recall that $H^r(K,H) \overset{df}{=} \varinjlim H^r(K,H_\nu)$.

Proposition 1.13. *Assume that* R *is complete, and let* H *be a* p-*divisible group over* X *such that the torsion subgroups of* H(R) *and* $H^t(R)$ *are finite. Then* $H^1(X,H) = 0$, *and there is an exact sequence*

$$0 \longrightarrow H(R)\otimes\mathbb{Q}/\mathbb{Z} \longrightarrow H^1(K,H) \longrightarrow H^t(R)^* \longrightarrow 0.$$

Proof: On applying Remark 1.12 to the isogeny $p^\nu\colon H \longrightarrow H$, we find that $H(R)^{(p^\nu)} \xrightarrow{\approx} H^1(X,H_\nu)$ and $H^1(X,H)_{p^\nu} = 0$. As $H^1(X,H)$ is p-primary, the equality shows that it is zero. From (1.4) we know there is an exact sequence

$$0 \longrightarrow H^1(X,H_\nu) \longrightarrow H^1(K,H_\nu) \longrightarrow H^1(X,H_\nu^t)^* \longrightarrow 0.$$

On using the isomorphism to replace the first and third terms in this sequence, we obtain an exact sequence

$$0 \longrightarrow H(R)^{(p^\nu)} \longrightarrow H^1(K,H_\nu) \longrightarrow H^t(R)^{(p^\nu)*} \longrightarrow 0.$$

Now one has only to pass to the direct limit to obtain the result.

Euler-Poincaré characteristics

If N is a finite flat group scheme over X, then the groups $H^r(X,N)$ are finite for all r and zero for $r > 1$. We define $\chi(X,N) = [H^0(X,N)]/[H^1(X,N)]$. Let N = Spec B. Recall that the *order* of N is defined to be the rank of B over R, and the *discriminant ideal* of N is the discriminant ideal of B over R.

Theorem 1.14. *Let N be a finite flat group scheme over X, and let n be its order and* $\mathfrak{d}$ *its discriminant ideal. Then* $(R:\mathfrak{d})$ *is an* n^{th} *power and*

$$\chi(X,N) = (R:\mathfrak{d})^{-1/n}.$$

Proof: When N is étale, $H^r(X,N) = H^r(g,N(R^{un}))$, and so both sides of the equation are 1. This allows us to assume that N is local. We can also assume that R is complete because passing to the completion does not change either side. Consider an exact sequence

$$0 \longrightarrow N \longrightarrow H \xrightarrow{\varphi} H' \longrightarrow 0$$

with H and H' connected p-divisible groups. As $H^1(X,H) = 0$,

$$\chi(X,N) = z(\varphi(R)) \overset{df}{=} [\mathrm{Ker}\ \varphi(R)]/[\mathrm{Coker}\ \varphi(R)].$$

Write H = Spf A and H' = Spf A'. Then A and A' are power series rings in d variables over R, where d is the common dimension of H and H'. The map φ corresponds to a homomorphism φ^a: $A' \longrightarrow A$ making A into a free A'-module of rank n. It also defines a map $d\varphi^a$: $\Omega^1_{A'/R} \longrightarrow \Omega^1_{A/R}$.

Lemma 1.15. *Choose bases for* $\Omega^1_{A'/R}$ *and* $\Omega^1_{A/R}$, *and let* θ' *and* θ *be the corresponding basis elements for* $\Lambda^d\Omega_{A'/R}$ *and* $\Lambda^d\Omega_{A/R}$ *over A' and A respectively. If* $\Lambda^d d\varphi^a$: $\Lambda^d\Omega_{A'/R} \longrightarrow \Lambda^d\Omega_{A/R}$ *maps* θ' *to* $a\theta$, $a \in A$, *then* $N_{A/A'}a$ *generates the discriminant ideal of A over A'.*

Proof: This follows from the existence of a trace map Tr: $\Lambda^d\Omega_{A/R} \longrightarrow \Lambda^d\Omega_{A'/R}$; see [Tate (1967b), p165].

Let $\omega_{A/R}$ and $\omega_{A'/R}$ be the R-modules of invariant differentials on H and H' respectively. The inclusion $\omega_{A/R} \longrightarrow \Omega^1_{A/R}$ induces isomorphisms $\omega_{A/R}\otimes_R A \xrightarrow{\approx} \Omega^1_{A/R}$ and $\omega_{A/R} \xrightarrow{\approx} \Omega^1_{A/R}\otimes_A R$. Let θ and θ' be basis elements for $\Lambda^d\omega_{A/R}$ and $\Lambda^d\omega_{A'/R}$. On taking θ and θ' to be $\theta\otimes 1$ and $\theta'\otimes 1$ in the lemma, we find that $d\varphi^a(\theta') = a\theta$, $a \in R$, and that the discriminant ideal of A over A' is generated by a^n. Since $A\otimes_{A'}R = B$, where $B = \Gamma(N,\mathcal{O}_N)$, this shows that the discriminant ideal $\mathfrak{d}$ of N is generated by a^n. It remains to show that $\chi(X,N) = (R:aR)$.

Let T(H) and T(H') be the tangent spaces to H and H' at zero. They are dual to $\omega_{A/R}$ and $\omega_{A'/R}$, and so aR is equal to the determinant ideal of the map of R-modules $d\varphi\colon T(H) \longrightarrow T(H')$. Recall [Tate (1967b), 2.4] that there exists a logarithm map $\log\colon H(R) \longrightarrow T(H)\otimes_R K$, and that if we choose an isomorphism $A \approx R[[X_1,\dots,X_d]]$, then for any c with $c^{p-1} < |p|$, log gives an isomorphism between

$$H(R)_c = \{x \in H(R) \mid |x_i| \le c \text{ all } i\}$$

and

$$T(H)_c = \{\tau \in T(H) \mid |\tau(X_i)| \le c \text{ all } i\} = p^c T(H).$$

From the commutative diagram

$$\begin{array}{ccc} H(R)_c & \xrightarrow{\varphi_c} & H'(R)_c \\ \approx\downarrow\log & & \approx\downarrow\log \\ T(H)_c & \xrightarrow{(d\varphi)_c} & T(H')_c \end{array}$$

we see that $(R:\det(d\varphi)) = (R:\det(d\varphi)_c) = [\mathrm{Coker}(\varphi_c)]$. But

$$[\mathrm{Coker}(\varphi_c)] = z(\varphi)(H(R):H(R)_c)(H'(R):H'(R)_c)^{-1}$$

and $(H(R):H(R)_c) = p^{d(c-1)} = (H'(R):H'(R)_c)$. Therefore

$$(R:aR) = (R:\det(d\varphi)) = z(\varphi) = \chi(X,N).$$

For a finite flat group scheme N over X, define

$$\chi_x(X,N) = \frac{[H^2_x(X,N)]}{[H^1_x(X,N)]}$$

Corollary 1.16. *Let* N *be a finite flat group scheme over* X, *and let* n *be its order and* $\mathfrak{d}$ *its discriminant ideal. Then*

$$\chi_x(X,N) = (R:nR)(R:\mathfrak{d})^{-1/n}.$$

Proof: From the cohomology sequence of $X \supset u$, we find that

$$\chi_x(X,N) = \chi(X,N)\chi(K,N)^{-1} = (R:\mathfrak{d}R)^{-1/n}(R:nR)$$

(see (1.14) and (I.2.8).)

As $H^r(X,N)$ is dual to $H_x^{3-r}(X,N^D)$, $\chi(X,N)\chi_x(X,N^D) = 1$. Therefore (1.14) and (1.16) imply that $\mathfrak{d}(N)\mathfrak{d}(N^D) = (n^n)$. This formula can also be directly deduced from the formulas

$$\mathrm{Norm}_{B/R}\mathfrak{D}(N) = \mathfrak{d}(N), \qquad \mathfrak{D}(N)\mathfrak{D}(N^D) = (n),$$

where $\mathfrak{D}(N)$ and $\mathfrak{D}(N^D)$ are the differents of N and N^D and $B = \Gamma(N,\mathcal{O}_N)$ (see [Raynaud (1974), Pptn 9] and [Mazur and Roberts (1970), A.2]).

Let N be a quasi-finite, flat, separated group scheme over X. Because X is Henselian, there is a finite flat group scheme $N^f \subset N$ having the same closed fibre as N; moreover, N/N^f is étale and it is the extension by zero of generic fibre (cf. [Milne (1980), I.4.2c]). In this case we write n^f and $\mathfrak{d}^f$ for the order of N^f and its discriminant ideal.

Corollary 1.17. *Let* N *be as above.*

(a) $\chi(X,N) = (R:\mathfrak{d}^f)^{-1/n^f}$.

(b) $\chi_x(X,N) = (R:nR)(R:\mathfrak{d}^f)^{-1/n^f}$.

Proof: (a) From the cohomology sequence of

$$0 \longrightarrow N^f \longrightarrow N \longrightarrow N/N^f \longrightarrow 0$$

we find that $\chi(X,N) = \chi(X,N^f)\chi(X,N/N^f)$. Therefore it suffices to prove the formula in the cases that $N = N^f$ or $N^f = 0$. In the first case, it becomes the formula in (1.14), and in the second $N = j_!N_K$, and so both sides are 1.

(b) Again, it suffices to prove the formula in the cases $N = N^f$ or $N^f = 0$. In the first case the formula becomes that in (1.16). In the second, $H^r(K,N_K) \xrightarrow{\approx} H_x^{r+1}(U,N)$, and so $\chi_x(X,N) = \chi(K,N_K)^{-1} = (R:nR)$ by (I.2.8).

Corollary 1.18. *Let* $H = (H_\upsilon)_{\upsilon\geq 1}$ *be a p-divisible group over* X; *the* $\chi(X,H_\upsilon) = (R:p^{d\upsilon}R)$, *where* d *is the dimension of* H.

Proof: According to [Tate (1967b), Pptn 2], the discriminant ideal of H_υ is generated by $p^{d\upsilon p^{h\upsilon}}$. Therefore $\chi(X,H_\upsilon) = ((R:p^{d\upsilon}R)^{p^{h\upsilon}})^{p^{-h\upsilon}} = (R:p^{d\upsilon}R)$.

Extensions of morphisms

For each finite group scheme N over X, define $h(N) = (N_K, H^1(X,N))$. A morphism $h(N) \longrightarrow h(N')$ is a K-morphism $\varphi_K: N_K \longrightarrow N'_K$ such that $H^1(\varphi_K): H^1(K,N'_K) \longrightarrow H^1(K,N'_K)$ maps $H^1(X,N)$ into $H^1(X,N')$.

Theorem 1.19. *The functor* $N \mapsto h(N)$ *of finite group schemes over* X *is fully faithful; that is, a homomorphism* $\varphi_K: N_K \longrightarrow N'_K$ *extends to a homomorphism* $\varphi: N \longrightarrow N'$ *if and only if* $H^1(\varphi_K)$ *maps* $H^1(X,N)$ *into* $H^1(X,N')$, *and the extension is unique when it exists.*

Proof: This is the main theorem of [Mazur (1970b)].

Examples

Assume that K contains the pth roots of 1. For each $a,b \in R$ with $ab = p$, there is a well-defined finite flat group scheme $N_{a,b}$ over R given by the classification of Oort and Tate (cf. 0.9). It is a finite group scheme of order p and discriminant a^p. Therefore $\chi(X,N_{a,b}) = (R:aR)$.

Because K contains a pth root of 1, $m \overset{df}{=} \mathrm{ord}(p)/(p-1)$ is an integer. We say that $N_{a,b}$ *splits generically* if its generic fibre is isomorphic to $\mathbb{Z}/p\mathbb{Z}$. This is equivalent to a being a nonzero $(p-1)^{st}$ power in R. Choose a uniformizing parameter π in R. Then the generically split group schemes of order p over R correspond to the pairs (a,b) with $a = \pi^{(p-1)i}$, $0 \leq i \leq m$, and $b = p/a$. For example, if $a = 1$, then $N = \mathbb{Z}/p\mathbb{Z}$, and if $a = \pi^{(p-1)m} = (\mathrm{unit})p$, then $N = \mu_p$. Let $U = R^{\times}$ and $U^{(i)} = \{a \in U \mid \mathrm{ord}(1-a) \geq i\}$.

Proposition 1.20. *Let N be a finite flat generically split group scheme of order p over X. Then there is a nonzero map* $\varphi\colon N \longrightarrow \mu_p$, *and for any choice of* φ, *the map* $H(\varphi)\colon H^1(X,N) \longrightarrow H^1(X,\mu_p)$ *identifies* $H^1(X,N)$ *with the subgroup* $U^{(i)}U^p/U^p$ *of* $H^1(X,\mu_p) = U/U^p$, *where* $i = pm - \mathrm{ord}(\mathrm{disc}\ N)/(p-1)$.

Proof: Let $N = N_{a,b}$, and suppose $a = \pi^{(p-1)i}$. Then, for any X-scheme Y, $N(Y) = \{y \in \Gamma(Y,\mathcal{O}_Y) \mid y^p = ay\}$, and so $y \mapsto y\pi^{(m-i)}$ defines a morphism of functors $N(Y) \longrightarrow \mu_p(Y)$ and hence a nonzero map $N \longrightarrow \mu_p$. For the proof of the proposition, one first shows that the image of $H^1(X,N)$ is contained in $U^{(i)}U^p/U^p$ and then uses (1.14) to show that it equals this group. See [Roberts (1973)].

More explicitly, if $N = N_{a,b}$ with $a = \pi^{(p-1)i}$, then $H^1(X,N) = U^{(j)}U^p/U^p$, where $j = p.\mathrm{ord}(p)/(p-1) - pi$. For example, if $a = 1$ so that $N = \mathbb{Z}/p\mathbb{Z}$, then $H^1(X,N) = U^{(pm)}U^p/U^p$, and if $N = \pi^{(p-1)m}$ so that $N = \mu_p$, then $H^1(X,N) = U^{(0)}U^p/U^p$.

Remark 1.21. In the examples in (1.20), the map $H^1(X,N) \longrightarrow H^1(X_i,N)$

is an isomorphism for i >> 0. This is true for any finite group scheme N, as can be easily deduced from the exact sequence

$$H(R) \longrightarrow H'(R) \longrightarrow H^1(X,N) \longrightarrow 0$$

arising from a resolution of N by p-divisible groups.

Notes: Theorems 1.3 and 1.14 are due to Mazur and Roberts ([Mazur and Roberts (1970)] and [Mazur (1970a)]). The second proof of (1.3) and the proof of (1.14) are taken from [Milne (1973)]. The first proof of (1.3) is new. Theorem 1.19 is due to Mazur [Mazur (1970b)].

§2 Local results: mixed characteristic, abelian varieties

The notations are the same as in §1. Except in the last two results, X will be endowed with its smooth topology.

Let A be an abelian variety over K, and let $\mathcal{A}$ be its Néron model over X. As in Appendix C, we write $\mathcal{A}^o$ for the open subgroup scheme of $\mathcal{A}$ whose closed fibre $\mathcal{A}^o_x$ is connected. There is an exact sequence of sheaves on X_{sm}

$$0 \longrightarrow \mathcal{A}^o \longrightarrow \mathcal{A} \longrightarrow i_*\Phi \longrightarrow 0.$$

We often regard Φ as a $Gal(k_s/k)$-module. Recall that for any submodule Γ of Φ, $\mathcal{A}^\Gamma$ denotes the inverse image of Γ in $\mathcal{A}$. There is an exact sequence

$$0 \longrightarrow \mathcal{A}^o \longrightarrow \mathcal{A}^\Gamma \longrightarrow i_*\Gamma \longrightarrow 0 \qquad (2.0.1).$$

Proposition 2.1. *The map* $\mathcal{A}^\Gamma(X) \longrightarrow \Gamma(x)$ *arising from* (2.0.1) *is surjective, and* $H^r(X,\mathcal{A}^\Gamma) \longrightarrow H^r(x,\Gamma)$ *is an isomorphism for* $r \geq 1$; *therefore,* $H^r(X,\mathcal{A}^\Gamma) = 0$ *for* $r \geq 2$.

Proof: According to [Milne (1980), III.3.11], $H^r(X,\mathcal{A}^\Gamma) = H^r(x,\mathcal{A}^\Gamma_x)$

for $r > 0$, and Lang's lemma implies that $H^r(x,\mathscr{A}_x^o) = 0$ for $r > 0$. Therefore the cohomology sequence of (2.0.1) leads immediately to the result.

Lemma 2.2. *For any Γ, there is an exact sequence*

$$\Phi(x) \longrightarrow (\Phi/\Gamma)(x) \longrightarrow H^1(X,\mathscr{A}^\Gamma) \longrightarrow H^1(K,A),$$

in which the last map is the restriction map; in particular, if $Gal(k_s/k)$ acts trivially on Φ, then $H^1(X,\mathscr{A}^\Gamma) \longrightarrow H^1(K,A)$ is injective.

Proof: We first consider the case that $\Gamma = \Phi$. Then $\mathscr{A}^\Gamma = \mathscr{A}$, and as $\mathscr{A} = j_*A$, the Leray spectral sequence for j shows immediately that the map $H^1(X,\mathscr{A}) \longrightarrow H^1(K,A)$ is injective. In the general case, the lemma can be deduced from the diagram

$$\begin{array}{ccccccc} & & & & H^1(X,\mathscr{A}^\Gamma) & \longrightarrow & H^1(X,\mathscr{A}) \\ & & & & \downarrow\approx & & \downarrow\approx \\ \Phi(x) & \longrightarrow & (\Phi/\Gamma)(x) & \longrightarrow & H^1(x,\Gamma) & \longrightarrow & H^1(x,\Phi). \end{array}$$

Lemma 2.3. *We have*

$$H_x^r(X,\mathscr{A}^\Gamma) = \begin{cases} 0 & \text{for } r \neq 1,2 \\ (\Phi/\Gamma)(x) & \text{for } r = 1, \end{cases}$$

and there is an exact sequence

$$0 \to \Gamma(x) \longrightarrow \Phi(x) \longrightarrow (\Phi/\Gamma)(x) \longrightarrow H^1(X,\mathscr{A}^\Gamma) \longrightarrow H^1(K,A) \longrightarrow H_x^2(X,\mathscr{A}^\Gamma) \to 0.$$

Proof: Consider the exact sequence

$$0 \longrightarrow H_x^0(X,\mathscr{A}^\Gamma) \longrightarrow H^0(X,\mathscr{A}^\Gamma) \longrightarrow H^0(K,A) \longrightarrow H_x^1(X,\mathscr{A}^\Gamma) \longrightarrow H^1(X,\mathscr{A}^\Gamma) \longrightarrow \dots .$$

Obviously $\mathscr{A}^\Gamma(X) \longrightarrow A(K)$ is injective, which shows that $H_x^0(X,\mathscr{A}^\Gamma) = 0$. As $H^r(X,\mathscr{A}^\Gamma)$ and $H^r(K,\mathscr{A}^\Gamma)$ are both zero for $r > 1$, the sequence shows that $H_x^r(X,\mathscr{A}^\Gamma) = 0$ for $r > 2$.

Take $\Gamma = \Phi$, so that $\mathscr{A}^\Gamma = \mathscr{A}$; then $\mathscr{A}(X) \longrightarrow A(K)$ is an isomorphism and (2.2) shows that $H^1(X,\mathscr{A}) \longrightarrow H^1(K,A)$ is injective. Therefore the sequence shows that $H^1_x(X,\mathscr{A}) = 0$. In the general case the exact sequence

$$0 \longrightarrow H^0_x(X,\Phi/\Gamma) \longrightarrow H^1_x(X,\mathscr{A}^\Gamma) \longrightarrow H^1_x(X,\mathscr{A})$$

gives an isomorphism $(\Phi/\Gamma)(x) \xrightarrow{\approx} H^1_x(X,\mathscr{A}^\Gamma)$. The existence of the required exact sequence follows from

$$\begin{array}{ccccccccc} H^1_x(X,\mathscr{A}^\Gamma) & \longrightarrow & H^1(X,\mathscr{A}^\Gamma) & \longrightarrow & H^1(K,A) & \longrightarrow & H^2_x(X,\mathscr{A}^\Gamma) \longrightarrow H^2(X,\mathscr{A}^\Gamma) \longrightarrow 0 \\ \downarrow\approx & & \| & & \| & & \\ \Phi(x) \longrightarrow (\Phi/\Gamma)(x) & \longrightarrow & H^1(X,\mathscr{A}^\Gamma) & \longrightarrow & H^1(K,A). & & \end{array}$$

We now consider an abelian variety A over K, its dual abelian variety B, and a Poincaré biextension W of (B,A) by $\mathbb{G}_m$. Recall (C.12) that W extends to a biextension of $(\mathscr{B}^{\Gamma'},\mathscr{A}^\Gamma)$ by $\mathbb{G}_m$ if and only if Γ' and Γ annihilate each other in the canonical pairing $\Phi' \times \Phi \longrightarrow \mathbb{Q}/\mathbb{Z}$.

Lemma 2.4. *If Γ' and Γ are subgroups of Φ' and Φ that annihilate each other, then the following diagrams commute:*

$$\begin{array}{ccccc} H^1(K,B) & \times & H^0(K,A) & \longrightarrow & H^2(K,\mathbb{G}_m); \\ \uparrow & & \downarrow & & \downarrow\approx \\ H^1(X,\mathscr{B}^{\Gamma'}) & \times & H^1_x(X,\mathscr{A}^\Gamma) & \longrightarrow & H^3_x(X,\mathbb{G}_m); \end{array} \qquad \begin{array}{ccccc} H^1(K,B) & \times & H^0(K,A) & \longrightarrow & H^2(K,\mathbb{G}_m) \\ \downarrow & & \uparrow & & \downarrow\approx \\ H^2_x(X,\mathscr{B}^{\Gamma'}) & \times & H^0(X,\mathscr{A}^\Gamma) & \longrightarrow & H^3_x(X,\mathbb{G}_m). \end{array}$$

Proof: In the first diagram, the first vertical arrow is the restriction map, and the second and third arrows are boundary maps $H^r(u,-) \longrightarrow H^{r+1}_x(X,-)$. Since the top pairing is defined by the restriction to u of the biextension defining the bottom pairing, the commutativity is obvious. The proof that the second diagram commutes is similar.

Theorem 2.5. *The canonical pairing $\Phi' \times \Phi \longrightarrow \mathbb{Q}/\mathbb{Z}$ is nondegenerate; that is, Conjecture C.13 holds in this case.*

Proof: The groups and the pairing are unchanged when we replace K with its completion. After making an unramified extension of R, we can assume that $\mathrm{Gal}(k_s/k)$ acts trivially on Φ and Φ'. By symmetry, it suffices to show that the pairing $\Phi' \times \Phi \longrightarrow \mathbb{G}_m$ is left non-degenerate, and for this, it suffices to show that the pairing $H^1(x,\Phi') \times H^0(x,\Phi) \longrightarrow H^1(x,\mathbb{Q}/\mathbb{Z}) \approx \mathbb{Q}/\mathbb{Z}$ is left nondegenerate.

The canonical pairing of Φ' with Φ is so defined that

$$\begin{array}{ccc} \mathcal{B} & \longrightarrow & \mathcal{E}xt^1_X(\mathcal{A}, j_*\mathbb{G}_m) \\ \downarrow & & \downarrow \\ i_*\Phi' & \longrightarrow & \mathcal{H}om_X(i_*\Phi, i_*\mathbb{Q}/\mathbb{Z}) \end{array}$$

commutes (see C.11). Alternatively, we can regard it as being the unique homomorphism $\Phi' \longrightarrow \mathcal{E}xt^1_x(\Phi,\mathbb{Z})$ making

$$\begin{array}{ccc} \mathcal{B} & \longrightarrow & \mathcal{E}xt^1_X(\mathcal{A}, j_*\mathbb{G}_m) \\ \downarrow & & \downarrow \\ i_*\Phi' & \longrightarrow & \mathcal{E}xt^1_X(i_*\Phi, i_*\mathbb{Z}) \end{array}$$

commute. From this we get a commutative diagram

$$\begin{array}{ccc} H^1(X,\mathcal{B}) & \longrightarrow & \mathrm{Ext}^2_X(\mathcal{A}, j_*\mathbb{G}_m) \\ \downarrow & & \downarrow \\ H^1(X, i_*\Phi') & \longrightarrow & \mathrm{Ext}^2_X(i_*\Phi, i_*\mathbb{Z}). \end{array}$$

These maps are used to define thc two lower pairings in the following diagram

$$\begin{array}{ccccc}
H^1(K,B) & \times H^0(K,A) & \longrightarrow H^2(K,\mathbb{G}_m) \approx \mathbb{Q}/\mathbb{Z} \\
\uparrow \text{inj} & \uparrow \approx & \uparrow \approx \\
H^1(X,\mathcal{B}) & \times H^0(X,\mathcal{A}) & \longrightarrow H^2(X,j_*\mathbb{G}_m) \\
\downarrow \approx & \downarrow \text{surj} & \downarrow \approx \\
H^1(X,i_*\Phi') & \times H^0(X,\Phi) & \longrightarrow H^2(X,i_*\mathbb{Z}),
\end{array}$$

and so the the diagram commutes (the upper arrows are all restriction maps). The top pairing is nondegenerate (I.3.4), and so the lower two pairings are left nondegenerate. This proves the theorem.

Corollary 2.6. *Suppose that Γ' and Γ are exact annihilators under the canonical pairing of Φ' and Φ. Then the map*

$$\mathcal{B}^{\Gamma'} \longrightarrow \mathcal{E}xt^1_{X_{sm}}(\mathcal{A}^{\Gamma},\mathbb{G}_m)$$

defined by the extension of W is an isomorphism (of sheaves on X_{sm}).

Proof: See (C.14).

Theorem 2.7. *Assume that Γ' and Γ are exact annihilators. Then the pairing*

$$H^r(X,\mathcal{B}^{\Gamma'}) \times H^{2-r}_x(X,\mathcal{A}^{\Gamma}) \longrightarrow H^3_x(X,\mathbb{G}_m) \approx \mathbb{Q}/\mathbb{Z}$$

defined by the canonical biextension of $(\mathcal{B}^{\Gamma'},\mathcal{A}^{\Gamma})$ by $\mathbb{G}_m$ induces an isomorphism $H^2_x(X,\mathcal{A}^{\Gamma}) \xrightarrow{\approx} \mathcal{B}^{\Gamma'}(X)^*$ of discrete groups for r = 0 and an isomorphism of finite groups $H^1(X,\mathcal{B}^{\Gamma'}) \xrightarrow{\approx} \mathcal{A}^{\Gamma}(X)^*$ for r = 1. For $r \neq 0,1$, both groups are zero.

Proof: Consider the diagram

$$\begin{array}{ccccccccc}
\Phi'(x) & \longrightarrow & (\Phi'/\Gamma')(x) & \longrightarrow & H^1(X,\mathcal{B}^{\Gamma'}) & \longrightarrow & H^1(K,B) & \longrightarrow & H^2_x(X,\mathcal{B}^{\Gamma'}) \longrightarrow 0 \\
\downarrow \approx & & \downarrow \approx & & \downarrow a & & \downarrow \approx & & \downarrow b \\
H^1(x,\Phi)^* & \longrightarrow & H^1(x,\Gamma)^* & \longrightarrow & H^1_x(X,\mathcal{A}^{\Gamma})^* & \longrightarrow & H^0(K,A)^* & \longrightarrow & H^0(X,\mathcal{A}^{\Gamma})^* \longrightarrow 0.
\end{array}$$

The top row is the exact sequence in (2.3), and the bottom row is the dual of the cohomology sequence of the pair $X \supset u$. That the last two squares commute is proved in (2.4). The first two vertical maps are the isomorphisms induced by the canonical pairing between Φ' and Φ. Thus the first square obviously commutes, and second was essentially shown to commute in the course of the proof of (2.5). It follows from the diagram that a is injective and b is surjective. But the two groups $H^1(X,\mathcal{B}^{\Gamma'})$ and $H^1_x(X,\mathcal{A}^{\Gamma})$ have the same order (see (2.1) and (2.3)), and so a is an isomorphism. This in turn shows that b is an isomorphism.

Remark 2.8. (a) Once (2.7) is acquired, it is easy to return and prove (2.5): the map $H^1(x,\Phi') \longrightarrow H^0(x,\Phi)^*$ can be identified with the isomorphism $H^1(X,\mathcal{B}') \longrightarrow H^1_x(X,\mathcal{A}^o)^*$ given by the (2.6).

(b) Let $\hat{X} = \operatorname{Spec} \hat{R}$. Then it follows from (I.3.10) that the maps $H^r_x(X,\mathcal{A}^{\Gamma}) \longrightarrow H^r_x(\hat{X},\mathcal{A}^{\Gamma})$ are isomorphisms for all r, and that $H^r(X,\mathcal{A}^{\Gamma}) \longrightarrow H^r(\hat{X},\mathcal{A}^{\Gamma})$ is an isomorphism for all $r > 0$. The map $A(X) \longrightarrow A(\hat{X})$ is injective and maps onto the torsion subgroup of $A(\hat{X})$; $A(\hat{X})$ is the completion of $A(X)$ for the topology of subgroups of finite index.

(c) When R is complete, $\mathcal{B}^{\Gamma'}(X)$ is compact. Therefore in this case the pairing induces dualities between:

the compact group $\mathcal{B}^{\Gamma'}(X)$ and the discrete group $H^2_x(X,\mathcal{A}^{\Gamma})$;
the finite group $H^1(X,\mathcal{B}^{\Gamma'})$ and the finite group $\mathcal{A}^{\Gamma}(X)$.

Write $\mathcal{B}\{n\}$ for the complex $\mathcal{B} \xrightarrow{n} \mathcal{B}^{n\Phi'}$ and $\mathcal{A}\{n\}$ for the complex $\mathcal{A}^{\Phi n} \xrightarrow{n} \mathcal{A}^o$. The pairings $\mathcal{B}\otimes^L\mathcal{A}^o \longrightarrow \mathbb{G}_m[1]$ and $\mathcal{B}^{n\Phi'}\otimes^L\mathcal{A}^{\Phi n} \longrightarrow \mathbb{G}_m[1]$ defined by a Poincaré biextension induce a pairing $\mathcal{B}\{n\}\otimes^L\mathcal{A}\{n\} \longrightarrow \mathbb{G}_m$ in the derived category of sheaves on X_{sm} (see [Grothendieck (1972), VIII.2]).

Theorem 2.9. *The map* $\mathcal{B}\{n\}\otimes^{L}\mathcal{A}\{n\} \longrightarrow \mathbb{G}_m$ *defines nondegenerate pairings*

$$H^r(X,\mathcal{B}\{n\}) \times H_x^{3-r}(X,\mathcal{A}\{n\}) \longrightarrow H_x^3(x,\mathbb{G}_m) \approx \mathbb{Q}/\mathbb{Z}$$

of finite groups for all r.

Proof: From the exact sequences of complexes

$$0 \longrightarrow \mathcal{B}^{n\Phi'}[-1] \longrightarrow \mathcal{B}\{n\} \longrightarrow \mathcal{B} \longrightarrow 0$$

$$0 \longrightarrow \mathcal{A}^{o}[-1] \longrightarrow \mathcal{A}\{n\} \longrightarrow \mathcal{A}^{\Phi n} \longrightarrow 0$$

we get the rows of the diagram

$$\begin{array}{ccccccc} \cdots \longrightarrow H^{r-1}(X,\mathcal{B}^{n\Phi'}) & \longrightarrow & H^r(X,\mathcal{B}(n)) & \longrightarrow & H^r(X,\mathcal{B}) & \longrightarrow & \cdots \\ \downarrow\approx & & \downarrow & & \downarrow\approx & & \\ \cdots \longrightarrow H_x^{3-r}(X,\mathcal{A}^{\Phi n})^* & \longrightarrow & H_x^{3-r}(X,\mathcal{A}(n))^* & \longrightarrow & H_x^{2-r}(X,\mathcal{A}^{o})^* & \longrightarrow & \cdots . \end{array}$$

Since the diagram obviously commutes, the theorem follows from (2.7).

Corollary 2.10. *Assume that* n *is prime to the characteristic of* k *or that* A *has semistable reduction. Then for all* r, *there is a canonical nondegenerate pairing of finite groups*

$$H_x^r(X_{fl},\mathcal{B}_n) \times H^{3-r}(X_{fl},\mathcal{A}_n) \longrightarrow H_x^3(X_{fl},\mathbb{G}_m) \approx \mathbb{Q}/\mathbb{Z}.$$

Proof: The hypothesis implies that $\mathcal{B} \xrightarrow{n} \mathcal{B}^{n\Phi'}$ and $\mathcal{A}^{\Phi n} \xrightarrow{n} \mathcal{A}^{o}$ are surjective when regarded as a maps of sheaves for the flat topology (see C.9). Hence $\mathcal{B}_n \approx \mathcal{B}\{n\}$ and $\mathcal{A}_n \approx \mathcal{A}\{n\}$, and so $H^r(X_{fl},\mathcal{B}_n) \approx H^r(X_{fl},\mathcal{B}\{n\}) \approx H^r(X_{sm},\mathcal{B}\{n\})$ and $H_x^r(X_{fl},\mathcal{A}_n) \approx H_x^r(X_{fl},\mathcal{A}\{n\}) \approx H_x^r(X_{sm},\mathcal{A}\{n\})$.

Curves over X

By exploiting the autoduality of the Jacobian, it is possible to use (2.9) to prove a duality theorem for a curve over X.

Theorem 2.11. *Let* $\pi\colon Y \longrightarrow X$ *be a proper flat map whose fibres are pure of dimension one. Assume that the generic fibre* Y_K *is smooth and connected, that the special fibre* Y_x *is connected, and that there is a section to* π. *Assume further that* $\mathbf{Pic}^{\tau}_{Y/X} = \mathcal{J}$, *where* $\mathcal{J}$ *is the Néron model of the Jacobian of* Y_K. *Then there is a canonical duality of finite groups*

$$H^r(Y,\mu_n) \times H^{5-r}_{Y_x}(Y,\mu_n) \longrightarrow H^3_x(X,\mathbb{G}_m) \approx \mathbb{Q}/\mathbb{Z}.$$

Proof: We use the Leray spectral sequence of π. Under the hypotheses $R^0\pi_*\mu_n \approx \mu_n$, $R^1\pi_*\mu_n \approx \operatorname{Ker}(\mathcal{J} \xrightarrow{n} \mathcal{J})$, and $R^2\pi_*\mu_n \approx \mathbb{Z}/n\mathbb{Z}$; for $r > 2$, $R^r\pi_*\mu_n = 0$. Moreover $\mathcal{J} = \mathcal{J}^o$. On taking $\mathcal{A} = \mathcal{J} = \mathcal{B}$ in the (2.9), we find that $H^r(X,R^1\pi_*\mu_n)$ is dual to $H^{3-r}_x(X,R^1\pi_*\mu_n)$ for all r. The result can be obtained by combining this duality with the duality of $H^r(X,R^0\pi_*\mu_n)$ and $H^{3-r}_x(X,R^2\pi_*\mu_n)$.

For conditions on Y/X ensuring that the hypotheses of the theorem hold, see the last few parapraphs of Appendix C. Our hypotheses are surely too stringent. Because of this, we make the following definition. Let X be the spectrum of an excellent Henselian discrete valuation ring (not necessarily of characteristic zero) with finite residue field, and let $\pi\colon Y \longrightarrow X$ be a proper flat morphism whose generic fibre is a smooth curve. If there is a canonical pairing

$$R\pi_*\mu_n \times R\pi_*\mu_n \longrightarrow \mathbb{G}_m[2].$$

extending that on the generic fibre and such that the resulting pairing

$$H^r(Y,\mu_n) \times H^{5-r}_{Y_x}(Y,\mu_n) \longrightarrow H^3(X,\mathbb{G}_m) \approx \mathbb{Q}/\mathbb{Z}$$

is nondegenerate, then we shall say that the *local duality theorem holds for Y/X and n.*

Notes: This section is based on [McCallum (1986)].

§3 Global results: number field case

Throughout this section, X will be the spectrum of the ring of integers $\mathcal{O}_K$ in a number field K. For an open subscheme U of X, U[1/n] denotes Spec $\Gamma(U,\mathcal{O}_X)[1/n]$, and $H^r_c(U,-)$ denotes the flat cohomology group with compact support as defined in (0.6a) (thus, it takes account of the infinite primes).

Finite sheaves

Let U be an open subscheme of X. As $\mathbb{G}_m$ is smooth, $H^r_c(U_{fl},\mathbb{G}_m) = H^r_c(U_{et},\mathbb{G}_m)$, and so (see II.3) there is a canonical trace map $H^3_c(U,\mathbb{G}_m) \xrightarrow{\approx} \mathbb{Q}/\mathbb{Z}$. Therefore, for any sheaf F on U, there is a canonical pairing

$$\mathrm{Ext}^r_U(F,\mathbb{G}_m) \times H^{3-r}_c(U,F) \longrightarrow H^3_c(U,\mathbb{G}_m) \approx \mathbb{Q}/\mathbb{Z}$$

(see 0.4e).

Let $f\colon U_{fl} \longrightarrow U_{et}$ be the morphism of sites defined by the identity map. Recall [Milne (1980), V.1] that the constructible sheaves on U_{et} are precisely those sheaves that are representable by étale algebraic spaces of finite-type over U; moreover, if $\tilde{F}$ represents F on U_{et}, then it represents f^*F on U_{fl}.

Theorem 3.1. *Let* U *be an open subscheme of* X, *and let* F *be a sheaf on* U_{fl} *such that* nF = 0 *for some integer* n. *Assume*

(i) *the restriction of* F *to* $U[1/n]_{fl}$ *is represented by an étale algebraic space of finite-type over* $U[1/n]$;

(ii) *for each* $v \in U - U[1/n]$, *the restriction of* F *to* $(\mathrm{Spec}\, \mathcal{O}_v)_{fl}$ *is represented by a finite flat group scheme.*

Let F^D *be the sheaf on* U *such that*

$$F^D|U[1/n] = f^*\mathbf{R}\mathcal{H}om_{U[1/n]_{et}}(F,\mathbb{G}_m)$$

$F^D|V = \mathcal{H}om_{V_{fl}}(F,\mathbb{G}_m)$ *for any open subscheme* V *of* U *where* $F|V$ *is represented by a finite flat group scheme.*

Then there are canonical maps $F^D \longrightarrow \mathbf{R}\mathcal{H}om_U(F,\mathbb{G}_m)$, *hence* $H^r(U,F^D) \longrightarrow \mathrm{Ext}^r_U(F,\mathbb{G}_m)$, *and the resulting pairing*

$$H^r(U,F^D) \times H^{3-r}_c(U,F) \longrightarrow H^3_c(U,\mathbb{G}_m) \approx \mathbb{Q}/\mathbb{Z}$$

is a nondegenerate pairing of finite groups.

Proof: We first note that on $V[1/n] = V \cap U[1/n]$, F is represented by a finite flat étale group scheme whose order is prime to the residue characteristics. Therefore

$$\mathbf{R}\mathcal{H}om_{V[1/n]_{et}}(F,\mathbb{G}_m) = \mathcal{H}om_{V[1/n]_{et}}(F,\mathbb{G}_m),$$

and so the requirements on F^D coincide on $V[1/n]$, which shows that F^D exists.

For all r, $H^r_c(U[1/n]_{et},F) = H^r_c(U[1/n]_{fl},F)$, and $H^r(U[1/n]_{fl},F^D) = H^r(U[1/n]_{et},F^D)$. Therefore, for the restriction of F to $U[1/n]$, the theorem becomes (II.3.3). To pass from $U[1/n]$ to the whole of U, one uses the diagram

$$\begin{array}{ccccccc}
\cdots \longrightarrow & H^r(U,F^D) & \longrightarrow & H^r(U[1/n],F^D) & \longrightarrow & \bigoplus_{v\in U-U[1/n]} H^{r+1}_v(\mathcal{O}^h_v,F^D) & \longrightarrow \cdots \\
& \downarrow & & \downarrow \approx & & \downarrow \approx & \\
\cdots \longrightarrow & H^{3-r}_c(U,F)^* & \longrightarrow & H^{3-r}_c(U[1/n],F)^* & \longrightarrow & \bigoplus_{v\in U-U[1/n]} H^{2-r}(\mathcal{O}^h_v,F)^* & \longrightarrow \cdots,
\end{array}$$

and (1.3).

Corollary 3.2. *Let* N *be a finite flat group scheme over* U, *and let* N^D *be its Cartier dual. Then*

$$H^r(U,N^D) \times H_c^{3-r}(U,N) \longrightarrow H_c^3(U,\mathbb{G}_m) \approx \mathbb{Q}/\mathbb{Z}$$

is a nondegenerate pairing of finite groups for all r.

Proof: When the sheaf F in (3.1) is taken to be that defined by N, then F^D is the sheaf defined by N^D.

Corollary 3.3. *Let* N *be a quasi-finite flat separated group scheme over* U, *and let* $nN = 0$. *Assume that there exists an open subscheme* V *of* U *such that*

(i) V *contains all points* v *of* U *whose residue characteristic divides* n;

(ii) $N|V$ *is finite;*

(iii) *if* j *denotes the inclusion of* $V[1/n]$ *into* $U[1/n]$, *then the canonical map* $N|U[1/n]_{et} \longrightarrow j_*j^*(N|V[1/n]_{et})$ *is an isomorphism.*

Let $N^D = \mathcal{H}om_{U_{fl}}(N,\mathbb{G}_m)$. *Then the canonical pairing*

$$H^r(U,N^D) \times H_c^{3-r}(U,N) \longrightarrow H_c^3(U,\mathbb{G}_m) \approx \mathbb{Q}/\mathbb{Z}$$

is a nondegenerate pairing of finite groups.

Proof: Because $N|V$ is finite, $N^D|V$ is the Cartier dual of $N|V$. Therefore the theorem shows that $H^r(V,N^D)$ is finite and dual to $H_c^{3-r}(V,N)$. The corollary therefore follows from

$$\begin{array}{ccccccc}
\cdots \longrightarrow & H^r(U,F^D) & \longrightarrow & H^r(V,F^D) & \longrightarrow & \bigoplus_{v\in U-V} H_v^{r+1}(\mathcal{O}_v^h,F^D) & \longrightarrow \cdots \\
& \downarrow & & \downarrow & & \downarrow\approx & \\
\cdots \longrightarrow & H_c^{3-r}(U,F)^* & \longrightarrow & H_c^{3-r}(V,F)^* & \longrightarrow & \bigoplus_{v\in U-V} H^{2-r}(\mathcal{O}_v^h,F)^* & \longrightarrow \cdots,
\end{array}$$

and (II.1.10b).

Let A be an abelian variety over K, and let $\mathcal{A}$ and $\mathcal{B}$ be the Néron minimal models over U of A and its dual B. Let n be an integer such that $\mathcal{A}$ has semistable reduction at all v dividing n. There are exact sequences

$$0 \longrightarrow \mathcal{B}_n \longrightarrow \mathcal{B} \longrightarrow \mathcal{B}^{n\Phi'} \longrightarrow 0$$

$$0 \longrightarrow \mathcal{A}_n \longrightarrow \mathcal{A}^{\Phi n} \longrightarrow \mathcal{A}^{o} \longrightarrow 0.$$

The Poincaré biextension of (B,A) by $\mathbb{G}_m$ extends uniquely to biextensions of $(\mathcal{B},\mathcal{A}^o)$ by $\mathbb{G}_m$ and of $(\mathcal{B}^{n\Phi'},\mathcal{A}^{\Phi n})$ by $\mathbb{G}_m$. Therefore (cf. §1), we get a canonical pairing

$$\mathcal{B}_n \times \mathcal{A}_n \longrightarrow \mathbb{G}_m.$$

Corollary 3.4. *Let* $\mathcal{B}_n$ *and* $\mathcal{A}_n$ *be as above. Then*

$$H^r(U,\mathcal{B}_n) \times H_c^{3-r}(U,\mathcal{A}_n) \longrightarrow H_c^3(U,\mathbb{G}_m) \approx \mathbb{Q}/\mathbb{Z}$$

is a nondegenerate pairing of finite groups for all r.

Proof: Over the open subset V where A has good reduction, $\mathcal{A}_n$ is a finite flat group scheme with Cartier dual $\mathcal{B}_n$, and so over V, the corollary is a special case of (3.2). To pass from V to U, use (2.10).

Euler-Poincaré characteristics

We extend (II.2.13) to the flat site. Let N be a quasi-finite flat separated group scheme over U. For each closed point v ∈ U, let n_v^f be the order of the maximal finite subgroup scheme N_v^f of $N \times_U \operatorname{Spec}(\mathcal{O}_v^h)$, and $\mathfrak{d}_v^f$ be the discriminant of N_v^f over $\mathcal{O}_v^h$. Also, we set

$$\chi(U,N) = \frac{[H^0(U,N)][H^2(U,N)]}{[H^1(U,N)][H^3(U,N)]}, \qquad \chi_c(U,N) = \frac{[H_c^0(U,N)][H_c^2(U,N)]}{[H_c^1(U,N)][H_c^3(U,N)]}.$$

Theorem 3.5. *Let* N *be quasi-finite, flat, and separated over* U. *Then*

$$\chi(U,N) = \prod_{v\in X-U} |[N(K_s)]|_v \times \prod_{v\in U} (\mathcal{O}_v^h : \mathfrak{d}_v^f)^{-1/n_v^f} \times \prod_{v \text{ arch}} \frac{[N(K_v)]}{[H^0(K_v,N)]}$$

and

$$\chi_c(U,N) = \prod_{v\in U} (R:\mathfrak{d}_v^f)^{-1/n^f} \times \prod_{v \text{ arch}} [N(K_v)].$$

Proof: Let V be an open subset of U such that N|V is finite and has order prime to the residue characteristics of V, so that, in particular, N|V is étale. The exact sequence

$$\cdots \longrightarrow \prod_{v\in U-V} H_v^r(U,N) \longrightarrow H^r(U,N) \longrightarrow H^r(V,N) \longrightarrow \cdots$$

shows that $\chi(U,N) = \chi(V,N) \times \prod_{v\in U-V} \chi_v(\mathcal{O}_v^h,N)$, and (II.2.13) and (1.17b) show respectively that

$$\chi(V,N) = \prod_{v \text{ arch}} [N(K_v)]/[H^0(K_v,N)]|[N(K_s)]|_v$$

and $\chi_v(\mathcal{O}_v^h,N) = |[N(K_s)]|_v^{-1}(R:\mathfrak{d}_v^f)^{-1/n_v^f}$. The formula in (a) follows immediately.

The exact sequence

$$\cdots \longrightarrow H_c^r(V,N) \longrightarrow H_c^r(U,N) \longrightarrow \bigoplus_{v\in U-V} H^r(\mathcal{O}_v^h,N) \longrightarrow \cdots$$

shows that $\chi_c(U,N) = \chi_c(V,N) \times \prod \chi_v(\mathcal{O}_v^h,N)$ and (II.2.13) and (1.17a) show respectively that $\chi_c(V,N) = \prod_{v \text{ arch}} [N(K_v)]$ and $\chi(\mathcal{O}_v^h,N) = (R:\mathfrak{d}_v^f)^{-1/n^f}$.

Néron models

Let A be an abelian variety over K, and let $\mathcal{A}$ be its Néron model. Then $\mathcal{A}/\mathcal{A}^o \overset{\text{df}}{=} \Phi = \oplus\, i_{v*}\Phi_v$ (finite sum) where $\Phi_v = i_v^*(\mathcal{A}_v/\mathcal{A}_v^o)$.

Proposition 3.6. *Let Γ be a subgroup of Φ, and let $\mathcal{A}^{\Gamma}$ be the corresponding subscheme of $\mathcal{A}$.*

(a) *The group $H^0(U,\mathcal{A}^{\Gamma})$ is finitely generated; for $r > 0$, $H^r(U,\mathcal{A}^{\Gamma})$ is torsion and of cofinite-type; the map $H^r(U,\mathcal{A}^{\Gamma}) \longrightarrow \bigoplus_{v \text{ arch}} H^r(K_v,A)$ is surjective for $r = 2$ and an isomorphism for $r > 2$.*

(b) *For $r < 0$, $\prod_{v \text{ arch}} H^r(K_v,\mathcal{A}^{\Gamma}) \longrightarrow H^r_c(U,\mathcal{A}^{\Gamma})$ is an isomorphism; $H^0_c(U,\mathcal{A}^{\Gamma})$ is finitely generated; $H^1_c(U,\mathcal{A}^{\Gamma})$ is an extension of a torsion group by a subgroup which has a natural compactification; $H^2_c(U,\mathcal{A}^{\Gamma})$ is torsion and of cofinite-type; for $r \geq 3$, $H^r_c(U,\mathcal{A}^{\Gamma}) = 0$.*

Proof: Fix an integer m, and let V be an open subscheme of U such that m is invertible on U and $\mathcal{A}$ is an abelian scheme over V. Then all statements are proved in (II.5.1) for $\mathcal{A}|V$ and m. The general case follows by writing down the usual exact sequences.

Let B be the dual abelian variety to A, and let $\mathcal{B}$ be its Néron model. Let $\mathcal{B}/\mathcal{B}^o \overset{df}{=} \Phi' = \oplus\, i_{v*}\Phi'_v$. For any subgroups $\Gamma = \oplus\, i_{v*}\Gamma_v$ and $\Gamma' = \oplus\, i_{v*}\Gamma'_v$ of Φ and Φ', the Poincaré biextension over K extends to a biextension over U if and only if each Γ_v annihilates each Γ'_v in the canonical pairing. In this case we get a map

$$\mathcal{B}^{\Gamma'} \otimes^L \mathcal{A}^{\Gamma} \longrightarrow \mathbb{G}_m[1].$$

Theorem 3.7. *Suppose that Γ_v and Γ'_v are exact annihilators at each closed point v.*

(a) *The group $H^0(U,\mathcal{B}^{\Gamma'})_{tors}$ is finite; the pairing*

$$H^0(U,\mathcal{B}^{\Gamma'}) \times H^2_c(U,\mathcal{A}^{\Gamma}) \longrightarrow \mathbb{Q}/\mathbb{Z}$$

is nondegenerate on the left and its right kernel is the divisible subgroup of $H^2_c(U,\mathcal{A}^{\Gamma})$.

(b) *The groups* $H^1(U,\mathcal{B}^{\Gamma'})$ and $H^1_c(U,\mathcal{A}^{\Gamma})_{tors}$ *are of cofinite-type, and the pairing*

$$H^1(U,\mathcal{B}^{\Gamma'}) \times H^1_c(U,\mathcal{A}^{\Gamma})_{tors} \longrightarrow \mathbb{Q}/\mathbb{Z}$$

annihilates exactly the divisible groups.

(c) *If the divisible subgroup of* $Ш^1(K,A)$ *is zero, then the compact group* $H^0(U,\mathcal{B}^{\Gamma'})^\wedge$ *(completion for the topology of subgroups of finite index) is dual to the discrete torsion group* $H^2_c(U,\mathcal{A}^{\Gamma})$.

Proof: Fix an integer m, and choose an open subscheme V of U on which m is invertible and A and B have good reduction. Theorem II.5.2 proves the result over V for the m-components of the groups. To pass from there to the m-components of the groups over U, use (2.7). As m is arbitrary, this completes the proof.

Curves over U

For a proper map $\pi\colon Y \longrightarrow U$ and sheaf F on Y_{fl}, we define $H^r_c(Y,F)$ to be $H^r_c(U,R\pi_*F)$.

Theorem 3.8. *Let* $\pi\colon Y \longrightarrow U$ *be a proper flat map whose fibres are pure of dimension one and whose generic fibre is a smooth geometrically connected curve. Assume that for all* $v \in U$, $Y\times_U \mathrm{Spec}\,\mathcal{O}^h_v \longrightarrow \mathrm{Spec}\,\mathcal{O}^h_v$ *satisfies the local duality theorem for* n *(see §2). Then there is a canonical nondegenerate pairing of finite groups*

$$H^r(Y,\mu_n) \times H^{5-r}_c(Y,\mu_n) \longrightarrow H^3_c(U,\mathbb{G}_m) \approx \mathbb{Q}/\mathbb{Z}.$$

Proof: Choose an open subscheme V of U such that n is invertible on V and $\pi|\pi^{-1}(V)$ is smooth. For $\pi|\pi^{-1}(V)$ the statement becomes that proved in Theorem II.7.7. Let $Z = Y - Y_V$. To pass from V to U, use the exact sequences

$$\cdots \longrightarrow H^r_Z(Y,\mu_n) \longrightarrow H^r(Y,\mu_n) \longrightarrow H^r(Y_V,\mu_n) \longrightarrow \cdots$$

and

$$\cdots \longrightarrow H^r_c(Y_V,\mu_n) \longrightarrow H^r_c(Y,\mu_n) \longrightarrow \bigoplus_{v\in U-V} H^r(Y\times_U \mathrm{Spec}(\mathcal{O}^h_v),\mu_n) \longrightarrow \cdots$$

and note that $H^r_Z(Y,\mu_n) = \bigoplus_{v\in U-V} H^r_v(Y\times_U \mathrm{Spec}(\mathcal{O}^h_v),\mu_n)$.

Notes: Theorem 3.1 was proved by the author in 1978. Earlier Artin and Mazur had announced the proof of a flat duality theorem over X (neither the statement of the theorem or its proof have been published, but two corollaries are stated in [Mazur (1972), 7.2, 7.3]; I believe that the original theorem is the special case of (3.3) in which U = X and n is odd).

§4 Local results: mixed characteristic, perfect residue field

In this section we summarize the results of [Bégeuri (1980)]. Throughout, X will be the spectrum of a complete discrete valuation ring R whose field of fractions K is of characteristic zero, and whose residue field k is perfect of characteristic $p \neq 0$. (Essentially the same results should hold if R is only Henselian.) We let m be the maximal ideal of R, and we let $X_i = \mathrm{Spec}\, R/m^{i+1}$.

Some cohomological properties of K

Proposition 4.1. *If* k *is algebraically closed, then for any torus* T *over* K, $H^r(K,T) = 0$ *all* $r \geq 0$.

Proof: Let L be a finite Galois extension of K with Galois group G. Then $H^1(G,L^\times) = 0$ by Hilbert's theorem 90, and $H^2(G,L^\times) = 0$ because the Brauer group of K is trivial (K is quasi-algebraically closed

[Shatz (1972), p116]). These two facts show that $L^{\times}$ is a cohomologically trivial G-module [Serre (1962), IX.5, Thm 8]. Choose L to split T. Then $\mathrm{Hom}(X^*(T),L^{\times}) = T(L)$, and (ibid. Thm 9) shows that $\mathrm{Hom}(X^*(T),L^{\times})$ is also cohomologically trivial because $\mathrm{Ext}^1(X^*(T),L^{\times}) = 0$.

Corollary 4.2. *Assume that k is algebraically closed, and let N be a finite group scheme over K.*

(a) *For all* $r \geq 2$, $H^r(K,N) = 0$.

(b) *Let K' be a finite Galois extension of K, and let* $G = \mathrm{Gal}(K'/K)$. *Then* $H_T^r(G,H^1(K',N))$ *is finite for all* $r \in \mathbb{Z}$, *and is isomorphic to* $H_T^{r+2}(G,N(K'))$. *The canonical homomorphism* $H_0(G,H^1(K',N)) \longrightarrow H^1(K,N)$ *deduced from the corestriction map is an isomorphism.*

Proof: (a) Resolve N by tori,

$$0 \longrightarrow N \longrightarrow T_0 \longrightarrow T_1 \longrightarrow 0,$$

and apply the proposition.

(b) From the above resolution, we get an exact sequence

$$0 \longrightarrow N(K') \longrightarrow T_0(K') \longrightarrow T_1(K') \longrightarrow H^1(K',N) \longrightarrow 0.$$

Since the middle two G-modules are cohomologically trivial, the iterated coboundary map is an isomorphism $H_T^r(G,H^1(K',N)) \longrightarrow H_T^{r+2}(G,N(K'))$. The last statement is proved similarly (see [Bégeuri (1980), p34]).

The algebraic structure on $H^r(X,N)$

For any k-algebra Λ, let $W_i(\Lambda)$ be the ring of Witt vectors over Λ of length i, and let $W(\Lambda)$ be the full Witt ring. For any scheme Y over W(k) and any i, the *Greenberg realization of level* i, $\mathrm{Green}_i(Y)$, of Y is the scheme over over k such that

$$\mathrm{Green}_i(Y)(\Lambda) = Y(W_i(\Lambda))$$

for all k-algebras Λ (see [Greenberg (1961)]). Note that R has a canonical structure as a W(k)-algebra, and so for any scheme Y over X, we can define $\mathcal{G}_i(Y)$ to be the Greenberg realization of level i of the restriction of scalars of Y, $\mathrm{Res}_{X/\mathrm{Spec}(W(k))}Y$. Then $\mathcal{G}_i(Y)$ is characterized by the following condition: for any k-algebra Λ,

$$\mathcal{G}_i(Y)(\Lambda) = Y(R\otimes_{W(k)}W_i(\Lambda)).$$

In particular, $\mathcal{G}_i(Y)(k) = Y(R/p^iR) = Y(X_{i-1})$. Note that $\mathcal{G}_1(Y) = Y\otimes_R k = Y_k$. For varying i, the $\mathcal{G}_i(Y)$ form a projective system $\mathcal{G}(Y) = (\mathcal{G}_i(Y))$. The perfect group scheme associated with $\mathcal{G}_i(Y)$ will be denoted by $\mathbf{G}_i(Y)$. Thus

$$\mathbf{G}_i(Y)(\Lambda) = Y(R\otimes_{W(k)}W_i(\Lambda))$$

for any perfect k-algebra Λ and $\mathbf{G}_i(Y)(k) = Y(R/p^iR)$. We let $\mathbf{G}(Y)$ be the perfect pro-group scheme $(\mathbf{G}_i(Y))$.

When G is a smooth group scheme over X, we let $\mathbf{V}(\omega_G)$ be the vector group associated with the R-module ω_G of invariant differentials on G.

Proposition 4.3. *Let G be a smooth group scheme over R. For all $i \geq 1$, $\mathcal{G}_i(G)$ is a smooth group scheme over k, and for all $i' \geq i$, there is an exact sequence of k-groups*

$$0 \longrightarrow \mathcal{G}_i(\mathbf{V}(\omega_G)) \longrightarrow \mathcal{G}_{i+i'}(G) \longrightarrow \mathcal{G}_{i'}(G) \longrightarrow 0.$$

In particular, $\mathcal{G}_{i+1}(G) \longrightarrow \mathcal{G}_i(G)$ is surjective with kernel $\omega_G\otimes_R k$, and $\mathcal{G}_i(G)$ is an extension of G_k by a smooth connected unipotent group. The group scheme $\mathcal{G}_i(G)$ is connected if and only if its special fibre is connected. The dimension of $\mathcal{G}_i(G)$ is $ei.\dim(G_k)$ where e is the *absolute ramification index of* R.

Proof: We may assume that k is algebraically closed and apply [Bégueri (1980), 4.1.1].

Lemma 4.4. *Let*

$$0 \longrightarrow N \longrightarrow G_0 \xrightarrow{\varphi} G_1 \longrightarrow 0$$

be an exact sequence of R*-groups with* G_0 *and* G_1 *smooth and connected and* N *finite. For all* $i \geq 1$, *the* k*-group* $\mathrm{Coker}(\mathcal{G}_i(\varphi))$ *is smooth, and when* k *is algebraically closed its group of* k*-points is* $H^1(X_{i-1},N)$.

Proof: The first statement follows from the fact that $\mathcal{G}_i(\varphi)$ is a homomorphism of smooth group schemes over k. For the second, note that $H^r(X_{i-1},G) = H^r(k,G_k) = 0$ for $r > 0$, and so we have a diagram

$$\begin{array}{ccccccc} \mathcal{G}_i(G_0)(k) & \longrightarrow & \mathcal{G}_i(G_1)(k) & \longrightarrow & \mathrm{Coker}(\mathcal{G}_i(\varphi)(k)) & \longrightarrow & 0 \\ \| & & \| & & \downarrow\approx & & \\ G_0(X_{i-1}) & \longrightarrow & G_1(X_{i-1}) & \longrightarrow & H^1(X_{i-1},N) & \longrightarrow & 0. \end{array}$$

Define $\tilde{H}^1(X_i,N)$ to be the sheaf on $\mathrm{Spec}(k)_{qf}$ associated with the presheaf $\Lambda \mapsto H^1(X_i \otimes_{W(k)} W(\Lambda),N)$. Then the lemma realizes $\tilde{H}^1(X_i,N)$ as an algebraic group, and the next lemma shows that this realization is essentially independent of the choice of the resolution.

Lemma 4.5. *Let*

$$0 \longrightarrow N \longrightarrow G_0' \xrightarrow{\varphi'} G_1' \longrightarrow 0$$

be a second resolution of N *by smooth algebraic groups. Then there is a canonical isomorphism* $\mathrm{Coker}(\mathcal{G}_i(\varphi)) \xrightarrow{\approx} \mathrm{Coker}(\mathcal{G}_i(\varphi'))$.

Proof: It is easy to construct a diagram

$$\begin{array}{ccccccccc}
 & & 0 & \longrightarrow & G_0' & \longrightarrow & G_0' & \longrightarrow & 0 \\
 & & \downarrow & & \downarrow & & \downarrow & & \\
0 & \longrightarrow & N & \longrightarrow & G_0 \oplus G_0' & \xrightarrow{\varphi''} & G & \longrightarrow & 0 \\
 & & \| & & \downarrow & & \downarrow & & \\
0 & \longrightarrow & N & \longrightarrow & G_0 & \xrightarrow{\varphi} & G_1 & \longrightarrow & 0
\end{array}$$

with G a smooth algebraic group. When we apply $\mathcal{G}_i$, the resulting diagram gives an isomorphism

$$\operatorname{Coker}(\mathcal{G}_i(\varphi'')) \xrightarrow{\approx} \operatorname{Coker}(\mathcal{G}_i(\varphi)),$$

and a similar construction gives an isomorphism

$$\operatorname{Coker}(\mathcal{G}_i(\varphi'')) \xrightarrow{\approx} \operatorname{Coker}(\mathcal{G}_i(\varphi')).$$

We now regard $\tilde{H}^1(X_i,N)$ as an algebraic group, and we write $\tilde{H}^1(X,N)$ for the pro-algebraic group $(\tilde{H}^1(X_i,N))_{i\geq 0}$.

For the definition of the *absolute different* $\mathfrak{D}$ of a finite group N scheme over R, we refer the reader to [Raynaud (1974), Appendice]. It is an ideal in R.

Theorem 4.6. *Let* N *be a finite flat group scheme of order a power of* p *over* X. *For all* $i \geq 0$, *the smooth algebraic* k*-group* $\tilde{H}^1(X_i,N)$ *is affine, connected, and unipotent. There exists an integer* i_0 *such that* $\tilde{H}^1(X,N) \longrightarrow \tilde{H}^1(X_i,N)$ *is an isomorphism for all* $i \geq i_0$. *The group scheme* $\tilde{H}^1(X,N)$ *has dimension* $\operatorname{ord}(\mathfrak{D})$ *where* $\mathfrak{D}$ *is the different of* N.

Proof: We may assume that the residue field is algebraically closed and apply [Bégeuri (1980), 4.2.2].

Proposition 4.7. *A short exact sequence*

$$0 \longrightarrow N' \longrightarrow N \longrightarrow N'' \longrightarrow 0$$

of finite flat p-*primary group schemes gives rise to an exact sequence of algebraic groups*

$$0 \longrightarrow \mathcal{G}(N') \longrightarrow \mathcal{G}(N) \longrightarrow \mathcal{G}(N'') \longrightarrow \tilde{H}^1(X,N') \longrightarrow \tilde{H}^1(X,N) \longrightarrow \tilde{H}^1(X,N'') \longrightarrow 0.$$

Proof: We may assume that the residue field is algebraically closed and apply [Bégeuri (1980), 4.2.3].

We write $H^1(X_i,N)$ and $H^1(X,N)$ for the perfect algebraic groups associated with $\tilde{H}^1(X_i,N)$ and $\tilde{H}^1(X,N)$. Suppose that k is algebraically closed. If

$$0 \longrightarrow N \longrightarrow G_0 \longrightarrow G_1 \longrightarrow 0$$

is a smooth resolution of N and i is so large that $N(R) \cap p^i G_0(R) = 0$, then the kernel and cokernel of the map

$$G(G_0)(R)^{(p^i)} \longrightarrow G(G_1)(R)^{(p^i)}$$

are N(R) and $H^1(X,N)(k)$ respectively (ibid. p44-45).

The algebraic structure on $H^1(K,N)$

Let T be a torus over K. According to [Raynaud (1966)], T admits a Néron model over X: this is a smooth group scheme $\mathcal{T}$ over X (not necessarily of finite type) such that $\mathcal{T}(Y) = T(Y_K)$ for all smooth X-schemes Y. Write G(T) for $G(\mathcal{T})$. It is a perfect pro-algebraic group over k whose set of connected components $\pi_0(G(T))$ is a finitely generated abelian group, equal to the set of connected components of the special fibre of $\mathcal{T}$.

Lemma 4.8. *Let* N *be a finite group scheme over* K, *and let*

$$0 \longrightarrow N \longrightarrow T_0 \xrightarrow{\varphi} T_1 \longrightarrow 0$$

be a resolution of N *by tori. The cokernel of* $G(\varphi): G(T_0) \longrightarrow G(T_1)$

is a pro-algebraic perfect group scheme, and when k *is algebraically closed it has* $H^1(X,N)$ *as its group of* k*-points.*

Proof: [Bégeuri (1980), 4.3.1].

The lemma allows us to define on $H^1(K,N)$ the structure of a perfect pro-algebraic group scheme. We write $H^1(K,N)$ for this group scheme. An argument as in the proof of (4.5) shows that $H^1(K,N)$ is independent of the resolution. The identity component of $H^1(K,N)$ is unipotent.

Proposition 4.9. *Let* N *be a finite flat* p*-primary group scheme over* X. *Then the standard resolution defines a closed immersion*

$$H^1(X,N) \longrightarrow H^1(K,N).$$

Proof: We may assume that the residue field is algebraically closed and apply [Bégeuri (1980), 4.4.4].

Theorem 4.10. *For any finite* K*-group* N, *the perfect group scheme* $H^1(K,N)$ *is affine and algebraic. Its dimension is* ord([N]), *where* [N] *is the order of* N.

Proof: The basic strategy of the proof is the same as that of the proof of (I.2.8); see [Bégeuri (1980), 4.3.3].

The reciprocity isomorphism

Assume first that k is algebraically closed. For any finite extension K'/K, let $U_{K'} = \mathcal{G}(\mathbb{G}_{m,R'})$ where R' is the ring of integers in K'. Then the norm map $N_{R'/R}$: $\mathrm{Res}_{R'/R}\mathbb{G}_{m,R'} \longrightarrow \mathbb{G}_{m,R}$ induces a surjective map $U_{K'} \longrightarrow U_K$ of affine k-groups. Let $V_{K'}$ be the kernel of this map, and let $V^o_{K'}$ be the identity component of $V_{K'}$. Then we have an exact sequence

$$0 \longrightarrow \pi_0(V_{K'}) \longrightarrow U_{K'}/V^o_{K'} \longrightarrow U_K \longrightarrow 0.$$

Assume that K' is Galois over K, and let t' be a uniformizing parameter in K'. The homomorphism

$$\mathrm{Gal}(K'/K)^{ab} = H^{-2}(\mathrm{Gal}(K'/K),\mathbb{Z}) \longrightarrow (U_{K'}/V^o_{K'})(k),$$

sending $\sigma \in \mathrm{Gal}(K'/K)$ to the class of $\sigma(t')/t'$ in $U_{K'}$ allows us to identify the preceding exact sequence with an exact sequence

$$0 \longrightarrow H^{-2}(\mathrm{Gal}(K'/K),\mathbb{Z}) \longrightarrow U_{K'}/V^o_{K'} \xrightarrow{N} U_K \longrightarrow 0.$$

On passing to the inverse limit over the fields K', we get an exact sequence

$$0 \longrightarrow \mathrm{Gal}(K_{ab}/K) \longrightarrow \varprojlim U_{K'}/V^o_{K'} \longrightarrow U_K \longrightarrow 0.$$

As U_K is connected, this sequence defines a continuous homomorphism

$$\mathrm{rec}_K\colon \pi_1(U_K) \longrightarrow \mathrm{Gal}(K_{ab}/K)$$

and the main result of [Serre (1961)] is that this map is an isomorphism.

It is also possible to show that $\pi_1(U) \overset{\mathrm{df}}{=} \varinjlim \pi_1(U_{K'})$ is a class formation, and so define rec_K as in (I.1).

Recall [Serre (1960), 5.4] that for any perfect algebraic group G and finite perfect group N, there is an exact sequence

$$0 \longrightarrow \mathrm{Ext}^1_k(\pi_0(G),N) \longrightarrow \mathrm{Ext}^1_k(G,N) \longrightarrow \mathrm{Hom}_k(\pi_1(G),N) \longrightarrow 0.$$

In particular, when G is connected $\mathrm{Ext}^1_k(G,N) \xrightarrow{\approx} \mathrm{Hom}(\pi_1(G),N)$. (We are still assuming that k is algebraically closed.) Therefore, rec_K gives rise to an isomorphism

$$\mathrm{Hom}(\mathrm{Gal}(K_{ab}/K),\mathbb{Z}/p^n\mathbb{Z}) = H^1(K,\mathbb{Z}/p^n\mathbb{Z})$$
$$\xrightarrow{\approx} \mathrm{Hom}(\pi_1(U_K),\mathbb{Z}/p^n\mathbb{Z}) = \mathrm{Ext}^1_K(U_K,\mathbb{Z}/p^n\mathbb{Z}).$$

If we assume that K contains the p^nth roots of 1, and we replace $\mathbb{Z}/p^n\mathbb{Z}$ with $\mu_{p^n}(K)$, then the isomorphism becomes

$$\Phi_n\colon H^1(K,\mu_{p^n}) \xrightarrow{\approx} \mathrm{Ext}^1_K(U_K,\mu_{p^n}).$$

Both groups have canonical structures of perfect algebraic groups.

Proposition 4.11. *The map Φ_n is a morphism of perfect algebraic groups.*

Proof: [Bégeuri (1980), 5.3.2].

When we drop the assumption that k is algebraically closed, we obtain an isomorphism

$$\mathrm{rec}_K\colon \eta(U_K) \longrightarrow \mathrm{Gal}(K_{ab}/K)$$

where $\eta(U_K)$ is the maximal constant quotient of $\pi_1(U_K)$. See [Hazewinkel (1969)].

Duality for finite group schemes over K

Theorem 4.12. *Let N be a finite group scheme over K; then there is a canonical isomorphism of connected perfect unipotent groups* $H^1(K,N)^o \longrightarrow (H^1(K,N^D)^o)^\vee$.

Proof: We can assume that k is algebraically closed and apply [Bégeuri (1980), 6.1.6].

This result can be improved by making use of derived categories (ibid. 6.2). Assume that k is algebraically closed, and let

$\mathbf{M}_n$ = category of finite group schemes over K killed by p^n,

$\mathbf{Q}_n$ = category of perfect pro-algebraic groups over k killed by p^n,

$\mathbf{S}_n$ = the category of sheaves on $(\mathrm{Spec}\ k)_{pf}$ killed by p^n

Let $C\colon D^b(\mathbf{M}_n) \longrightarrow D^b(\mathbf{M}_n)$, $S\colon D^b(\mathbb{Q}_n) \longrightarrow D^b(\mathbb{Q}_n)$, and $B\colon D^b(\mathbf{S}_n) \longrightarrow D^b(\mathbf{S}_n)$ be the functors defined respectively by Cartier duality, Serre duality, and Breen-Serre duality (see §0; here $D^b(*)$ denotes the derived category obtained from the category $K^b(*)$ of bounded complexes and homotopy classes of maps). Then $H^1\colon \mathbf{M}_n \longrightarrow \mathbb{Q}_n$ admits a left derived functor, and we have a commutative diagram (up to an isomorphism of functors):

$$\begin{array}{ccccc} D^b(\mathbf{M}_n) & \xrightarrow{LH^1} & D^b(M_n) & \xrightarrow{\text{can}} & D^b(M_n) \\ \downarrow C & & \downarrow S & & \downarrow B \\ D^b(\mathbf{M}_n) & \xrightarrow{LH^1} & D^b(M_n) & \xrightarrow{\text{can}} & D^b(M_n); \end{array} \tag{4.12.1}$$

moreover, $(\text{can}\circ LH^1)(N) \xrightarrow{\approx} RH^0(N)[1]$. See [Bégeuri (1980), 6.2.4].

Duality for finite group schemes over R

Theorem 4.13. *For any finite flat* p*-primary group scheme* N *over* X, *there is a canonical isomorphism of* k*-groups*

$$H^1(X,N) \xrightarrow{\approx} (H^1(K,N^D)^{\circ}/H^1(X,N^D))^t.$$

Proof: Ibid. 6.3.2..

Duality for tori

Let T be a torus over K, and let $\mathcal{T}$ be its Néron model over X.

Theorem 4.14. *The pairing* $H^0(K,X^*(T)) \times T(K) \longrightarrow \mathbb{Z}$ *defines isomorphisms*

$$H^0(K,X^*(T)) \xrightarrow{\approx} \text{Hom}(\pi_0(\mathcal{T}_k),\mathbb{Z})$$

$$H^1(K,X^*(T)) \xrightarrow{\approx} \text{Ext}^1_k(\pi_0(\mathcal{T}_k),\mathbb{Z}) \text{ (finite groups).}$$

$$H^2(K,X^*(T)) \xrightarrow{\approx} \mathrm{Hom}_{cts}(\pi_1(T(K)),\mathbb{Q}/\mathbb{Z}).$$

Proof: Ibid. 7.2..

Duality for abelian varieties

Let A be an abelian variety over K, and let $\mathcal{A}$ be its Néron model over X. We write G(A) for $G(\mathcal{A})$ and $\pi_i(A)$ for $\pi_i(G(A))$.

Theorem 4.15. *Let* A *be an abelian variety over* K.

(a) *The pairing* $\pi_0(\mathcal{A}_k) \times \pi_0(\mathcal{A}_k^t) \longrightarrow \mathbb{Q}/\mathbb{Z}$ *defined in* (C.11) *is non-degenerate.*

(b) *There is a canonical isomorphism*

$$H^1(K,A^t) \xrightarrow{\approx} \mathrm{Ext}_k^1(G(A),\mathbb{Q}/\mathbb{Z}).$$

Proof: (a) We can assume that k is algebraically closed, and in this case the result is proved in [Bégeuri (1980), 8.3.3].

(b) From (ibid. 8.3.6) we know that the result holds if k is algebraically closed; to deduce the result in the general case, apply the Hochschild-Serre spectral sequence to the left hand side and the spectral sequence (I.0.17) to the right hand side.

Corollary 4.16. Assume *that* $\mathcal{A}_k$ *is connected. Then there is a non-degenerate pairing of* $\mathrm{Gal}(K_{un}/K)$*-modules*

$$H^1(K_{un},A^t) \times \pi_1(A) \longrightarrow \mathbb{Q}/\mathbb{Z}.$$

Proof: Again we can assume that k is algebraically closed. As we noted above, for any connected perfect group scheme G over k and finite perfect group scheme N, $\mathrm{Ext}_k^1(G,N) = \mathrm{Hom}_k(\pi_1(G),N)$. This shows that $\mathrm{Ext}_k^1(G(A),\mathbb{Q}/\mathbb{Z}) = \mathrm{Hom}_k(\pi_1(G(A),\mathbb{Q}/\mathbb{Z})$, and so the result follows from the theorem.

Notes: This results in this section are due to [Bégeuri (1980)]. Partial results in the same direction were obtained earlier by Vvedens'kii (see [Vvedens'kii (1973), (1976)] and earlier papers).

§5 Two exact sequences

We write down two canonical short resolutions that are of great value in the proof of duality theorems in characteristic p. Throughout, X will be a scheme of characteristic $p \neq 0$.

The first exact sequence

The first sequence generalizes the sequences

$$0 \longrightarrow \mathbb{Z}/p\mathbb{Z} \longrightarrow \mathbb{G}_a \xrightarrow{1-F} \mathbb{G}_a \longrightarrow 0$$

$$0 \longrightarrow \alpha_p \longrightarrow \mathbb{G}_a \xrightarrow{F} \mathbb{G}_a \longrightarrow 0$$

to any group scheme N^D that is the Cartier dual of a finite group scheme of height one. Note that in each sequence, $\mathbb{G}_a$ is the cotangent space to N.

Let N be a finite flat group scheme over X of height 1, and let $e: X \longrightarrow N$ be the zero section. Let $\mathcal{I} \subset \mathcal{O}_N$ be the ideal defining the closed immersion e (so that $(\mathcal{O}_N/\mathcal{I})|e(X) = \mathcal{O}_X$), and let $\mathcal{I}n\ell^1_X(N) \overset{\mathrm{df}}{=} \mathcal{S}pec(\mathcal{O}_N/\mathcal{I}^2)$ be the first order infinitesimal neighbourhood of the zero section. Then $\mathcal{I}/\mathcal{I}^2$ is the cotangent space ω_N of N over X. Locally on X, there is an isomorphism of pointed schemes

$$N \approx \mathcal{S}pec(\mathcal{O}_X[T_1,\ldots,T_m]/(T_1^p,\ldots,T_m^p)),$$

and therefore

$$\mathcal{I}/\mathcal{I}^2 \approx (T_1,\ldots,T_m)/(T_1^2,\ldots,T_m^2)$$

(see [Messing (1972), II.2.1.2]). In particular, ω_N is a locally free $\mathcal{O}_X$-module of finite rank, and hence it defines a vector group $V(\omega_N)$ over X. We shall almost always write ω_N for $V(\omega_N)$. This vector group represents $\mathcal{M}or_{X\text{-ptd}}(\mathcal{I}n\ell_X^1(N), \mathbb{G}_m)$ viewed as a functor of schemes over X. (The notation X-ptd means that the morphisms are required to respect the canonical X-valued points of the two schemes.) The Cartier dual N^D of N represents $\mathcal{H}om_X(N, \mathbb{G}_m)$, and so the inclusion $\mathcal{I}n\ell_X^1(N) \hookrightarrow N$ defines a canonical homomorphism $\iota(N)\colon N^D \longrightarrow \omega_N$.

Recall from §0 that the Verschiebung is a map $V\colon N^{(p)} \longrightarrow N$. It induces a map $\omega_N \longrightarrow \omega_{N^{(p)}}$, and on combining this with the canonical isomorphism $\omega_{N^{(p)}} \approx \omega_N^{(p)}$, we obtain a homomorphism $\varphi_0\colon \omega_N \longrightarrow \omega_N^{(p)}$. The relative Frobenius morphism for the vector group ω_A over X is also a homomorphism $\varphi_1\colon \omega_N \longrightarrow \omega_N^{(p)}$, and we define $\varphi = \varphi_0 - \varphi_1$.

Theorem 5.1. *For any finite flat group scheme* N *of height one over* X, *the sequence*

$$0 \longrightarrow N^D \xrightarrow{\iota} \omega_N \xrightarrow{\varphi} \omega_N^{(p)} \longrightarrow 0$$

is exact.

Proof: For $N = \mu_p$ and $N = \alpha_p$ the sequence becomes one of those listed above. In the case that X is an algebraically closed field, every N has a composition series whose quotients are isomorphic to μ_p or to α_p, and the theorem can be proved in this case by induction on the length of N (see [Artin and Milne (1976), p115]).

Lemma 5.2. *The sequence is a complex, that is,* $\varphi \circ \iota = 0$.

Proof: This can be proved by direct calculation (ibid., p 114).

The next lemma shows that, when X in Noetherian, the theorem follows from the case that X is an algebraically closed field.

Lemma 5.3. *Let*

$$0 \longrightarrow G' \xrightarrow{\iota} G \xrightarrow{\varphi} G'' \longrightarrow 0$$

be a complex of flat group schemes of finite type over a Noetherian scheme X. Assume that for all geometric points x of X, the sequence of fibres

$$0 \longrightarrow G'_x \longrightarrow G_x \longrightarrow G''_x \longrightarrow 0$$

is exact. Then the original sequence is exact.

Proof: The faithful flatness of the φ_x, combined with the local criterion for flatness [Grothendieck (1971), IV.5.9], implies that φ is faithfully flat. Thus Ker(φ) is flat and of finite type, and by assumption ι factors through it. Now the same argument shows that ι: N $\longrightarrow$ Ker(φ) is faithfully flat. Finally, the kernel of ι a is group scheme over X whose geometric fibres are all zero, and hence is itself zero.

We now complete the proof of (5.1). It suffices to check the exactness of the sequence locally on X, and so we can assume that X is quasi-compact. Then there will exist a Noetherian scheme X_0, a finite flat group scheme N_0 over X_0, and a map X $\longrightarrow$ X_0 such that N = $N_0 \times_{X_0} X$. We know that the sequence for N_0 is exact, but since the construction of the sequence commutes with base change, this proves that the sequence for N is exact.

Example 5.4. Suppose N has order p. Then it can be written N = $N_{0,a}^{\mathcal{L}}$ in the Oort-Tate classification (0.9) with $\mathcal{L}$ an invertible sheaf on X and $a \in \mathcal{L}^{\otimes 1-p}$. Its dual $N^D = N^{\mathcal{L}^\vee}_{a,0}$. The cotangent sheaf ω_N is equal to $\mathcal{L}^\vee$, and when we identify $\mathcal{L}^{\vee(p)}$ with $\mathcal{L}^{\vee\otimes p}$, the map φ in the sequence in the theorem becomes

$$z \mapsto z^{\otimes p} - a\otimes z\colon \mathcal{L}^\vee \longrightarrow \mathcal{L}^{\vee\otimes p}.$$

In this case the exactness of the sequence is obvious from the description Oort and Tate give of the points of N^D (see 0.9d).

Remark 5.5. The theorem shows that every finite flat group scheme N over X whose dual has height one gives rise to a locally free $\mathcal{O}_X$-module $\mathcal{V}$ of finite rank and to a linear map $\varphi_0\colon \mathcal{V} \longrightarrow \mathcal{V}^{(p)}$. To recover N from the pair $(\mathcal{V},\varphi_0)$, simply form the kernel of $\varphi_0 - \varphi_1$ where φ_1 is the relative Frobenius of $\mathcal{V}$ (regarded as a vector group). These remarks lead to a classification of finite group schemes of this type that is similar, but dual, to the classification of finite flat group schemes of height one by their p-Lie algebras (see [Demazure and Gabriel (1970), II, §7]).

The second exact sequence

We now let $\pi\colon X \longrightarrow S$ be a smooth map of schemes of characteristic p, and we assume that S is perfect. Write X' for X regarded as an S-scheme by means of $\pi' = F_{abs}\circ\pi$. Because S is perfect, we can identify (X',π') with $(X^{(1/p)},\pi^{(1/p)})$ (see §0), and when we do this, the relative Frobenius map $F_{X^{(1/p)}/S}\colon X^{(1/p)} \longrightarrow X$ becomes identified with the absolute Frobenius map $F = F_{abs}$. For example, if S = Spec R and X = Spec A for some R-algebra $i\colon R \longrightarrow A$, then X' = Spec A, with A regarded as an R-algebra by means of $a \mapsto i(a)^p$, and the Frobenius map $X \longleftarrow X'$ corresponds to $a \mapsto a^p\colon A \longrightarrow A$. We write $\Omega^1_{X/S,cl}$ for the sheaf of closed differential forms on X relative to S, that is,

$\Omega^1_{X/S,cl} = \mathrm{Ker}(d\colon \Omega^1_{X/S} \longrightarrow \Omega^1_{X/S})$. We regard $\Omega^1_{X/S}$ and $\Omega^1_{X/S,cl}$ as sheaves on X_{et}.

Again N is a finite flat group scheme of height one on X, and we let $\mathfrak{n}$ be the Lie algebra of N (equal to the tangent sheaf of N over X).

Theorem 5.6. *Let* $f\colon X_{fl} \longrightarrow X_{et}$ *be the morphism of sites defined by the identity map. Then* $R^r f_* N = 0$ *for* $r \neq 1$, *and there is an exact sequence*

$$0 \longrightarrow R^1 f_* N \longrightarrow \mathfrak{n} \otimes_{\mathcal{O}_X} \Omega^1_{X'/S,cl} \xrightarrow{\psi} \mathfrak{n} \otimes_{\mathcal{O}_X} \Omega^1_{X/S} \longrightarrow 0.$$

Proof: We first show that $R^r f_* N = 0$ for $r \neq 1$. According to (A.5), there is an exact sequence

$$0 \longrightarrow N \longrightarrow G_0 \longrightarrow G_1 \longrightarrow 0$$

with G_0 and G_1 smooth. As $R^r f_* G = 0$ for $r > 0$ if G is a smooth group scheme (see [Milne (1980), III.3.9]), it is clear that $R^r f_* N = 0$ for $r > 1$ (this part of the argument works for a finite flat group scheme over any scheme). The sheaf $f_* N$ is the sheaf defined by N on X_{et}, which is zero because N is infinitesimal and all connected schemes étale over X are integral.

As far as the sequence is concerned, we confine ourselves to defining the maps; for the proof of the exactness, see [Artin and Milne (1976), §2]. First we need a lemma.

Lemma 5.7. *The map* $H^1(X,N) \longrightarrow H^1(X',N)$ *is zero.*

Proof: Since N has height one, $F_{N/X}\colon N \longrightarrow N^{(p)}$ factors through $e^{(p)}(X)$ where $e\colon X \longrightarrow N$ is the zero section. Therefore, $F_{abs}\colon N \longrightarrow N$ factors through $e(X)$, which means that the image of $a \mapsto a^p\colon \mathcal{O}_N \longrightarrow \mathcal{O}_N$ is contained in $\mathcal{O}_X \subset \mathcal{O}_N$. By descent theory, this last statement holds for any principal homogeneous space P of N over X: there is a map $\epsilon\colon \mathcal{O}_P \longrightarrow \mathcal{O}_X$ whose composite with the inclusion $\mathcal{O}_X \hookrightarrow \mathcal{O}_P$ is the pth power map. The composite $\mathcal{O}_P \xrightarrow{\epsilon} \mathcal{O}_X \xrightarrow{id} \mathcal{O}_{X'}$ is an $\mathcal{O}_X$-morphism, and therefore defines an X-morphism $X' \longrightarrow P$. This shows that P becomes trivial over X'. Since all elements of $H^1(X,N)$ are represented by principal homogeneous spaces, this proves the lemma.

Let P be a principal homogeneous space for N over X. Then the lemma shows that there is a trivialization $\varphi'\colon X' \longrightarrow P$, and the fact that N is purely infinitesimal implies that φ' is unique. It gives rise to two maps $X'\times_X X' = X'' \longrightarrow P$, namely, $\varphi'\circ p_1$ and $\varphi'\circ p_2$. Their difference is an element α'' of $N(X'')$ that is zero if and only if φ' is arises from N(X). Now if $\mathcal{N}$ denotes the nilradical of $\mathcal{O}_{X''}$, then $\mathcal{O}_{X''}/\mathcal{N} \approx \mathcal{O}_{X'}$ and $\mathcal{N}/\mathcal{N}^2 \approx \Omega^1_{X'/S}$. Since α'' is trivial on X', the restriction of α'' to $\mathcal{S}pec(\mathcal{O}_{X''}/\mathcal{N}^2)$ defines a map $\omega_N \longrightarrow \Omega^1_{X'/S}$, whose image can be shown to lie in $\Omega^1_{X'/S,cl}$. This map can be identified with an element of $n\otimes\Omega^1_{X'/S}$.

The same construction works for any U étale over X; for such a U, we get a map $H^1(U_{fl},N) \longrightarrow n\otimes_{\mathcal{O}_{U'}}\Omega^1_{U'/S,cl}$. As R^1f_*N is the sheaf associated with $U \mapsto H^1(U_{fl},N)$, this defines a map

$$R^1f_*N \longrightarrow n\otimes_{\mathcal{O}_{U'}}\Omega^1_{U'/S,cl},$$

which we take to be the first map in the sequence.

Lemma 5.8. *There exists a unique map* $C\colon \Omega^1_{X/S,cl} \longrightarrow \Omega^1_{X/S}$ *with the properties:*

(i) $C(f^p\omega) = fC(\omega)$, $f \in \mathcal{O}_X$, $\omega \in \Omega^1_{X/S,cl}$;

(ii) $C(\omega) = 0$ *if and only if* ω *is exact;*

(iii) $C(f^{p-1}df) = df$.

Proof: Every closed differential 1-form is locally a sum of exact differentials and differentials of the form $f^{p-1}df$, and so (ii) and (iii) completely describe C. For the proof that C exists, see [Milne (1976), 1.1] and [Katz (1970), 7.2].

Note that (ii) says that C is p^{-1}-linear. According to our conventions, $\Omega^1_{X'/S} = \Omega^1_{X/S}$ as sheaves of abelian groups on X' = X, but $f \in \mathcal{O}_X$ acts as f^p on $\Omega^1_{X'/S}$. Therefore, when regarded as a map $\Omega^1_{X'/S,cl} \longrightarrow \Omega^1_{X/S}$, C is $\mathcal{O}_X$-linear. Define $\psi_0\colon n\otimes\Omega^1_{X'/S,cl} \longrightarrow n\otimes\Omega^1_{X/S}$ to be $1\otimes C$.

Recall [Demazure and Gabriel (1970), II, §7] that n has the structure of a p-Lie algebra, that is, there is a map $n \mapsto n^{(p)}\colon n \to n$ such that $(fx)^{(p)} = f^p x^{(p)}$. Also we have a canonical inclusion $\Omega^1_{X'/S,cl} \longrightarrow \Omega^1_{X/S}$ (because X' = X). Define $\psi_1\colon n\otimes\Omega^1_{X'/S,cl} \longrightarrow n\otimes\Omega^1_{X/S}$ to be $n\otimes\omega \mapsto n^{(p)}\otimes\omega$, and ψ to be $\psi_0 - \psi_1$; thus $\psi(n\otimes\omega) = n\otimes C\omega - n^{(p)}\otimes\omega$.

Example 5.9. Let $N = N^{\mathcal{L}}_{0,b}$, $b \in \Gamma(X,\mathcal{L}^{\otimes 1-p})$ (in the Oort-Tate classification (0.9)). Then we can describe ψ explicitly. It is the map

$$\mathcal{L}\otimes\Omega^1_{X'/S,cl} \longrightarrow \mathcal{L}\otimes\Omega^1_{X/S},\quad x\otimes\omega \mapsto x\otimes C\omega - (b\otimes x)\otimes\omega.$$

If $\mathcal{L} = \mathcal{O}_X$, then the $\mathcal{O}_X$-structure on $\Omega^1_{X'/S,cl}$ is irrelevant, and so we can identify it with $\Omega^1_{X/S,cl}$. The map ψ then becomes

$$(\omega \mapsto C\omega - b\omega)\colon \Omega^1_{X/S,cl} \longrightarrow \Omega^1_{X/S}, \ .$$

For example, if b = 1, then $N = \mu_p$ and $R^1f_*\mu_p = \mathcal{O}_X^\times/\mathcal{O}_X^{\times p}$; the sequence is

$$0 \longrightarrow \mathcal{O}_X^\times/\mathcal{O}_X^{\times p} \xrightarrow{\text{dlog}} \Omega^1_{X/S,cl} \xrightarrow{C-1} \Omega^1_{X/S} \longrightarrow 0, \qquad \text{dlog}(f) = \frac{df}{f}.$$

If b = 0, then $N = \alpha_p$ and $R^1f_*\alpha_p = \mathcal{O}_X/\mathcal{O}_X^p$; the sequence is

$$0 \longrightarrow \mathcal{O}_X/\mathcal{O}_X^p \xrightarrow{d} \Omega^1_{X/S,cl} \xrightarrow{C} \Omega^1_{X/S} \longrightarrow 0.$$

The canonical pairing of the complexes

We continue with the notations of the last subsection. Write $U^{\cdot}(N)$ for the complex $n\otimes\Omega^1_{X'/S,cl} \xrightarrow{\psi} n\otimes\Omega^1_{X/S}$, and write $V^{\cdot}(N^D)$ for the complex $\omega_N \xrightarrow{\varphi} \omega_N^{(p)}$ (both supported in degrees zero and one).

Proposition 5.10. *There is a canonical pairing of complexes*

$$V^{\cdot}(N^D) \times U^{\cdot}(N) \longrightarrow U^{\cdot}(\mu_p).$$

Proof: In order to define a pairing of complexes, we have to define pairings

$$(\ ,\ )_{0,0}\colon V^0(N^D) \times U^0(N) \longrightarrow U^0(\mu_p)$$

$$(\ ,\ )_{1,0}\colon V^1(N^D) \times U^0(N) \longrightarrow U^1(\mu_p)$$

$$(\ ,\ )_{0,1}\colon V^0(N^D) \times U^1(N) \longrightarrow U^1(\mu_p)$$

such that

$$\psi(v,u)_{0,0} = (\varphi v,u)_{1,0} + (v,\psi u)_{0,1}$$

for all $(v,u) \in V^0(N^D) \times U^0(N)$. If we set

$$(\alpha, n\otimes\omega')_{0,0} = \alpha(n)\omega'$$

$$(\beta, n\otimes\omega')_{1,0} = \beta(n)\omega'$$

$$(\alpha, n\otimes\omega)_{0,1} = \alpha(n)\omega,$$

then it is routine matter to verify that these pairings satisfy the conditions (ibid., §3).

Note that Theorems 5.1 and 5.6 give us quasi-isomorphisms $Rf_*N^D \xrightarrow{\sim} V^{\cdot}(N^D)$ and $Rf_*N \xrightarrow{\sim} U^{\cdot}(N)[-1]$. Also, that the pairing $N^D \times N \longrightarrow \mu_p$ gives a pairing $Rf_*N^D \otimes^L Rf_*N \longrightarrow Rf_*\mu_p$.

Proposition 5.11. *The following diagram commutes (in the derived category)*

$$\begin{array}{ccc} Rf_*N^D \otimes^L Rf_*N & \longrightarrow & Rf_*\mu_p \\ \downarrow \sim & & \downarrow \sim \\ V^{\cdot}(N^D)\otimes^L U^{\cdot}(N)[-1] & \longrightarrow & U^{\cdot}(\mu_p)[-1]. \end{array}$$

Proof: This is a restatement of [Artin and Milne (1976), 4.6].

Notes: This section summarizes [Artin and Milne (1976)].

§6 Local fields of characteristic p

Throughout this section, K will be a local field of characteristic $p \neq 0$ with finite residue field k. Let R be the ring of integers in K. A choice of a uniformizing parameter t for R determines isomorphisms $R \xrightarrow{\approx} k[[t]]$ and $K \xrightarrow{\approx} k((t))$.

We shall frequently use that any group scheme G of finite-type over K has a composition series with quotients of the following types: a smooth connected group scheme; a finite étale group scheme; a finite group scheme that is local with étale Cartier dual; a finite group scheme that is local with local Cartier dual. We shall refer to the last two group schemes as being local-étale and local-local respectively. A finite local group scheme has a composition series whose quotients are all of height one, and a finite local-local group scheme has a composition series whose quotients are all isomorphic to α_p [Demazure and Gabriel (1970), IV,§3,5].

Čech cohomology

Fix an algebraic closure of K_a of K. For any sheaf F on

$(\mathrm{Spec}\ K)_{fl}$ and finite extension L of K, we write $\check{H}^r(L/K,F)$ for the rth cohomology group of the complex of abelian groups

$$0 \longrightarrow F(L) \longrightarrow F(L\otimes_K L) \longrightarrow \dots \longrightarrow F(\otimes_K^r L) \longrightarrow F(\otimes_K^{r+1} L) \longrightarrow \dots \quad \mathbf{(6.0.1)}.$$

In the case that L is Galois over K with Galois group G, this complex can be identified with the complex of inhomogeneous cochains of the G-module F(L) (see [Shatz (1972), p207] or [Milne (1980), III.2.6]). We define $\check{H}^r(K,F)$ to be $\varinjlim \check{H}^r(L/K,F)$ where L runs over the finite field extensions of K contained in K_a.

Proposition 6.1. *For any group scheme* G *of finite-type over* K *and any* $r \geq 0$, *the canonical map* $\check{H}^r(K,G) \longrightarrow H^r(K,G)$ *is an isomorphism.*

Proof: The proof uses only that K is a field of characteristic p. The first step is to show that a short exact sequence of group schemes leads to a long exact sequence of Čech cohomology groups. This is an immediate consequence of the following lemma.

Lemma 6.2. *For any short exact sequence*

$$0 \longrightarrow G' \longrightarrow G \longrightarrow G'' \longrightarrow 0$$

of group schemes of finite type over K *and any* r, *the sequence*

$$0 \longrightarrow G'(\otimes_K^r K_a) \longrightarrow G(\otimes_K^r K_a) \longrightarrow G''(\otimes_K^r K_a) \longrightarrow 0$$

is exact.

Proof: We show that $H^1(\otimes_K^r K_a,G)$ $(= \varinjlim H^1(\otimes_K^r L,G))$ is zero. For any finite extension L of K, $\otimes_K^r L$ is an Artin ring. It is therefore a finite product of local rings whose residue fields L_i are finite extensions of L. If G is smooth, $H^1(\otimes_K^r L,G) = \prod H^1(L_i,G)$ (see [Milne (1980), III.3.11]), and so obviously $\varinjlim H^1(\otimes_K^r L,G) = 0$. It remains to treat the case of a finite group scheme N of height one. Denote

Spec $\otimes_K^r L$ by X. Then (5.7) shows that the restriction map $H^1(X,N) \longrightarrow H^1(X^{(p^{-1})},N)$ is zero. Since $X^{(p^{-1})} = L^{p^{-1}} \otimes_K \cdots \otimes_K L^{p^{-1}}$, again we see that $\varinjlim H^1(\otimes_K^r L,G) = 0$.

We next need to know that $\check{H}^r(K,G)$ and $H^r(K,G)$ are effaceable in the category of group schemes of finite-type over K.

Lemma 6.3. *Let G be a group scheme of finite-type over* K.

(a) *For any* $c \in \check{H}^r(K,G)$, *there exists an embedding* $G \hookrightarrow G'$ *of* G *into a group scheme* G' *of finite-type over* K *such that* c *maps to zero in* $\check{H}^r(K,G')$.

(b) *Same statement with* $\check{H}^r(K,G)$ *replaced by* $H^r(K,G)$.

Proof: In both cases, there exists a finite extension L of K, $L \subset K_a$, such that c maps to zero in $\check{H}^r(L,G)$ (or $H^r(L,G)$). Take G' to be $\mathrm{Res}_{L/K}G$ and the map to be the canonical inclusion $G \hookrightarrow \mathrm{Res}_{L/K}G$. In the case of Čech cohomology, a simple direct calculation shows that c maps to zero in $\check{H}^r(K,G')$, and in the case of derived-functor cohomology, $H^r(K,G') = H^r(L,G)$.

We now prove the proposition by induction on r. For r = 0 it is obvious, and so assume that it holds for all r less than some r_0. For any embedding $G \hookrightarrow G'$, G'/G is again a group scheme of finite-type over K, and we have a commutative diagram

$$\begin{array}{ccccccc}
\cdots \longrightarrow \check{H}^{r_0-1}(K,G') & \longrightarrow & \check{H}^{r_0-1}(K,G'/G) & \longrightarrow & \check{H}^{r_0}(K,G) & \longrightarrow & \check{H}^{r_0}(K,G') \\
\downarrow \approx & & \downarrow \approx & & \downarrow & & \downarrow \\
\cdots \longrightarrow H^{r_0-1}(K,G') & \longrightarrow & H^{r_0-1}(K,G'/G) & \longrightarrow & H^{r_0}(K,G) & \longrightarrow & H^{r_0}(K,G').
\end{array}$$

Let c be a nonzero element of $\check{H}^{r_0}(K,G)$, and choose $G \hookrightarrow G'$ to be the embedding given by (6.3a); then a diagram chase shows that the image of c in $H^{r_0}(K,G)$ is nonzero. Let $c' \in H^{r_0}(K,G)$, and choose $G \hookrightarrow G'$ to be the embedding given by (6.3b); then a diagram chase shows that c' is in the image of $\check{H}^{r_0}(K,G) \longrightarrow H^{r_0}(K,G)$. As c and c' are arbitrary elements, this shows that $\check{H}^{r_0}(X,G) \longrightarrow H^{r_0}(X,G)$ is an isomorphism and so completes the proof.

First calculations

Note that for any finite group scheme N over K, $H^0(K,N)$ is finite.

Proposition 6.4. *Let* N *be a finite group scheme over* K.

(a) *If* N *is étale-local, then* $H^r(K,N) = 0$ *for* $r \neq 0,1$.

(b) *If* N *is local-étale, then* $H^r(K,N) = 0$ *for* $r \neq 1,2$.

(c) *If* N *is local-local, then* $H^r(K,N) = 0$ *for* $r \neq 1$.

Proof: (a) Since N is étale, $H^1(K,N) = H^1(\mathrm{Gal}(K_s/K),N(K_s))$, and because its Cartier dual is local, N must have p-power order. Therefore the assertion follows from the fact that K has Galois p-cohomological dimension 1.

(b) We can assume that N has height one, and then (5.6) shows that $R^r f_* N = 0$ for $r \neq 1$. Therefore $H^r(K,N) = H^{r-1}(K_{et},R^1 f_* N)$, and $R^1 f_* N$ is a p-torsion sheaf.

(c) We can assume that $N = \alpha_p$. In this case the statement follows directly from the cohomology sequence of

$$0 \longrightarrow \alpha_p \longrightarrow \mathbb{G}_a \xrightarrow{F} \mathbb{G}_a \longrightarrow 0.$$

The topology on the cohomology groups

The ring $\otimes_K^r L$ has a natural topology. If G is an affine group scheme of finite type over K, then $G(\otimes_K^r L)$ also has a natural topology: immerse G into some affine space $\mathbb{A}^n$ and give $G(\otimes_K^r L) \subset \mathbb{A}^n(\otimes_K^r L)$ the subspace topology. One checks easily that the topology is independent of the immersion chosen. Therefore, for any group scheme G of finite type over K, $G(\otimes_K^r L)$ has a natural topology, and the boundary maps in (6.0.1) are continuous because they are given by polynomials. Endow $Z^r(L/K,G) \subset C^r(L/K,G)$ $(= G(\otimes_K^{r+1} L))$ with the subspace topology, and $\check{H}^r(L/K,G) = Z^r(L/K,G)/B^r(L/K,G)$ with the quotient topology. We can then give $H^r(K,G)$ the direct limit topology: a map $H^r(K,G) \longrightarrow T$ is continuous if and only if it defines continous maps on $\check{H}^r(L/K,G)$ for all L.

Lemma 6.5. *Let G be a group scheme of finite type over K, and let $L \subset K_a$ be a finite extension of K.*

(a) *The group $H^r(X,G)$ is Hausdorff, locally compact, and σ-compact (that is, a countable union of compact subspaces).*

(b) *The maps in the cohomology sequence arising from a short exact sequence of group schemes are continuous.*

(c) *The restriction maps $H^r(K,G) \longrightarrow H^r(L,G)$ are continous.*

(d) *When G is finite, the inflation map Inf: $\check{H}^1(L/K,G) \longrightarrow H^1(K,G)$ has closed image and defines a homeomorphism of $\check{H}^1(L/K,G)$ onto its image.*

(e) *Cup-product is continuous.*

Proof: (a) The groups $C^r(L/K,G)$ are Hausdorff, σ-compact, and locally compact. As $Z^r(L/K,G)$ is a closed subspace of $C^r(L/K,G)$, it has the same properties. Also the image $B^r(L/K,G)$ of $C^{r-1}(L/K,G)$ is a locally compact subgroup of a Hausdorff group, and so is closed. Hence $H^r(L/K,G)$ is the quotient of a Hausdorff, σ-compact, locally compact space by a closed subspace, and it therefore inherits the same properties.

(b) Obvious.

(c) It suffices to note that, for any $L' \supset L$, the maps $C^r(L'/K,G) \longrightarrow C^r(L'/L,G)$ are continuous.

(d) It suffices to show that for any $L' \supset L$, the inflation map $\check{H}^1(L/K,G) \longrightarrow \check{H}^1(L'/K,G)$ is closed. The map $G(L\otimes_K L) \longrightarrow G(L'\otimes_K L')$ is closed, and therefore its restriction to $Z^1(L/K,G) \longrightarrow Z^1(L'/K,G)$ is also closed. Since $B^1(L'/K,G)$ is compact (it is finite), the map $Z^1(L'/K,G) \longrightarrow \check{H}^1(L'/K,G)$ is closed, and the assertion follows.

(e) Obvious.

Remark 6.6. (a) Let

$$0 \longrightarrow N \longrightarrow G \longrightarrow G' \longrightarrow 0$$

be an exact sequence, and suppose that $H^1(K,G) = 0$. Then $H^0(K,N)$ has the subspace toplogy induced from $H^0(K,N) \hookrightarrow G(K)$ and $H^1(K,N)$ has the quotient topology induced from $G'(K) \twoheadrightarrow H^1(K,N)$.

(b) Our definition of the topology on $H^r(K,G)$ differs from, but is equivalent to, that of Shatz [Shatz (1972), VI].

Examples 6.7. (a) The cohomology sequence of

$$0 \longrightarrow \mathbb{Z}/p\mathbb{Z} \longrightarrow \mathbb{G}_a \xrightarrow{\wp} \mathbb{G}_a \longrightarrow 0$$

shows that $H^1(K,\mathbb{Z}/p\mathbb{Z}) = K/\wp K$. This group is infinite, and it has the discrete topology because $R/\wp R$ is a finite open subgroup of $K/\wp K$.

(b) From the Kummer sequence we find that $H^1(K,\mu_p) = K^\times/K^{\times p}$. This group is compact because $R^\times/R^{\times p}$ is a compact subgroup of finite index.

(c) The group $H^1(K,\alpha_p) = K/K^p$. The subgroup R/R^p of K/K^p is compact and open, and the quotient K/RK^p is an infinite discrete group.

(d) The group $H^0(K,\mathbb{G}_m)$ equals $K^\times$ with its locally compact topology. The group $H^2(K,\mathbb{G}_m)$ equals $\mathbb{Q}/\mathbb{Z}$ with the discrete topology because $H^2(K,\mathbb{G}_m) = \cup H^2(L/K,\mathbb{G}_m)$, and $H^2(L/K,\mathbb{G}_m)$ is finite and Hausdorff.

Table 6.8.

G	$H^0(K,G)$	$H^1(K,G)$	$H^2(K,G)$
étale-local	finite, discrete	torsion, discrete	0
local-étale	0	compact	finite, discrete
local-local	0	locally compact	0
torus	locally compact	finite, discrete	discrete

To verify the statements in the table, first note that, because the topologies on $\check{H}^r(L/K,G)$ and $H^r(K,G)$ are Hausdorff, they are discrete when the groups are finite. Next note that $\check{H}^r(L/K,G)$ contains $\check{H}^r(R_L/R_K,G)$ as an open subgroup. Moreover, if G is $\mathbb{Z}/p\mathbb{Z}$, μ_p, α_p, or $\mathbb{G}_m$, each assertion follows from (6.7). It is not difficult now to deduce that they are true in the general case.

Duality for tori

Let M be a finitely generated torsion-free module for $Gal(K_s/K)$. As M becomes a module with trivial action over some finite separable extension L of K, it is represented by an étale group scheme locally of finite-type over K. The method used above for group schemes of finite-type defines the discrete topology on the groups $H^r(K,M)$.

Theorem 6.9. *Let T be a torus over K and let $X^*(T)$ be its group of characters. Then the cup-product pairing*

$$H^r(K,T) \times H^{2-r}(K,X^*(T)) \longrightarrow H^2(K,\mathbb{G}_m) \approx \mathbb{Q}/\mathbb{Z}$$

defines dualities between:

the compact group $H^0(K,T)^\wedge$ *(completion relative to the topology of open subgroups of finite index) and the discrete group* $H^2(K,X^*(T))$;

the finite groups $H^1(K,T)$ *and* $H^1(K,X^*(T))$;

the discrete group $H^2(K,T)$ *and the compact group* $H^0(K,X^*(T))^\wedge$ *(completion relative to the topology of subgroups of finite index).*

Proof: As T is smooth, $H^r(K,T) = H^r(\mathrm{Gal}(K_s/K),T(K_s))$, and the group $H^r(K,X^*(T)) = H^r(\mathrm{Gal}(K_s/K),X^*(T))$. All the groups have the discrete topology except $H^0(K,T) = T(K)$, which has the topology induced by that on K. The theorem therefore simply restates (I.2.4).

Finite group schemes

We now let N be a finite group scheme over K.

Theorem 6.10. *For any finite group scheme N over K, the cup-product pairing*

$$H^r(K,N^D) \times H^{2-r}(K,N) \longrightarrow H^2(K,\mathbb{G}_m) \approx \mathbb{Q}/\mathbb{Z}$$

identifies each group with the Pontryagin dual of the other.

Proof: After (I.2.3) we may assume that N is a p-primary group scheme. Suppose first that N is étale. Then there is a short exact sequence of discrete $\mathrm{Gal}(K_s/K)$-modules

$$0 \longrightarrow M_1 \longrightarrow M_0 \longrightarrow N(K_s) \longrightarrow 0$$

with M_0 and M_1 finitely generated and torsion-free (as abelian groups). Dually there is an exact sequence

$$0 \longrightarrow N^D \longrightarrow T^0 \longrightarrow T^1 \longrightarrow 0$$

with T^0 and T^1 tori. Because the cohomology groups of the modules in the first sequence are all discrete, the dual of its cohomology sequence is exact. Therefore we get an exact commutative diagram

$$\begin{array}{ccccccc} \cdots \longrightarrow & H^r(K,N^D) & \longrightarrow & H^r(K,T^0) & \longrightarrow & H^r(K,T^1) & \longrightarrow \cdots \\ & \downarrow & & \downarrow & & \downarrow & \\ \cdots \longrightarrow & H^r(K,N)^* & \longrightarrow & H^r(K,M_0)^* & \longrightarrow & H^r(K,M_1)^* & \longrightarrow \cdots . \end{array}$$

For $r \geq 1$, the second two vertical arrows in the diagram are isomorphisms, and this shows that $H^r(K,N^D) \xrightarrow{\approx} H^r(K,N)^*$ for $r \geq 2$. The diagram also shows that the image of $H^1(K,N^D)$ in $H^1(K,N)^*$ is dense. As $H^1(K,N^D)$ is compact, its image is closed and so equals $H^1(K,N)^*$. If $H^1(K,N^D) \longrightarrow H^1(K,N)$ were not injective, then there would exist an element $b \in T^1(K)$ that is in the image of $T^0(K)^\wedge \longrightarrow T^1(K)^\wedge$, but which is not in the image of $T^0(K) \longrightarrow T^1(K)$. I claim that the image of $T^0(K)$ is closed in $T^1(K)$. Let L be a splitting field for T^0 and T^1, and consider the diagram

$$\begin{array}{ccccccc} 0 \longrightarrow & \mathrm{Hom}(X^*(T^0),R_L^\times) & \longrightarrow & T^0(K) & \longrightarrow & \mathrm{Hom}(X^*(T^0),\mathbb{Z}) & \longrightarrow \cdots \\ & \downarrow & & \downarrow & & \downarrow & \\ 0 \longrightarrow & \mathrm{Hom}(X^*(T^1),R_L^\times) & \longrightarrow & T^1(K) & \longrightarrow & \mathrm{Hom}(X^*(T^1),\mathbb{Z}) & \longrightarrow \cdots \end{array}$$

(Hom's as G-modules). Let $b \in T^1(K)$. If the image of b in $\mathrm{Hom}(X^*(T^1),\mathbb{Z})$ is not in the image from T^0, then $b.\mathrm{Hom}(X^*(T),R_L^\times)$ is an open neighbourhood of b that is disjoint from the image of $T^0(K)$. Therefore we may assume $b \in \mathrm{Hom}(X^*(T^1),R_L^\times)$; but $\mathrm{Hom}(X^*(T^0),R_L^\times)$ is compact snd so its complement in $\mathrm{Hom}(X^*(T^1),R_L^\times)$ is an open neighbourhood of b. This proves the claim and completes the proof of the theorem in the case that N or its dual is étale.

Next assume that $N = \alpha_p$. Here one shows that the pairing

$$H^1(K,\alpha_p) \times H^1(K,\alpha_p) \longrightarrow \mathbb{Q}/\mathbb{Z}$$

can be identified with the pairing

$$K/K^p \times K/K^p \longrightarrow p^{-1}\mathbb{Z}/\mathbb{Z} \subset \mathbb{Q}/\mathbb{Z},\ (f,g) \mapsto p^{-1}\mathrm{Tr}_{k/\mathbb{F}_p}(\mathrm{res}(fdg)),$$

(see [Shatz (1972), p240-243]; it also follows immediately from the elementary case of (5.11) in which $N = \alpha_p$), and this last pairing is a duality.

Finally, to complete the proof of the theorem we need the following lemma.

Lemma 6.11. *Let*

$$0 \longrightarrow N' \longrightarrow N \longrightarrow N'' \longrightarrow 0$$

be an exact sequence of finite group schemes over K. *If the theorem is true for* N' *and* N", *then it is true for N.*

Proof: Proposition I.0.22 and the discussion preceding it show that the bottom row of the following diagram is exact, and so this follows from the five-lemma:

$$\begin{array}{ccccccccc} \cdots \longrightarrow & H^r(K,N''^D) & \longrightarrow & H^r(K,N^D) & \longrightarrow & H^r(K,N'^D) & \longrightarrow \cdots \\ & \downarrow \approx & & \downarrow & & \downarrow \approx & \\ \cdots \longrightarrow & H^{2-r}(K,N'')^* & \longrightarrow & H^{2-r}(K,N)^* & \longrightarrow & H^{2-r}(K,N')^* & \longrightarrow \cdots . \end{array}$$

Remark 6.12. (a) It is possible to give an alternative proof of Theorem 6.10 using the sequences in §5. After (6.11) (and by symmetry), it suffices to prove the theorem for a group N of height one. Then the exact sequences in §5 yield cohomology sequences

$$0 \longrightarrow H^0(K,N^D) \longrightarrow V^0(N^D) \longrightarrow V^1(N^D) \longrightarrow H^1(K,N^D) \longrightarrow 0$$

$$0 \longrightarrow H^1(K,N) \longrightarrow U^0(N) \longrightarrow U^1(N) \longrightarrow H^2(K,N) \longrightarrow 0.$$

Here $V^0(N^D)$ is the K-vector space ω_N and $U^1(N)$ is the K-vector space $n\otimes\Omega^1_{K/k}$. The pairing $(\alpha, n\otimes\omega) \mapsto \mathrm{res}(\alpha(n)\omega)$ identifies ω_N with the k-linear dual of $n\otimes\Omega^1_{K/k}$, and so the pairing

$$(\alpha, n\otimes\omega) \mapsto p^{-1}\mathrm{Tr}_{k/\mathbb{F}_p}\mathrm{res}(\alpha(n)\omega)$$

identifies ω_N with the Pontryagin dual of $n\otimes\Omega^1_{K/k}$ (see 0.7). Similarly $V^1(N^D)$ is the Pontryagin dual of $U^0(N)$. Therefore the pairing of complexes in (5.10) shows that the dual of the first of the above sequences can be identified with an exact sequence

$$0 \longrightarrow H^1(K,N^D)^* \longrightarrow U^0(N) \longrightarrow U^1(N) \longrightarrow H^0(K,N^D)^* \longrightarrow 0.$$

Thus there are canonical isomorphisms $H^1(K,N) \xrightarrow{\approx} H^1(K,N^D)^*$ and $H^2(K,N) \xrightarrow{\approx} H^0(K,N^D)^*$, and (5.11) shows that these are the maps given by cup-product.

(b) It is also possible to deduce a major part of (6.10) from (I.2.1). Let N be étale over K. Then $H^r(K_{fl},N) = H^r(K_{et},N)$, and so we have to show that $\mathrm{Ext}^r_{K_{et}}(N,\mathbb{G}_m) = H^r(K_{fl},N^D)$. Note [Milne (1980), II.3.1d] that $f^*(N|K_{et}) = N$, where $f\colon (\mathrm{Spec}\ K)_{fl} \longrightarrow (\mathrm{Spec}\ K)_{et}$ is defined by the identity map. From the spectral sequence $\mathrm{Ext}^r_{K_{et}}(N,R^sf_*\mathbb{G}_m) \Rightarrow \mathrm{Ext}^{r+s}_{K_{fl}}(f^*N,\mathbb{G}_m)$ and the vanishing of the higher direct images of $\mathbb{G}_m$, we see that $\mathrm{Ext}^r_{K_{et}}(N,\mathbb{G}_m) = \mathrm{Ext}^r_{K_{fl}}(N,\mathbb{G}_m)$ for all r. But N is locally constant on $(\mathrm{Spec}\ K)_{fl}$, and the exact sequence

$$\dots \longrightarrow H^r(X,\mathbb{G}_m) \xrightarrow{n} H^r(X,\mathbb{G}_m) \longrightarrow \mathrm{Ext}^r_{K_{fl}}(\mathbb{Z}/n\mathbb{Z},\mathbb{G}_m) \longrightarrow \dots$$

and the divisibity of $\mathbb{G}_m$ on the flat site show that $\mathcal{E}xt^r_{K_{fl}}(\mathbb{Z}/n\mathbb{Z},\mathbb{G}_m) = 0$ for $r > 0$. Therefore $\mathcal{E}xt^r_{K_{fl}}(N,\mathbb{G}_m) = 0$ for $r > 0$, and the local-global spectral sequence for Exts shows that $\mathrm{Ext}^r_{K_{et}}(N,\mathbb{G}_m) = H^r(K_{fl},N^D)$.

Remark 6.13. Much of the above discussion continues to hold if K is the field of fractions of an excellent Henselian discrete valuation ring with finite residue field. For example, if N is étale, then $H^r(K,N) = H^r(\hat{K},N)$ for all r because K and $\hat{K}$ have the same absolute Galois group, and $N(K_s) = N(\hat{K}_s)$; if N is local-étale, then $H^1(K,N)$ is dense in $H^1(\hat{K},N)$ and $H^2(K,N) = H^2(\hat{K},N)$; if N is local-local, then $H^1(K,N)$ is dense in $H^1(\hat{K},N)$. The map $H^r(K,N^D) \longrightarrow H^{2-r}(K,N)^*$ given by cup-product is an isomorphism for $r \neq 1$, and is injective with dense image for $r = 1$.

Notes: The main results in this section are taken from [Shatz (1964)]; see also [Shatz (1972)]. Theorem 6.10 was the first duality theorem to be proved for the flat topology and so can be regarded as the forerunner of the rest of the results in this chapter.

§7 Local results: equicharacteristic, finite residue field

Throughout this section, R will be a complete discrete valuation ring of characteristic $p \neq 0$ with finite residue field k. As usual, we use the notations X = Spec R and

$$\operatorname{Spec} k = x \xrightarrow{i} X \xleftarrow{j} u = \operatorname{Spec} K.$$

Finite group schemes

Let N be a finite flat group scheme over X. As in (1.1), we find that

$$H^0(X,N) = N(X) = N(K) = H^0(K,N_K),$$
$$H^1(X,N) \hookrightarrow H^1(K,N),$$
$$H^r(X,N) = 0,\ r > 1,$$

and

$$H^2_x(X,N) = H^1(K,N)/H^1(X,N),$$
$$H^3_x(X,N) = H^2(K,N),$$
$$H^r_x(X,N) = 0, \quad r \neq 2,3.$$

In the last section we defined topologies on the groups $H^r(K,N)$. We endow $H^r(X,N)$ with its topology as a subspace of $H^r(K,N)$, and we endow $H^r_x(X,N)$ with its topology as a quotient of $H^r(K,N)$. With respect to these topologies $H^0(X,N)$ is discrete (and finite), and $H^1(X,N)$ is compact and Hausdorff; $H^2_x(X,N)$ and $H^3_x(X,N)$ are both discrete (and $H^3_x(X,N)$ is finite).

Theorem 7.1. *For any finite flat group scheme N over X, the canonical pairings*

$$H^r(X,N^D) \times H^{3-r}_x(X,N) \longrightarrow H^3_x(X,\mathbb{G}_m) \approx \mathbb{Q}/\mathbb{Z}$$

define dualities between:

the finite groups $H^0(K,N^D)$ *and* $H^3_x(X,N)$;

the compact group $H^1(X,N^D)$ *and the discrete torsion group* $H^2_x(X,N)$.

Before giving the proof, we list some corollaries.

Corollary 7.2. *For any finite flat group scheme N over X,* $H^1(X,N^D)$ *is the exact annihilator of* $H^1(X,N)$ *in the pairing*

$$H^1(K,N^D) \times H^1(K,N) \longrightarrow H^2(K,\mathbb{G}_m) = \mathbb{Q}/\mathbb{Z}$$

of (6.10).

Proof: As in the proof of (1.4), one sees easily that the corollary is equivalent to the case r = 1 of the theorem.

Corollary 7.3. *Let N be a finite flat group scheme over X. For all* $r < 2p-2$,

$$\mathrm{Ext}^r_X(N,\mathbb{G}_m) \times H^{3-r}_x(X,N) \longrightarrow H^3_x(X,\mathbb{G}_m) = \mathbb{Q}/\mathbb{Z}$$

is a nondegenerate pairing of finite groups.

Proof: The proof is the same as that of (1.5).

Write $f\colon X_{fl} \longrightarrow X_{et}$ for the morphism of sites defined by the identity map.

Corollary 7.4. *Let* N *be a quasi-finite flat group scheme over* X *whose* p*-primary component* N(p) *is finite over* X. *Let* N^D *be the complex of sheaves such that*

$$N^D(\ell) = \begin{cases} \mathcal{H}om_{X_{fl}}(N(\ell),\mathbb{G}_m) & \ell = p \\ f^*R\mathcal{H}om_{X_{et}}(N(\ell),\mathbb{G}_m) & \ell \neq p. \end{cases}$$

Then

$$H^r(X,N^D) \times H^{3-r}_x(X,N) \longrightarrow H^3_x(X,\mathbb{G}_m) = \mathbb{Q}/\mathbb{Z}$$

is a nondegenerate pairing of finite groups.

Proof: For the prime-to-p components of the groups, the corollary follows from (II.1.8); for the p-primary component, it follows immediately from the theorem.

Proof of 7.1: Assume first that N has height one. Then the first exact sequence in §5 leads to a cohomology sequence

$$0 \longrightarrow H^0(X,N^D) \longrightarrow H^0(X,\omega_N) \longrightarrow H^0(X,\omega_N^{(p)}) \longrightarrow H^1(X,N^D) \longrightarrow 0,$$

and the second leads to a cohomology sequence

$$0 \longrightarrow H^2_x(X,N) \longrightarrow H^1_x(X,n\otimes\Omega^1_{X'}) \longrightarrow H^1_x(X,n\otimes\Omega^1_X) \longrightarrow H^3_x(X,N) \longrightarrow 0.$$

The pairing $(\alpha,n\otimes w) \mapsto \alpha(n)w\colon \omega_N \times n\otimes\Omega^1_X \longrightarrow \Omega^1_X \approx \mathcal{O}_X$ realizes $H^0(X,\omega_N)$ as the R-linear dual of $H^0(X,n\otimes\Omega^1_X)$; therefore (0.8) shows that the compact group $H^0(X,\omega_N)$ is the Pontryagin dual of the discrete group

$H^1_x(X, n\otimes\Omega^1_X) = H^0(X, n\otimes\Omega^1_X)\otimes K/H^0(X, n\otimes\Omega^1_X)$. Similarly the compact group $H^0(X, \omega_N^{(p)})$ is the Pontryagin dual of $H^1_x(X, n\otimes\Omega^1_{X'})$, and so the pairing in (5.10) gives a commutative diagram

$$\begin{array}{ccccccccc} 0 \longrightarrow & H^0(X,N^D) & \longrightarrow & H^0(X,\omega_N) & \longrightarrow & H^0(X,\omega_N^{(p)}) & \longrightarrow & H^1(X,N^D) & \longrightarrow 0 \\ & & & \downarrow \approx & & \downarrow \approx & & & \\ 0 \longrightarrow & H^3_x(X,N)^* & \longrightarrow & H^1_x(X,n\otimes\Omega^1_X)^* & \longrightarrow & H^1_x(X,n\otimes\Omega^1_{X'})^* & \longrightarrow & H^2_x(X,N)^* & \longrightarrow 0. \end{array}$$

The diagram provides isomorphisms $H^0(X,N^D) \xrightarrow{\approx} H^3_x(X,N)$ and $H^1(X,N^D) \xrightarrow{\approx} H^2(X,N)$, which we must show are those given by the pairing in the theorem. For this we retreat to the derived category. In (5.11) we saw that there is a commutative diagram:

$$\begin{array}{ccc} Rf_*N^D\otimes^L Rf_*N & \longrightarrow & Rf_*\mu_p \\ \downarrow \sim & & \downarrow \sim \\ V^\cdot(N^D)\otimes^L U^\cdot(N)[-1] & \longrightarrow & U^\cdot(\mu_p)[-1]. \end{array}$$

From this we get a commutative diagram

$$\begin{array}{ccccc} H^r(X,N^D) & \times & H^{3-r}_x(X,N) & \longrightarrow & H^3_x(X,\mu_p) \\ \downarrow & & \downarrow & & \downarrow \approx \\ H^r(X,V^\cdot(N^D)) & \times & H^{2-r}_x(X,U^\cdot(N)) & \longrightarrow & H^2_x(X,U^\cdot(\mu_p)), \end{array}$$

which exactly says that the two pairs of maps agree. This completes the proof of the theorem when N has height one.

Next we note that for $r = 0$ the theorem follows from (6.10), and that for $r = 1$ it is equivalent to (7.2). Since this last statement is symmetric between N and N^D, we see that the theorem is also true if N is the dual of a finite group scheme of height one. Every finite group scheme over X has a composition series each of whose quotients is of height one or is the dual of a height one group (this is obvious over K, and one can apply (B.1) to obtain it over X), and so the theorem follows from the obvious fact that it is true for any extension of groups for which it is true.

Remark 7.5. The original proof of the Theorem 7.1 [Milne (1970/72), (1973)] was more explicit. We include a sketch. It clearly suffices to prove (7.2). Write dK ($\approx K/K^p$) and dR ($\approx R/R^p$) for the images of the maps $d\colon K \longrightarrow \Omega^1_{K/k}$ and $d\colon R \longrightarrow \Omega^1_{R/k}$.

Assume first that $N = \alpha_p$. Then the diagram

$$\begin{array}{ccccc} H^1(X,N^D) & \times & H^1(X,N) & & \\ \downarrow & & \downarrow & & \\ H^1(K,N^D) & \times & H^1(K,N) & \longrightarrow & H^2(K,\mathbb{G}_m) \approx \mathbb{Q}/\mathbb{Z} \end{array} \tag{7.5.1}$$

can be identified with

$$\begin{array}{ccccc} R/R^p & \times & dR & & \\ \downarrow & & \downarrow & & \\ K/K^p & \times & dK & \longrightarrow & p^{-1}\mathbb{Z}/\mathbb{Z} \subset \mathbb{Q}/\mathbb{Z}, \end{array}$$

where the bottom pairing is $(f,\omega) \mapsto p^{-1}\mathrm{Tr}_{k/\mathbb{F}_p}\,\mathrm{res}(f\omega)$. It is obvious that the upper groups are exact annihlators in the lower pairing.

Let $N = N_{a,0}$ in the Oort-Tate classification (0.9) with $a = t^{(p-1)c}$. Then the diagram (7.5.1) can be identified with

$$\begin{array}{ccccc} t^{-cp}R/((t^{-cp}R)\cap\wp R) & \times & \mathrm{dlog}(K^\times)\cap t^{cp}R\,dt & & \\ \downarrow & & \downarrow & & \\ K/\wp K & \times & \mathrm{dlog}(K^\times) & \longrightarrow & p^{-1}\mathbb{Z}/p\mathbb{Z} \subset \mathbb{Q}/\mathbb{Z}, \end{array}$$

where the lower pairing is $(f,\omega) \mapsto \mathrm{Tr}_{k/\mathbb{F}_p}(\mathrm{res}(f\omega))$. It is easy to check that the upper groups are exact annihilators in the lower pairing.

These calculations prove the theorem whenever $N = N_{0,0}$ (that is, $N = \alpha_p$), $N = N_{a,0}$ where $a = t^{c(p-1)}$, or $N = N_{0,b}$ where $b = t^{c(p-1)}$. The general case can be reduced to these special cases by means of the following statements [Milne (1973), p84-85]:

(i) for any finite group scheme of p-power order N over R, there exists a finite extension K' of K of degree prime to p such that over R', N has a composition series whose quotients have the above form;

(ii) the theorem is true over K if it is true over some finite extension K' of K of degree prime to p.

Remark 7.6. Let N be a finite flat group scheme over X. We define $\chi(X,N) = [H^0(X,N)]/[H^1(X,N)]$ when both groups are finite. Then the same proof as in (1.14) shows that when N_K is étale,

$$\chi(X,N) = (R:\mathfrak{d})^{-1/n},$$

where n is the order of N and $\mathfrak{d}$ is the discriminant ideal of N over R. Note that

$$H^1(X,N) \text{ is infinite} \iff N_K \text{ is not étale} \iff \mathfrak{d} = 0,$$

and so, if we interprete $1/\infty$ as 0, then the assertion continues to hold when N_K is not étale.

It is possible to prove a weak form of (1.19).

Exercise 7.7. (a) Let N and N' be finite flat group schemes over R; then a homomorphism $\varphi: N_K \to N'_K$ extends to a homomorphism $N \to N'$ if and only if, for all finite field extensions L of K, $H^1(\varphi_L): H^1(L,N_L) \to H^1(L,N_L)$ maps $H^1(R_L,N_L)$ into $H^1(R_L,N'_L)$.

(b)[1] Use (a) to prove the characteristic p analogue of the main theorem of [Tate (1967b)]: if G and G' are p-divisible groups over R, then every homomorphism $\varphi: G_K \to G'_K$ extends to a unique homomorphism

---

[1]The author does not pretend to be able to do part (b) of the problem; the question is still open in general.

$G \longrightarrow G'$.

Abelian varieties and Néron models

We extend (I.3.4) and the results of §2 to characteristic p. Note that in the main results, the coefficient groups are smooth, and so the cohomology groups can be computed using the étale topology (or even using Galois cohomology). The proofs however necessarily involve the flat site.

Theorem 7.8. *Let A be an abelian variety over K, and let B be its dual. The pairings*

$$H^r(K,B) \times H^{1-r}(K,A) \longrightarrow H^2(K,\mathbb{G}_m) \approx \mathbb{Q}/\mathbb{Z}$$

induced by the Poincaré biextension W of (B,A) by $\mathbb{G}_m$ define dualities between:

the compact group B(K) and the discrete group $H^1(K,A)$;

the discrete group $H^1(K,B)$ and the compact group A(K).

For $r \neq 0,1$, all the groups are zero.

Proof: We note first that the pairing is symmetric in the sense that the pairing defined by W and by its transpose W^t are the same (up to sign) (see C.4). Therefore it suffices to show that the map $\alpha^1(K,A)\colon H^1(K,B) \longrightarrow A(K)^*$ is an isomorphism and that $H^r(K,B) = 0$ for $r \geq 2$. Consider the diagram

$$\begin{array}{ccccccccc} 0 & \longrightarrow & B(K)^{(n)} & \longrightarrow & H^1(K,B_n) & \longrightarrow & H^1(K,B)_n & \longrightarrow & 0 \\ & & \downarrow & & \downarrow\approx & & \downarrow & & \\ 0 & \longrightarrow & H^1(K,A)_n^* & \longrightarrow & H^1(K,A_n)^* & \longrightarrow & A(K)^{(n)*} & \longrightarrow & 0. \end{array}$$

Theorem 6.10 shows that the middle vertical arrow is an isomorphism, and it follows that $\alpha^1(K,A)_n$ is surjective for all n. As A(K) is a profinite group, its dual $A(K)^*$ is torsion, and so this proves that

$\alpha^1(K,A)$ is surjective. To show that it is injective, it suffices to prove that $H^1(K,B)_\ell \longrightarrow A(K)^{(\ell)*}$ is injective for all primes. For $\ell \neq p$, we saw in (I.3) that this can be done by a counting argument. Unfortunately, for $\ell = p$ the groups involved are not finite (nor even compact), and so we must work more directly with the cohomology groups of finite group schemes.

We dispose of the statement that $H^2(K,B) = 0$ (note that $H^r(K,B) = 0$ for $r > 2$ because K has strict (Galois) cohomological dimension 2). Consider the diagram

$$\begin{array}{ccccccc} H^1(K,B) & \longrightarrow & H^2(K,B_n) & \longrightarrow & H^2(K,B)_n & \longrightarrow & 0 \\ \downarrow & & \downarrow & & \downarrow & & \\ A(K)^* & \longrightarrow & A_n(K)^* & \longrightarrow & 0 & & \end{array}$$

which is just a continuation to the right of the previous diagram. We have seen that the first vertical arrow is surjective, and the second is an isomorphism by (6.10). A diagram chase now shows that $H^2(K,B)_n = 0$ for all n, and so $H^2(K,B)$ is zero.

The next lemma will allow us to replace K by a larger field.

Lemma 7.9. *If for some finite Galois extension* L *of* K, $\alpha^1(L,A_L)$ *is injective, then* $\alpha^1(K,A)$ *is injective.*

Proof: Since K is local, the Galois group G of L over K is solvable, and so we may assume it to be cyclic. There is an exact commutative diagram

$$\begin{array}{ccccccccc} 0 \to H^1(G,B(L)) & \longrightarrow & H^1(K,B) & \longrightarrow & H^1(L,B)^G & \longrightarrow & H^0_T(G,B(L)) & \longrightarrow & H^2(K,B) = 0 \\ \downarrow & & \downarrow \text{surj} & & \downarrow \approx & & \downarrow & & \\ 0 \to H^0_T(G,A(L))^* & \longrightarrow & A(K)^* & \longrightarrow & A(L)^{*G} & \longrightarrow & H^1(G,A(L))^* & \longrightarrow & 0. \end{array}$$

in which the top row is part of the sequence coming from the

Hochschild-Serre spectral sequence (except that we have replaced $H_T^2(G,B(L))$ with $H_T^0(G,B(L))$), and the bottom row is the dual of the sequence that explicitly describes H_T^0 and H_T^1 for a cyclic group. The second and third vertical arrows are $\alpha^1(K,A)$ and $\alpha^1(L,A)$, and the first and fourth are induced by $\alpha^1(K,A)$ and by the dual of $\alpha^1(K,B)$ respectively. From the right hand end of the diagram we see that

$$H_T^0(G,B(L)) \longrightarrow H^1(G,A(L))^*$$

is an isomorphism, and by interchanging A and B we see that

$$H_T^0(G,A(L)) \longrightarrow H^1(G,B(L))^*$$

is an isomorphism. Thus all vertical maps but the second are isomorphisms, and the five-lemma shows that it also is an isomorphism.

To proceed further, we need to consider the Néron models $\mathcal{A}$ and $\mathcal{B}$ of A and B. Let $i_*\Phi = \mathcal{A}/\mathcal{A}^o$, and write $\mathcal{A}^\Gamma$ for the subscheme of $\mathcal{A}$ corresponding to a subgroup Γ of Φ.

Lemma 7.10. (a) *The map* $\mathcal{A}^\Gamma(X) \longrightarrow \Gamma(x)$ *is surjective, and the map* $H^r(X,\mathcal{A}^\Gamma) \longrightarrow H^r(x,\Gamma)$ *is an isomorphism for all* $r \geq 1$; *therefore* $H^r(X,\mathcal{A}^\Gamma) = 0$ *for* $r \geq 2$.

(b) *There is an exact sequence*

$$\Phi(x) \longrightarrow (\Phi/\Gamma)(x) \longrightarrow H^1(X,\mathcal{A}^\Gamma) \longrightarrow H^1(K,A).$$

(c) *We have*

$$H_x^r(X,\mathcal{A}^\Gamma) = \begin{cases} 0 & \textit{for } r \neq 1,2 \\ (\Phi/\Gamma)(x) & \textit{for } r = 1, \end{cases}$$

and there is an exact sequence

$$0 \to \Gamma(x) \longrightarrow \Phi(x) \longrightarrow (\Phi/\Gamma)(x) \longrightarrow H^1(X,\mathcal{A}^\Gamma) \longrightarrow H^1(K,A) \longrightarrow H_x^2(X,\mathcal{A}^\Gamma) \to 0.$$

Proof: The proofs of (2.1 - 2.3) apply also in characteristic p.

After we make a finite separable field extension, A (and B) will have semistable reduction and $\mathcal{A}_p$ and $\mathcal{B}_p$ will extend to finite group schemes over X. The group $\Phi(k)_p$ then has order p^{μ} where μ is the dimension of the toriodal part of the reduction $\mathcal{A}_0^o$ of $\mathcal{A}^o$ (equal to the dimension of the toroidal part of the reduction of $\mathcal{B}^o$). The extension of the Poincaré biextension to $(\mathcal{B}^o, \mathcal{A}^o)$ defines a pairing

$$H_x^2(X,\mathcal{B}^o) \times H^0(X,\mathcal{A}^o) \longrightarrow H_x^3(X,\mathbb{G}_m) \approx \mathbb{Q}/\mathbb{Z},$$

which the proposition allows us to identify with a map $H^1(K,B) \longrightarrow \mathcal{A}^o(X)^*$. Clearly this map is the composite of $\alpha^1(K,A)\colon H^1(K,B) \twoheadrightarrow A(K)^*$ with $A(K)^* \twoheadrightarrow \mathcal{A}^o(X)^*$. In order to complete the proof of the theorem, it suffices therefore to show that the kernel of $H^1(K,B)_p \longrightarrow \mathcal{A}^o(X)^{(p)*}$ has order $[A(K)^{(p)}/\mathcal{A}^o(X)^{(p)}] = [\Phi(k)^{(p)}]$.

From the diagram

$$\begin{array}{ccccccc} 0 \longrightarrow & \Phi'(k)^{(p)} & \longrightarrow H_x^2(X,\mathcal{B}_p^o) & \longrightarrow & H_x^2(X,\mathcal{B}^o)_p & \longrightarrow 0 \\ & \downarrow & \downarrow & & \downarrow & \\ & 0 & \longrightarrow H^0(X,\mathcal{A}_p^o)^* & \longrightarrow & \mathcal{A}^o(X)^{(p)*} & \longrightarrow 0, \end{array}$$

we see that it suffices to show that the kernel of $H_x^2(X,\mathcal{B}_p^o) \longrightarrow H^0(X,\mathcal{A}_p^o)^*$ has order $[\Phi'(k)^{(p)}][\Phi(k)^{(p)}]$.

The map $H_x^2(X,\mathcal{B}_p^o) \longrightarrow H^0(X,\mathcal{A}_p^o)^*$ is the composite of the maps

$$H_x^2(X,\mathcal{B}_p^o) \longrightarrow H_x^2(X,\mathcal{B}_p) \longrightarrow H^0(X,\mathcal{A}_p)^* \longrightarrow H^0(X,\mathcal{A}_p^o)^*.$$

The middle map is an isomorphism (6.1) and the remaining two maps are surjective with kernels respectively $H_x^1(X,\Phi_p') = H^1(x,\Phi_p')$ and $H^0(x,\Phi_p)^*$. The shows that the kernel of $H_x^2(X,\mathcal{B}_p^o) \longrightarrow H^0(X,\mathcal{A}_p^o)^*$ has the required order. (See also [Milne (1970/72)].)

Once (7.8) is acquired, the proofs of (2.5) to (2.10) apply when the base ring has characteristic p. We merely list the results.

Theorem 7.11. *The canonical pairing* $\Phi' \times \Phi \longrightarrow \mathbb{Q}/\mathbb{Z}$ *is nondegenerate; that is, Conjecture C.13 holds in this case.*

Corollary 7.12. *Suppose that* Γ' *and* Γ *are exact annihilators under the canonical pairing on* Φ' *and* Φ. *Then the map*

$$\mathcal{B}^{\Gamma'} \longrightarrow \mathcal{E}xt^1_{X_{sm}}(\mathcal{A}^{\Gamma},\mathbb{G}_m)$$

defined by the extension of the Poincaré biextension is an isomorphism (of sheaves on X_{sm}*).*

Theorem 7.13. *Assume that* Γ' *and* Γ *are exact annihilators. Then the pairing*

$$H^r(X,\mathcal{B}^{\Gamma'}) \times H^{2-r}_x(X,\mathcal{A}^{\Gamma}) \longrightarrow H^3_x(X,\mathbb{G}_m) \approx \mathbb{Q}/\mathbb{Z}$$

defined by the canonical biextension of $(\mathcal{B}^{\Gamma'},\mathcal{A}^{\Gamma})$ *by* $\mathbb{G}_m$ *induces an isomorphism* $H^2_x(X,\mathcal{A}^{\Gamma}) \xrightarrow{\approx} \mathcal{B}^{\Gamma'}(X)^*$ *of discrete groups for* $r = 0$ *and an isomorphism of finite groups* $H^1(X,\mathcal{B}^{\Gamma'}) \xrightarrow{\approx} \mathcal{A}^{\Gamma}(X)^*$ *for* $r = 1$. *For* $r \neq 0,1$, *both groups are zero.*

Remark 7.14. Assume that R is an excellent Henselian discrete valuation ring, and let $\hat{X} = \operatorname{Spec} \hat{R}$. Then it follows from (I.3.10) that the maps $H^r_x(X,\mathcal{A}^{\Gamma}) \longrightarrow H^r_x(\hat{X},\mathcal{A}^{\Gamma})$ are isomorphisms for all r, and $H^r(X,\mathcal{A}^{\Gamma}) \longrightarrow H^r(\hat{X},\mathcal{A}^{\Gamma})$ is an isomorphism for all $r > 0$. The map $A(X) \longrightarrow A(\hat{X})$ is injective and maps onto the torsion subgroup of $A(\hat{X})$; $A(\hat{X})$ is the completion of $A(X)$ for the topology of open subgroups of finite index.

Write $\mathcal{B}\{n\}$ for the complex $\mathcal{B} \xrightarrow{n} \mathcal{B}^{n\Phi'}$ and $\mathcal{A}\{n\}$ for the complex $\mathcal{A}^{\Phi n} \xrightarrow{n} \mathcal{A}^{o}$. The pairings $\mathcal{B}\otimes^{L}\mathcal{A}^{o} \longrightarrow \mathbb{G}_m[1]$ and $\mathcal{B}^{n\Phi'}\otimes^{L}\mathcal{A}^{\Phi n} \longrightarrow \mathbb{G}_m[1]$ defined by the Poincaré biextension induce a pairing $\mathcal{B}\{n\}\otimes^{L}\mathcal{A}\{n\} \longrightarrow \mathbb{G}_m$ in the derived category of sheaves on X_{sm}.

Theorem 7.15. *The map* $\mathcal{B}\{n\}\otimes^{L}\mathcal{A}\{n\} \longrightarrow \mathbb{G}_m$ *defines nondegenerate pairings*

$$H^r(X,\mathcal{B}\{n\}) \times H_x^{3-r}(X,\mathcal{A}\{n\}) \longrightarrow H_x^3(x,\mathbb{G}_m) \approx \mathbb{Q}/\mathbb{Z}$$

for all r.

Corollary 7.16. *Assume that* n *is prime to* p *or that* A *has semistable reduction. Then for all* r, *there is a canonical nondegenerate pairing*

$$H_x^r(X_{fl},\mathcal{B}_n) \times H^{3-r}(X_{fl},\mathcal{A}_n) \longrightarrow H_x^3(X_{fl},\mathbb{G}_m) \approx \mathbb{Q}/\mathbb{Z}.$$

Curves over X

It is possible to prove an analogue of Theorem 2.11. Note that the methods in [Artin and Milne (1976), §5] can be used to prove a more general result.

Notes: Theorem 7.1 was proved independently by the author [Milne (1970/72), (1973)] and by Artin and Mazur (unpublished). The above proof is new. Theorem 7.8 was also proved by the author [Milne (1970/72)] ([Shatz (1967)] contains a proof for elliptic curves with Tate parametrizations). The stronger forms of it are due to McCallum [McCallum (1986)].

§8 Global results: curves over finite fields, finite sheaves

Throughout this section, X will be a complete smooth curve over a finite field k. The function field of X is denoted by K, and p is the characteristic of k. For a sheaf F on an open subscheme U of X, $H^r_c(U,F)$ denotes the cohomology group with compact support as defined in (0.6b). Thus, there exist exact sequences

$$\cdots \longrightarrow H^r_c(U,F) \longrightarrow H^r(U,F) \longrightarrow \bigoplus_{v\in X-U} H^r(K_v,F) \longrightarrow \cdots$$

$$\cdots \longrightarrow H^r_c(V,F) \longrightarrow H^r_c(U,F) \longrightarrow \bigoplus_{v\in U-V} H^r(\hat{\mathcal{O}}_v,F) \longrightarrow \cdots$$

with K_v and $\hat{\mathcal{O}}_v$ the *completions* of K and $\mathcal{O}_v$ at v. With this definition, a short exact sequence of sheaves gives rise to a long exact cohomology sequence, and there is a pairing between Ext groups and cohomology groups with compact support (see 0.4b and 0.4e), but in general the flat cohomology groups with compact support will not agree with the étale groups even for a sheaf arising from an étale sheaf or a smooth group scheme (contrast 0.4d).

The duality theorem

When N is a quasi-finite flat group scheme on U, we endow $H^r(U,N)$ with the discrete topology.

Lemma 8.1. *For any quasi-finite étale group scheme* N *on an open subscheme* U *of* X, $H^r_c(U,N) = H^r_c(U_{et},N)$ *all* r.

Proof: Let $\tilde{K}_v$ be the field of fractions of $\mathcal{O}^h_v$. Then, as we observed in (6.13), $H^r(\tilde{K}_v,N) \xrightarrow{\approx} H^r(K_v,N)$ for all r, and as $H^r(\tilde{K}_{v,et},N) \xrightarrow{\approx} H^r(\tilde{K}_v,N)$, the lemma follows from comparing the two sequences

$$\begin{array}{ccccccc}
\cdots \longrightarrow H^r_c(U_{et},N) & \longrightarrow & H^r(U_{et},N) & \longrightarrow & \bigoplus_{v\in X-U} H^r(\tilde{K}_{v,et},N) & \longrightarrow & \cdots \\
\downarrow & & \downarrow \approx & & \downarrow \approx & & \\
\cdots \longrightarrow H^r_c(U,N) & \longrightarrow & H^r(U,N) & \longrightarrow & \bigoplus_{v\in X-U} H^r(K_v,N) & \longrightarrow & \cdots .
\end{array}$$

Theorem 8.2. *Let N be a finite flat group scheme over an open subscheme U of X. For all r, the canonical pairing*

$$H^r(U,N^D) \times H^{3-r}_c(U,N) \longrightarrow H^3_c(U,\mathbb{G}_m) \approx \mathbb{Q}/\mathbb{Z}$$

defines isomorphisms $H^{3-r}_c(U,N) \xrightarrow{\approx} H^r(U,N^D)^*$.

Proof: After Lemma 8.1 and Theorem II.3.1, it suffices to prove the theorem for a group scheme killed by a power of p. We first need some lemmas.

Lemma 8.3. *Let*

$$0 \longrightarrow N' \longrightarrow N \longrightarrow N'' \longrightarrow 0$$

be an exact sequence of finite flat group schemes on U. If the theorem is true for N' and N'', then it is true for N.

Proof: Because the groups are discrete, the Pontryagin dual of

$$\cdots \longrightarrow H^r(U,N''^D) \longrightarrow H^r(U,N^D) \longrightarrow H^r(U,N'^D) \longrightarrow \cdots$$

is exact. Therefore one can apply the five-lemma to the obvious diagram.

Lemma 8.4. *Let V be an open subscheme of U. The theorem is true for N on U if and only if it is true for* $N|V$ *on V.*

Proof: This follows from (7.1) and the diagram

$$\begin{array}{ccccccc} \cdots \longrightarrow H_c^{3-r}(V,N) & \longrightarrow & H_c^{3-r}(U,N) & \longrightarrow & \bigoplus_{v\in U-V} H^{3-r}(\hat{\mathcal{O}}_v,N) & \longrightarrow \cdots \\ \downarrow & & \downarrow & & \downarrow \approx & \\ \cdots \longrightarrow H^r(V,N^D)^* & \longrightarrow & H^r(U,N^D)^* & \longrightarrow & \bigoplus_{v\in U-V} H_v^r(\hat{\mathcal{O}}_v,N^D)^* & \longrightarrow \cdots . \end{array}$$

Lemma 8.5. *The theorem is true if* U = X *and* N *or its dual have height one; moreover, the groups involved are finite.*

Proof: Assume first that N has height one. The first exact sequence in §5 yields a cohomology sequence

$$\cdots \longrightarrow H^r(X,N^D) \longrightarrow H^r(X,\omega_N) \longrightarrow H^r(X,\omega_N^{(p)}) \longrightarrow \cdots$$

and the second a sequence

$$\cdots \longrightarrow H^{r+1}(X,N) \longrightarrow H^r(X,n\otimes\Omega^1_{X'}) \longrightarrow H^r(X,n\otimes\Omega^1_X) \longrightarrow \cdots .$$

The canonical pairing of complexes (5.10) together with the usual duality theorem for coherent sheaves on a curve show that the finite-dimensional k-vector spaces $H^r(X,\omega_N)$ and $H^r(X,\omega_N^{(p)})$ are the k-linear (hence Pontryagin) duals of the k-vector spaces $H^{1-r}(X,n\otimes\Omega^1_X)$ and $H^{1-r}(X,n\otimes\Omega^1_X)$, and moreover that there is a commutative diagram

$$\begin{array}{ccccccc} \cdots \longrightarrow H^{3-r}(X,N) & \longrightarrow & H^{2-r}(X,n\otimes\Omega^1_{X'}) & \longrightarrow & H^{2-r}(X,n\otimes\Omega^1_X) & \longrightarrow \cdots \\ \downarrow & & \downarrow \approx & & \downarrow \approx & \\ \cdots \longrightarrow H^r(X,N^D)^* & \longrightarrow & H^{r-1}(X,\omega_N^{(p)})^* & \longrightarrow & H^{r-1}(X,\omega_N)^* & \longrightarrow \cdots . \end{array}$$

The diagram gives an isomorphism $H^{3-r}(X,N) \xrightarrow{\approx} H^r(X,N^D)^*$ for all r, and (5.11) shows that this is the map in the statement of the theorem.

In this case the groups $H^r(X,N)$ and $H^r(X,N^D)$ are finite, and the statement of the theorem is symmetric between N and N^D. Therefore, the theorem is proved also if the dual of N has height one.

We now prove the theorem. Let N be a finite group scheme over

U. Lemma 8.4 allows us to replace U by a smaller open subset, and so we can assume that N has a composition series all of whose quotients have height 1 or are the Cartier duals of groups of height 1. Now Lemma 8.3 allows us to assume that N (or its dual) has height 1. According to Proposition B.4, N_K extends to a finite flat group scheme $\mathcal{N}$ on X which is of height one (or has a dual of height one). After again replacing U by a smaller open set, we can assume that $\mathcal{N}|U = N$. According to (8.5), the theorem is true for $\mathcal{N}$ on X, and (8.4) shows that this implies the same result for $\mathcal{N}|U = N$.

Corollary 8.6. *Let* N *be a finite flat group scheme on* U. *For all* $r < 2p-2$, *the pairing*

$$\mathrm{Ext}^r_U(N,\mathbb{G}_m) \times H^{3-r}_c(U,N) \longrightarrow H^3_c(U,\mathbb{G}_m) \approx \mathbb{Q}/\mathbb{Z}$$

defines isomorphisms $H^{3-r}_c(U,N) \longrightarrow \mathrm{Ext}^r_U(N,\mathbb{G}_m)^*$.

Proof: Under the hypotheses, $H^r(U,N^D) = \mathrm{Ext}^r_U(N,\mathbb{G}_m)$ (see the proof of (1.5)), and so this follows immediately from the theorem.

Corollary 8.7. *Let* N *be a quasi-finite flat group scheme over* U *whose* p*-primary component* N(p) *is finite over* U. *Let* N^D *be the complex of sheaves such that*

$$N^D(\ell) = \begin{cases} \mathcal{H}om_{U_{fl}}(N(\ell),\mathbb{G}_m) & \ell = p \\ f^*R\mathcal{H}om_{U_{et}}(N(\ell),\mathbb{G}_m) & \ell \neq p. \end{cases}$$

Then the pairing

$$H^r(U,N^D) \times H^{3-r}_c(U,N) \longrightarrow H^3_c(U,\mathbb{G}_m) \approx \mathbb{Q}/\mathbb{Z}$$

defines isomorphisms $H^{3-r}_c(U,N) \xrightarrow{\approx} H^r(U,N^D)^*$ *for all* r.

Proof: For the p-primary component of N, this follows directly from the theorem; for the ℓ-primary component, $\ell \neq p$, Lemma 8.1 shows that it follows from (II.1.11b).

Problem 8.8. The group $H^r(U,N^D)$ is torsion, and so $H^r(U,N^D)^*$ is a compact topological group. The isomorphism in the theorem therefore gives $H^r_c(U,N)$ a natural topology as a compact group. Find a direct description of this topology.

Euler–Poincaré characteristics

When $U \neq X$, the groups $H^r(U,N)$ will usually be infinite, even when N is a finite étale group scheme over U (for example, $H^1(\mathbb{A}^1,\mathbb{Z}/p\mathbb{Z}) = k[T]/\wp k[T]$, which is infinite). This restricts us to considering the case $U = X$.

Lemma 8.9. *For any finite flat group scheme* N *over* X, *the groups* $H^r(X,N)$ *are finite.*

Proof: When N or its dual have height one, we saw that the groups are finite in (8.5). In the general case, N_K will have a filtration all of whose quotients are of height one or have duals that are of height one, and by taking the closures of the groups in the filtration, we get a similar filtration for N (cf. B.1). The lemma now follows by induction on the length of the filtration.

When N is a finite flat group scheme on X, we define

$$\chi(X,N) = \frac{[H^0(X,N)][H^2(X,N)]}{[H^1(X,N)][H^3(X,N)]}.$$

Problem 8.10. Find a formula for $\chi(X,N)$.

Let $\bar{X} = X\otimes_k k_s$, and let $\Gamma = \mathrm{Gal}(k_s/k)$. If the groups $H^r(\bar{X},N)$ are finite, then it follows immediately from the exact sequences

$$0 \longrightarrow H^{r-1}(\bar{X},N)_{\Gamma} \longrightarrow H^{r}(X,N) \longrightarrow H^{r}(\bar{X},N)^{\Gamma} \longrightarrow 0$$

given by the Hochschild-Serre spectral sequence for $\bar{X}$ over X, that $\chi(X,N) = 1$. When N is étale, the finiteness theorem in étale cohomology [Milne (1980), VI.2.1] shows that $H^r(\bar{X},N)$ is finite, and a duality theorem (see §11) shows that the same is true when N is the dual of an étale group. Otherwise the groups are often infinite. For example,

$$H^1(\bar{X},\alpha_p) = \mathrm{Ker}(F\colon H^1(\bar{X},\mathcal{O}_{\bar{X}}) \longrightarrow H^1(\bar{X},\mathcal{O}_{\bar{X}})),$$

which is finite if and only if the curve $\bar{X}$ has an invertible Hasse-Witt matrix. Nevertheless, $\chi(X,\alpha_p) = 1$.

Let $N = N^{\mathcal{L}}_{a,0}$ in the Oort-Tate classification. Then (cf. 5.4), we have an exact sequence

$$0 \longrightarrow N \longrightarrow \mathcal{L} \xrightarrow{\ \varphi\ } \mathcal{L}^{\otimes p} \longrightarrow 0$$

with $\varphi(z) = z^{\otimes p} - a\otimes z$. Therefore $\chi(X,N) = q^{\chi(\mathcal{L}) - \chi(\mathcal{L}^{\otimes p})}$ where q is the order of k. But the Riemann-Roch theorem shows that

$$\chi(\mathcal{L}) = \deg(\mathcal{L}) + 1 - g$$
$$\chi(\mathcal{L}^{\otimes p}) = p\deg(\mathcal{L}) + 1 - g,$$

and so

$$\chi(X,N) = p^{(p-1)\deg(\mathcal{L})}.$$

It is easy to construct N for which $\deg(\mathcal{L}) \neq 0$: take $\mathcal{L}_0$ to be any invertible sheaf of degree > 0; then for some $r > 0$, $\Gamma(X,\mathcal{L}_0^{\otimes r(p-1)}) \neq 0$, and so we can take $N = N^{\mathcal{L}}_{a,0}$ with $\mathcal{L} = \mathcal{L}_0^{\otimes r}$ and a any element of $\Gamma(X,\mathcal{L}^{\otimes(p-1)})$.

Problem 8.11. As we mentioned above, the groups $H^r_c(U,N)$ have canonical compact topologies. Is it possible extend the above discussion to $\chi_c(U,N)$ by using Haar measures?

Remark 8.12. We show that, for any scheme Y proper and smooth over a finite field k of characteristic p, the groups $H^r(Y,\mu_p)$ are finite for all r and $\chi(Y,\mu_p) \stackrel{\mathrm{df}}{=} \prod [H^r(Y,\mu_p)]^{(-1)^r} = 1$. From the exact sequence (special case of (5.6))

$$0 \longrightarrow R^1 f_* \mu_p \longrightarrow \Omega^1_{Y/k,cl} \xrightarrow{C-1} \Omega^1_{Y/k} \longrightarrow 0$$

we see that it suffices to show that the groups $H^r(Y,\Omega^1_{Y/k,cl})$ and $H^r(Y,\Omega^1_{Y/k})$ are all finite and that $\chi(Y,\Omega^1_{Y/k,cl}) = \chi(Y,\Omega^1_{Y/k})$. Consider the exact sequences of sheaves on Y_{et},

$$0 \longrightarrow \mathcal{O}_Y \xrightarrow{F} \mathcal{O}_Y \longrightarrow d\mathcal{O}_Y \longrightarrow 0$$

and

$$0 \longrightarrow d\mathcal{O}_Y \longrightarrow \Omega^1_{Y/k,cl} \xrightarrow{C} \Omega^1_{Y/k} \longrightarrow 0.$$

From the cohomology sequence of the first sequence, we find that $H^r(Y_{et},d\mathcal{O}_Y)$ is finite for all r and is zero for $r > \dim(Y)$; moreover $\chi(Y,d\mathcal{O}_Y) = 1$. From the cohomology sequence of the second sequence, we find that $H^r(Y_{et},\Omega^1_{Y/k,cl})$ is also finite for all r and zero for $r > \dim(Y)$, and that $\chi(Y,\Omega^1_{Y/k,cl}) = \chi(Y,\Omega^1_{Y/k})$, which is what we had to prove.

Remark 8.13. It has been conjectured that for any scheme proper Y over Spec $\mathbb{Z}$, the cohomological Brauer group $H^2(Y,\mathbb{G}_m)$ is finite. The last remark shows that when the image of the structure map of Y is a single point (p) in Spec $\mathbb{Z}$, then $H^r(Y,\mathbb{G}_m)_p$ is finite for all r.

Notes: In the very special case that U = X and N is constant without local-local factors, Theorem 8.2 can be found in [Milne (1977), Thm A.2] with a similar proof. The case U = X and a general N is implicitly contained in [Artin and Milne (1975)].

§9 Global results: curves over finite fields, Néron models

The notations are the same as those in the last section. In particular, X is a complete smooth curve over a finite field, U is an open subscheme of X, and $H^r_c(U,F)$ is defined so that the sequence

$$\dots \longrightarrow H^r_c(U,F) \longrightarrow H^r(U,F) \longrightarrow \bigoplus_{v \notin U} H^r(K_v,F) \longrightarrow \dots$$

is exact with K_v the *completion* of K.

Let A be an abelian variety over K, and let $\mathcal{A}$ and $\mathcal{B}$ be the Néron models over U of A and its dual B. If either A has semistable reduction or n is prime to p, there are exact sequences

$$0 \longrightarrow \mathcal{B}_n \longrightarrow \mathcal{B} \longrightarrow \mathcal{B}^{n\Phi'} \longrightarrow 0$$

$$0 \longrightarrow \mathcal{A}_n \longrightarrow \mathcal{A}^{\Phi n} \longrightarrow \mathcal{A}^{o} \longrightarrow 0.$$

The Poincaré biextension of (B,A) by $\mathbb{G}_m$ extends uniquely to biextensions of $(\mathcal{B},\mathcal{A}^o)$ by $\mathbb{G}_m$ and of $(\mathcal{B}^{n\Phi'},\mathcal{A}^{\Phi n})$ by $\mathbb{G}_m$. Therefore (cf. §1), we get a canonical pairing

$$\mathcal{B}_n \times \mathcal{A}_n \longrightarrow \mathbb{G}_m$$

in this case.

Proposition 9.1. *Assume that* n *is prime to* p *or that* $\mathcal{A}$ *has semistable reduction at all primes of* K. *Then the pairing*

$$H^r(U,\mathcal{B}_n) \times H^{3-r}_c(U,\mathcal{A}_n) \longrightarrow H^3_c(U,\mathbb{G}_m) \approx \mathbb{Q}/\mathbb{Z}$$

induces an isomorphism $H^{3-r}_c(U,\mathcal{A}_n) \longrightarrow H^r(U,\mathcal{B}_n)^*$ *for all* r.

Proof: Let V be an open subset where A (hence also B) has good reduction. Over V, $\mathcal{A}_n$ is a finite flat group scheme with Cartier dual $\mathcal{B}_n$, and so the proposition is a special case of (8.2). To pass from V to U, we use the diagram

$$\begin{array}{ccccccc} \cdots \to & H_c^{3-r}(V,\mathcal{A}_n) & \to & H_c^{3-r}(U,\mathcal{A}_n) & \to & \bigoplus_{v\in U-V} H^{3-r}(\hat{\mathcal{O}}_v,\mathcal{A}_n) & \to \cdots \\ & \downarrow\approx & & \downarrow & & \downarrow & \\ \cdots \to & H^r(V,\mathcal{B}_n)^* & \to & H^r(U,\mathcal{B}_n)^* & \to & \bigoplus_{v\in U-V} H_v^r(\mathcal{O}_v^h,\mathcal{B}_n)^* & \to \cdots . \end{array}$$

Obviously $H_v^r(\mathcal{O}_v^h,\mathcal{B}_n) \xrightarrow{\approx} H_v^r(\hat{\mathcal{O}}_v,\mathcal{B}_n)$, and Corollary 7.16 shows that $H^{3-r}(\hat{\mathcal{O}}_v,\mathcal{A}_n) \to H_v^r(\hat{\mathcal{O}}_v,\mathcal{B}_n)^*$ is an isomorphism. Therefore the proposition follows from the five-lemma.

As usual, we write $\mathcal{A}/\mathcal{A}^o = \Phi = \oplus\, i_{v*}\Phi_v$ (finite sum).

Proposition 9.2. *There are exact sequences*

$$0 \to \mathcal{A}^o(X) \to A(K) \to \oplus\, \Phi_v(k(v)) \to H^1(X,\mathcal{A}^o) \to Ш(K,A) \to 0$$

and

$$0 \to Ш(K,A) \to H^1(X,\mathcal{A}) \to \oplus\, H^1(v,\Phi_v).$$

Proof: Let U be an open subscheme of X such that $\mathcal{A}|U$ is an abelian scheme; in particular, $\mathcal{A}|U = \mathcal{A}^o|U$. As in (II.5.5), we have an exact sequence

$$0 \to H^1(U,\mathcal{A}) \to H^1(K,A) \to \prod_{v\in U} H^1(K_v,A).$$

There is an exact sequence

$$\begin{array}{ccccccccccccc} 0 \to & \mathcal{A}^o(X) & \to & \mathcal{A}(U) & \to & \bigoplus_{X-U} H_v^1(\mathcal{O}_v^h,\mathcal{A}^o) & \to & H^1(X,\mathcal{A}^o) & \to & H^1(U,\mathcal{A}) & \to & \bigoplus_{X-U} H_v^2(\mathcal{O}_v^h,\mathcal{A}^o) \\ & & & \| & & \| & & & & & & \| \\ & & & A(K) & & \oplus\, \Phi_v(k(v)) & & & & & & \oplus\, H^1(\tilde{K}_v,A) \ ; \end{array}$$

(for the second two equalities, see (7.10)). According to (I.3.10), the field of fractions $\tilde{K}_v$ of $\mathcal{O}_v^h$ can be replaced by K_v in $H^1(\tilde{K}_v,A)$. The kernel-cokernel exact sequence of

$$H^1(K,A) \to \prod_{v\in X} H^1(K_v,A) \to \prod_{v\in U} H^1(K_v,A)$$

is an exact sequence

$$0 \longrightarrow Ш(K,A) \longrightarrow H^1(U,\mathcal{A}) \longrightarrow \prod_{v\in X-U} H^1(K_v,A),$$

and it follows from this and the six-term sequence that $Ш(K,A)$ is the image of $H^1(X,\mathcal{A}^o)$ in $H^1(U,\mathcal{A})$. The first exact sequence can now be obtained by truncating the six-term exact sequence, and the second sequence can be obtained by comparing the last sequence above with

$$0 \longrightarrow H^1(X,A) \longrightarrow H^1(U,\mathcal{A}) \longrightarrow \prod_{v\in X-U} H^2_v(\mathcal{O}^h_v,\mathcal{A}).$$

Corollary 9.3. *For any $\Gamma \subset \Phi$, $H^1(X,\mathcal{A}^\Gamma)$ is torsion and of cofinite-type.*

Proof: It suffices to prove this with $\Gamma = \emptyset$. Then the group equals $Ш(K,A)$, which is obviously torsion. It remains to show that $Ш(K,A)_p$ is finite. There is an elementary proof of this in [Milne (1970b)]. It can also be proved by using (8.12) in the case of a surface to show that $Ш(K,A)_p$ is finite when A is a Jacobian variety, and then embedding an arbitrary abelian variety into a Jacobian to deduce the general case.

Let $\mathcal{B}/\mathcal{B}^o \overset{\text{df}}{=} \Phi' = \oplus\, i_{v*}\Phi'_v$. For any subgroups $\Gamma = \oplus\, i_{v*}\Gamma_v$ and $\Gamma' = \oplus\, i_{v*}\Gamma'_v$ of Φ and Φ', the Poincaré biextension over K extends to a biextension over U if and only if each Γ_v annihilates each Γ'_v. In this case we get a map

$$\mathcal{B}^{\Gamma'} \otimes^L \mathcal{A}^\Gamma \longrightarrow \mathbb{G}_m[1].$$

Theorem 9.4. *Suppose that Γ_v and Γ'_v are exact annihilators at each closed point v.*

(a) *The the kernels of the pairing*

$$H^1(U,\mathcal{B}^{\Gamma'}) \times H^1_c(U,\mathcal{A}^{\Gamma})_{\mathrm{tors}} \longrightarrow \mathbb{Q}/\mathbb{Z}$$

are exactly the divisible groups.

(b) *If* Ш$^1(K,A)$ *is finite, then* $H^0(U,\mathcal{A}^{\Gamma})^\wedge$ *is dual to* $H^2_c(U,\mathcal{B}^{\Gamma'})$.

Proof: If A has good reduction on U, this can be proved by the same argument as in (II.5.2) (using 8.2). This remark shows that the theorem is true for some $V \subset U$, and to pass from V to U one uses (7.13).

Corollary 9.5. *The Cassels-Tate pairing* (II.5.7a)

$$Ш^1(K,B) \times Ш^1(K,A) \longrightarrow \mathbb{Q}/\mathbb{Z}$$

annihilates only the divisible subgroups.

Proof: This follows from (9.4) and the diagram

$$\begin{array}{ccccccc} \oplus\, \Phi'_v(k(v)) & \longrightarrow & H^1(X,\mathcal{B}^o) & \longrightarrow & Ш^1(K,B) & \longrightarrow & 0 \\ \downarrow & & \downarrow & & \downarrow & & \\ \oplus\, H^1(v,\Phi_v)^* & \longrightarrow & H^1(X,\mathcal{A})^* & \longrightarrow & Ш^1(K,A)^* & \longrightarrow & 0 \end{array}$$

because (7.11) shows that the first vertical map is an isomorphism.

Application to the conjecture (B-S/D) for Jacobians

Recall that the index of a curve C over a field F is the greatest common divisor of the degrees of the fields F' over F such that C has a rational point in F'. Equivalently, it is the least positive degree of a divisor on C.

In this subsection, we let Y be a regular connected surface over k, and we let $\pi\colon Y \longrightarrow X$ be a proper morphism such that

(i) the generic fibre π is a smooth geometrically connected curve over K;

(ii) for all $v \in X$, the curve Y_{K_v} has index one.

We write A for the Jacobian of the generic fibre of π.

Proposition 9.6. (a) *The orders of the Brauer group of* Y *and the Tate-Shafarevich group of* A *are related by*

$$\delta^2[\mathrm{Br}(Y)] = [Ш(K,A)]$$

where δ *is the index of* Y_K.

(b) *The conjecture of Artin and Tate* [Tate (1965/66), Conjecture C] *holds for* Y *if and only if the conjecture of Birch and Swinnerton-Dyer* (I.7, B-S/D) *holds for* A.

Proof: (a) Once (9.5) is acquired, the proof in [Milne (1981)] applies.

(b) It is proved in [Gordon (1979)] that (a) implies (b).

Corollary 9.7. *Let* A *be a Jacobian variety over* K *arisirg as above. The following statements are equivalent:*

(i) *for some prime* ℓ (ℓ = p *is allowed*), *the* ℓ*-primary component of* Ш(K,A) *is finite;*

(ii) *the* L-*series* L(s,A) *of* A *has a zero at* s = 1 *of order equal to the rank of* A(K);

(iii) *the Tate-Shafarevich group* Ш(K,A) *is finite, and the conjecture of Birch and Swinnerton-Dyer is true for* A.

Proof: After part (b) of the theorem, the equivalence of the three statements (i), (ii), and (iii) for A follows from the equivalence of the corresponding statements for Y [Milne (1975)].

The behaviour of conjecture (B-S/D) with respect to p-isogenies.

We partially extend Theorem I.7.3 to the case of p-isogenies.

Theorem 9.8. *Let $f\colon A \longrightarrow B$ be an isogeny of abelian varieties over K, and let N be the kernel of its extension $f\colon \mathcal{A} \longrightarrow \mathcal{B}$ to the Néron models of A and B over X. Assume that either the degree of f is prime to p or that A and B have semistable reduction at all points of X and $H^r(X\otimes k_s, N)$ is finite for all r. Then the conjecture of Birch and Swinnerton-Dyer is true for A if and only if it is true for B.*

Proof: After (I.7.3), we may assume that the degree of f is a power of p. The initial calculations in (I.7) show that in order to prove the equivalence, one must show that

$$z(f(K)) = \Big(\prod_{v\in X} \frac{\mu_v(A,\omega_A)}{\mu_v(B,\omega_B)}\Big).z(f^t(K)).z(Ш^1(f)).$$

(Note that we can not replace the local terms with $z(f(K_v))$ because they cokernel of $f(K_v)$ may not be finite.) Let $N^o = \mathrm{Ker}(f^o\colon \mathcal{A}^o \longrightarrow \mathcal{B}^o)$. From the exact sequence

$$0 \longrightarrow N^o \longrightarrow \mathcal{A}^o \longrightarrow \mathcal{B}^o \longrightarrow 0$$

we get an exact sequence

$$0 \longrightarrow N^o(X) \longrightarrow \mathcal{A}^o(X) \longrightarrow \mathcal{B}^o(X) \longrightarrow H^1(X,N^o) \longrightarrow H^1(X,\mathcal{A}^o) \longrightarrow H^1(X,\mathcal{B}^o)$$
$$\longrightarrow H^2(X,N^o) \longrightarrow H^2(X,\mathcal{A}^o) \longrightarrow H^2(X,\mathcal{B}^o) \longrightarrow H^3(X,N^o) \longrightarrow 0.$$

The sequence shows that

$$z(H^0(f^o)) = z(H^1(f^o)).z(H^2(f^o))^{-1}.\chi(X,N^o).$$

But $\chi(X,N^o) = 1$ (because we have assumed that the groups $H^r(X\otimes k_s,N)$ are finite; cf. the discussion following (8.10)), and (9.3) shows that $z(H^2(f^o))^{-1} = z(H^0(f^t))$. Therefore it remains to show that

$$z(f(K))/z(H^0(f^o)) = z(Ш(f))/z(H^1(f^o)).\Big(\prod_{v\in X} \frac{\mu_v(A,\omega_A)}{\mu_v(B,\omega_B)}\Big).$$

From (9.2), we get a diagram

$$\begin{array}{ccccccccccc} 0 \longrightarrow & \mathcal{A}^{\circ}(X) & \longrightarrow & A(K) & \longrightarrow & \oplus\, \Phi_v(k(v)) & \longrightarrow & H^1(X,\mathcal{A}^{\circ}) & \longrightarrow & Ш(K,A) & \longrightarrow 0 \\ & \downarrow & & \downarrow & & \downarrow & & \downarrow & & \downarrow & \\ 0 \longrightarrow & \mathcal{B}^{\circ}(X) & \longrightarrow & B(K) & \longrightarrow & \oplus\, \Phi'_v(k(v)) & \longrightarrow & H^1(X,\mathcal{B}^{\circ}) & \longrightarrow & Ш(K,B) & \longrightarrow 0, \end{array}$$

which shows that

$$z(f(K))/z(H^0(f^{\circ})) = z(Ш(f))/z(H^1(f^{\circ})).\prod \frac{[\Phi_v(k(v))]}{[\Phi'_v(k(v))]}.$$

It remains to show that $\dfrac{\mu_v(A,\omega_A)}{\mu_v(B,\omega_B)} = \dfrac{[\Phi_v(k(v))]}{[\Phi'_v(k(v))]}$, but this follows from the formula

$$\mu_v(A,\omega_v) = [\Phi_v(k(v))]/L_v(1,A),$$

ω_v a Néron differential on $A\otimes K_v$, for which we have no reference to offer the reader.

Remark 9.9. It was first pointed out in [Milne (1970), p296] that, because the group $H^1(X\otimes k_s,N)$ may be infinite, $Ш^1(K,A)_p$ may be infinite when K is a function field with algebraically closed field of constants. This phenomenon has been studied in the papers [Vvedens'kii (1979a), (1979b), (1980/81)], which give criteria for the finiteness $Ш_p$ (and hence of the groups $H^r(X\otimes k_s,N)$ in the above theorem).

Problem 9.10. Prove the above result for every isogeny $f\colon A \longrightarrow B$.

There seems to be some hope that the method used in (II.5) may be effective in the general case; the groups are no longer finite, but they are compact.

Duality for surfaces

It is possible to prove a similar result to (3.8) (see also

[Artin and Milne (1976)]).

§10 Local results: equicharacteristic, perfect residue field

Throughout this section, X = Spec R where R is a complete discrete valuation ring with algebraically closed residue field k. We let X_i = Spec R_i, where $R_i = R/m^{i+1}$.

Finite group schemes

In the equicharacteristic case, the Greenberg construction becomes a special case of Weil restriction of scalars: for each $i \geq 0$, the k-algebra structure on R_i defines a map $\alpha_i\colon X_i \longrightarrow$ Spec k, and for any group scheme G over X, $\mathcal{G}_i(G) = \mathrm{Res}_{X_i/k}G$. We write $\mathcal{G}(G)$ for the pro-algebraic group $(\mathcal{G}_i(G))_{i\geq 0}$ on X.

Lemma 10.1. (a) *The functor* $G \mapsto \mathcal{G}(G)$ *from smooth group schemes on* X *to pro-algebraic groups on* Spec(k) *is exact.*

(b) *If* G *is smooth and has connected fibres, then* $\mathcal{G}_i(G)$ *is smooth and connected for all* i.

(c) *If* N *is a finite flat group scheme of height one, then* $\mathcal{G}_i(N)$ *is connected for all* i.

Proof: See [Bester (1978), 1.1, 1.2].

Let π_0 be the functor sending a pro-algebraic group scheme over k to its maximal étale quotient, and let π_r be the rth left derived functor of π_0. Write $\pi_r(G)$ for $\pi_r(\mathcal{G}(G))$. When N is a finite flat group scheme X, we choose a resolution of N by smooth connected formal groups

$$0 \longrightarrow N_i \longrightarrow G_i \longrightarrow H_i \longrightarrow 0, \quad i \geq 0,$$

and define $\mathcal{F}_i(N) = \mathrm{Coker}(\pi_1(G_i) \longrightarrow \pi_1(H_i))$. It is an pro-étale group scheme on X, and we let $\mathcal{F}(N)$ be the pro-étale group scheme $(\mathcal{F}_i(N))_{i\geq 0}$.

Write α_i for the morphism $X_i \longrightarrow \mathrm{Spec}\ k$, and $R^r\hat{\alpha}_*$ for the functor $F \mapsto \varprojlim R^r\alpha_{i,*}(F|X_i)$. If G is a group scheme of finite-type over X, then $\hat{\alpha}_*G = \mathcal{G}(G)$, and if N is a finite flat group scheme over X, then $R^1\hat{\alpha}_*N$ is representable by a pro-algebraic group scheme over k.

Lemma 10.2. *Let N be a finite flat group scheme over X.*

(a) *The group scheme $\mathcal{F}(N)$ is independent of the choice of the resolution.*

(b) *A short exact sequence*

$$0 \longrightarrow N' \longrightarrow N \longrightarrow N'' \longrightarrow 0$$

of finite group schemes defines an exact sequence

$$0 \longrightarrow \pi_1(N') \longrightarrow \pi_1(N) \longrightarrow \pi_1(N'') \longrightarrow \mathcal{F}(N') \longrightarrow \mathcal{F}(N) \longrightarrow \mathcal{F}(N'') \longrightarrow 0.$$

(c) *There is an exact sequence*

$$0 \longrightarrow \pi_0(N) \longrightarrow \mathcal{F}(N) \longrightarrow \pi_1(R^1\hat{\alpha}_*N) \longrightarrow 0$$

of pro-sheaves on X.

Proof: See [Bester (1978), 3.2, 3.7, 3.9].

Write μ_∞ for the direct system of finite group schemes (μ_{p^n}).

Lemma 10.3. *There is a canonical isomorphism* $H^2_x(X,\mathcal{F}(\mu_\infty)) \xrightarrow{\approx} \mathbb{Q}_p/\mathbb{Z}_p$.

Proof: From the Kummer sequence

$$0 \longrightarrow \mu_{p^n} \longrightarrow \mathbb{G}_m \xrightarrow{p^n} \mathbb{G}_m \longrightarrow 0$$

we obtain an exact sequence

$$0 \longrightarrow \pi_1(\mathbb{G}_{m,X}) \longrightarrow \pi_1(\mathbb{G}_{m,X}) \longrightarrow \mathcal{F}(\mu_{p^n}) \longrightarrow 0.$$

This yields a cohomology sequence

$$0 \longrightarrow H^2_Z(X,\pi_1(\mathbb{G}_{m,X})) \longrightarrow H^2_Z(X,\pi_1(\mathbb{G}_{m,X})) \longrightarrow H^2_Z(\mathcal{F}(\mu_{p^n})) \longrightarrow 0.$$

But the higher cohomology groups of the universal covering group of $\mathcal{G}(\mathbb{G}_{m,X})$ are zero, and so

$$H^2_x(X,\pi_1(\mathbb{G}_{m,X})) = H^1_x(X,\mathbb{G}_{m,X}) = K^\times/R^\times = \mathbb{Z},$$

and so $H^2_Z(\mathcal{F}(\mu_{p^n})) \approx \mathbb{Z}/p^n\mathbb{Z}$.

Assume that N is killed by p^n. Then the pairing

$$N^D \times N \longrightarrow \mu_{p^n}$$

induces a pairing

$$N^D \times \mathcal{F}(N) \longrightarrow \mathcal{F}(\mu_{p^n}),$$

and hence a pairing

$$H^2_x(X,N^D) \times \mathcal{F}(N) \longrightarrow H^2_x(X,\mathcal{F}(\mu_{p^n})) \approx \mathbb{Z}/p^n\mathbb{Z}.$$

Theorem 10.4. *The above pairing defines an isomorphism*

$$H^2_x(X,N^D) \longrightarrow \mathrm{Hom}_k(\mathcal{F}(N),\mathbb{Z}/p^n\mathbb{Z}).$$

Proof: If N (or its dual) has height one, this can be proved using the exact sequencec in §5. The general case follows by induction on the length of N_K. See [Bester (1978), §2.6].

As in §4, we can endow $H^1(X,N)$ and $H^1(K,N)$ with the structures of perfect pro-algebraic group schemes over k. We write $H^1(X,N)$ and $H^1(K,N)$ for these group schemes. Note that $H^1(X,N)$ is the perfect group scheme associated with $R^1\hat{\alpha}_*N$. For any finite group scheme N over X, the map $H^1(X,N) \hookrightarrow H^1(K,N)$ is a closed immersion, and we write $H^2_x(X,N)$ for the quotient group.

Theorem 10.5. *For any finite flat* p*-primary group scheme over* X, *there is a canonical isomorphism*

$$H^1(X,N) \xrightarrow{\approx} (H^2_x(X,N^D)^\circ)^t.$$

Proof: This follows from the commutative diagram (see 10.2c)

$$\begin{array}{ccccccccc} 0 \to & \mathrm{Hom}_k(\pi_1(H^1(X,N)),\mathbb{Q}/\mathbb{Z}) & \to & \mathrm{Hom}_k(\mathcal{F}(N),\mathbb{Q}/\mathbb{Z}) & \to & \mathrm{Hom}_k(\pi_0(N),\mathbb{Q}/\mathbb{Z}) & \to 0 \\ & \downarrow & & \downarrow \approx & & \downarrow & \\ 0 \to & H^2_x(X,N^D)^\circ & \to & H^2_x(X,N^D) & \to & \pi_0(H^2_x(X,N^D)) & \to 0, \end{array}$$

and the isomorphism $\mathrm{Hom}_k(\pi_1(H^1(X,N)),\mathbb{Q}_p/\mathbb{Z}_p)) \xrightarrow{\approx} \mathrm{Ext}^1_k(H^1(X,N),\mathbb{Q}_p/\mathbb{Z}_p)$.

Remark 10.6. The above proof shows that the dual of the continuous part $H^2_x(X,N)^\circ$ of $H^2_x(X,N)$ is $H^1(X,N)$ and the dual of its finite part $\pi_0(H^2_x(X,N))$ is the finite part $\pi_0(N)$ of $H^0(X,N)$. Write $H_x(X,N)$ and $H(X,N)$ for the canonical objects in the derived category such that $H^r(H_x(X,N)) = H^r_x(X,N)$ and $H^r(H(X,N)) = H^r(X,N)$. Then the correct way to state the above results is that there is an isomorphism

$$H_x(X,N^D) \xrightarrow{\approx} H(X,N)^t[2]$$

where the t denotes Breen-Serre dual (0.14).

Abelian varieties

Let A be an abelian variety over K, and let $\mathcal{A}$ be its Néron model over X. We write $\pi_r(A)$ for $\pi_r(\mathcal{A})$ and G(A) for $G(\mathcal{A})$.

Conjecture 10.7. There is a canonical isomorphism

$$H^1(K,A^t) \longrightarrow \mathrm{Ext}^1_k(G(A),\mathbb{Q}/\mathbb{Z}).$$

In particular, if $\mathcal{A}_k$ is connected, then $H^1(K,A^t) \xrightarrow{\approx} \mathrm{Hom}_k(\pi_1(\mathcal{A}),\mathbb{Q}/\mathbb{Z})$.

The second part of the statement follows from the first, as in (4.16). For the components of the groups prime to p, the conjecture is proved in [Ogg (1962)] and [Shafarevich (1962)].

In the case that A has good reduction, $\mathcal{F}(A_{p^n}) = \pi_1(A)^{(p^n)}$ and $H^2_x(X,A^t)_{p^n} \approx H^2_x(X,A^t_{p^n})$ and so the conjecture can be obtained by passing to the limit in (10.4). See [Bester (1978), 7.1].

One can also show by a similar argument to that in (7.9) that it suffices to prove the result after passing to a finite separable extension of K.

Finally, one can show that the result is true if A is an elliptic curve with a Tate parametrization (cf. [Shatz (1967)]. In this case there is an exact sequence

$$0 \longrightarrow \mathbb{Z} \longrightarrow L^\times \longrightarrow A(L) \longrightarrow 0,$$
$$n \mapsto q^n$$

for all fields L finite over K. Therefore

$$H^1(K,A) \xrightarrow{\approx} H^2(K,\mathbb{Z}) \xrightarrow{\approx} \mathrm{Hom}_{cts}(\mathrm{Gal}(K_s/K),\mathbb{Q}/\mathbb{Z}).$$

On the other hand, $A(K) = R^\times \times (\mathbb{Z}/\mathrm{ord}(q))$, and so

$$\mathrm{Ext}^1_k(G(A),\mathbb{Q}_p/\mathbb{Z}_p) \xrightarrow{\approx} \mathrm{Ext}^1_k(G(\mathbb{G}_{m,R}),\mathbb{Q}_p/\mathbb{Z}_p).$$

As $\mathrm{Ext}^1_k(G(\mathbb{G}_{m,R}),\mathbb{Q}_p/\mathbb{Z}_p) = \mathrm{Hom}_k(\pi_1(\mathbb{G}_{m,R}),\mathbb{Q}_p/\mathbb{Z}_p)$ (see [Serre (1960), 5.4]), the duality in this case follows from the class field theory of [Serre (1961)].

It is to be hoped that the general case can be proved by the methods of §7.

Remark 10.8. The discussion in this section holds with only minor changes when the residue field k is an arbitrary perfect field.

Notes: This section is based on [Bester (1978)]. Some partial results in the same direction were obtained earlier by Vvedens'kii (see [Vvedens'kii (1973) (1976)] and earlier papers).

§11 Global results: curves over perfect fields

Throughout this section, S = Spec k with k a perfect field of characteristic $p \neq 0$, and $\pi\colon X \longrightarrow S$ is a complete smooth curve over S. Again we define $H^r_c(U,F)$ so that the sequence

$$\dots \longrightarrow H^r_c(U,F) \longrightarrow H^r(U,F) \longrightarrow \bigoplus_{v \notin U} H^r(K_v,F) \longrightarrow \dots$$

is exact with K_v the completion of K at v.

Let N be a finite flat group scheme over $U \subset X$, and write $R^r\pi_* N$ and $R^r\pi_! N$ for the sheaves on the perfect site S_{pf} associated with $S' \mapsto H^r(U_{S'},N)$ and $S' \mapsto H^r_c(U_{S'},N)$.

Theorem 11.1. (a) *The sheaves* $R^r\pi_* N$ *and* $R^r\pi_! N$ *are representable by perfect group schemes on* S.

(b) *The canonical pairing*

$$R\pi_* N^D \times R\pi_! N \longrightarrow R\pi_! \mathbb{G}_m \approx \mathbb{Q}/\mathbb{Z}[-2]$$

induces an isomorphism

$$R\pi_* N^D \longrightarrow R\mathcal{H}om_S(R\pi_! N, \mathbb{Q}/\mathbb{Z}[-2]).$$

Proof: We begin with the case that X = U.

Lemma 11.2. *The theorem is true if* U = X *and* N *or its dual have* *height one.*

Proof: Assume first that N has height one. The first exact sequence in §5 yields an exact sequence

$$\cdots \longrightarrow R^r\pi_*N^D \longrightarrow R^r\pi_*\omega_N \longrightarrow R^r\pi_*\omega_N^{(p)} \longrightarrow \cdots$$

and the second a sequence

$$\cdots \longrightarrow R^{r+1}\pi_*N \longrightarrow R^r\pi_*(n\otimes\Omega^1_{X'}) \longrightarrow R^r\pi_*(n\otimes\Omega^1_X) \longrightarrow \cdots .$$

Since two out of three terms in these sequences are vector groups, it is clear that $R^r\pi_*N$ and $R^r\pi_*N^D$ are represented by perfect algebraic groups. The usual duality theorem for coherent sheaves on a curve show that the k-vector spaces $H^r(X,\omega_N)$ and $H^r(X,\omega_N^{(p)})$ are the k-linear duals (hence Breen-Serre duals) of the k-vector spaces $H^{1-r}(X,n\otimes\Omega^1_X)$ and $H^{1-r}(X,n\otimes\Omega^1_X)$. The pairing (5.10) induces an isomorphism $R\pi_*V^\cdot(N^D) \xrightarrow{\approx} R\pi_*U^\cdot(N)^t[1]$. Since $R\pi_*N^D \xrightarrow{\approx} R\pi_*V^\cdot(N^D)$ and $R\pi_*N \xrightarrow{\approx} R\pi_*U^\cdot(N)[-1]$, this (together with (5.11)) shows that $R\pi_*N^D \xrightarrow{\approx} (R\pi_*N)^t[2]$, as required. (For more details, see [Artin-Milne (1976)].)

Lemma 11.3. *Let*

$$0 \longrightarrow N' \longrightarrow N \longrightarrow N'' \longrightarrow 0$$

be an exact sequence of finite flat group schemes on U. *If the theorem is true for* N' *and* N'', *then it is true for* N.

Proof: This is obvious from (0.14).

Lemma 11.4. *Let* V *be an open subscheme of* U. *The theorem is true for* N *on* U *if and only if it is true for* N|V *on* V.

Proof: This follows from the distinguished triangles

$$\bigoplus_{v\in U-V} H_v(\hat{\mathcal{O}}_v,N^D) \longrightarrow R\pi_*(N^D) \longrightarrow R\pi_*(N|V)^D \longrightarrow \bigoplus_{v\in U-V} H_v(\hat{\mathcal{O}}_v,N^D)[1]$$

$$R\pi_!(N|V) \longrightarrow R\pi_!N \longrightarrow \bigoplus_{v\in U-V} H(\hat{\mathcal{O}}_v, N) \longrightarrow R\pi_!(N|V)[1],$$

and (10.6).

We now prove the theorem. Let N be a finite group scheme over U. After replacing U by a smaller open subset we can assume that N has a composition series all of whose quotients have height 1 or are the Cartier duals of groups of height 1. Now Lemma 11.3 shows that we can assume that N (or its dual) has height 1. According to Appendix B, N_K extends to a finite flat group scheme $\mathcal{N}$ on X which is of height one (or has a dual of height one). After again replacing U by a smaller open set, we can assume that $\mathcal{N}|U = N$. According to (11.2), the theorem is true for $\mathcal{N}$ on X, and (11.4) shows that this implies the same result for $\mathcal{N}|U = N$.

Néron models

We now assume the ground field k to be algebraically closed. Let A be an abelian variety over K, and let $\mathcal{A}$ be its Néron model over X.

Lemma 11.5. *The restriction map* $H^1(X,\mathcal{A}) \longrightarrow H^1(K,A)$ *identifies* $H^1(X,\mathcal{A})$ *with* $Ш^1(K,A)$.

Proof: The argument in the proof of Proposition (9.2) shows again that there is an exact sequence

$$0 \longrightarrow Ш^1(K,A) \longrightarrow H^1(X,\mathcal{A}) \longrightarrow \bigoplus H^1(v,\Phi_v),$$

but in the present case, the final term is zero.

In the proof of the next theorem, we shall use without proof that $R^2\pi_*\mathcal{A}$ has no connected part. The argument that the tangent space to $R^2\pi_*\mathcal{A}$ should equal $R^2\pi_*$(tangent sheaf to $\mathcal{A}$), which is zero because X is a curve, makes this plausible. This assumption is not needed if $\mathcal{A}$ has good reduction everywhere.

Theorem 11.6. *There is an exact sequence*

$$0 \longrightarrow H^1(X,\mathcal{A}) \longrightarrow H^1(K,A) \longrightarrow \bigoplus_{\text{all } v} H^1(K_v,A) \longrightarrow (T_pA^t(K))^*$$

with A^t *the dual abelian variety to* A; *in particular, if* A *has no constant part (that is, the* K/k*-trace of* A *is zero), then* $H^1(K,A) \longrightarrow \oplus H^1(K_v,A)$ *is surjective.*

Proof: Let U be an open subscheme of X. The cohomology sequence of the pair $X \supset U$

$$0 \longrightarrow H^1(X,\mathcal{A}) \longrightarrow H^1(U,\mathcal{A}) \longrightarrow \prod_{v\notin U} H^2_v(\mathcal{O}_v,\mathcal{A}) \longrightarrow H^2(X,\mathcal{A}) \longrightarrow \dots$$

can be rewritten as

$$0 \longrightarrow Ш(K,A) \longrightarrow H^1(U,\mathcal{A}) \longrightarrow \prod_{v\notin U} H^1(K_v,A) \longrightarrow H^2(X,\mathcal{A}) \longrightarrow \dots .$$

Because the residue fields at closed points are algebraically closed, for any open $V \subset X$, $H^2_c(V,\mathcal{A}) \xrightarrow{\approx} H^2(X,\mathcal{A})$. Choose a V such that $\mathcal{A}|V$ is an abelian scheme. There is an exact sequence

$$0 \longrightarrow H^1_c(V,\mathcal{A})\otimes\mathbb{Q}_p/\mathbb{Z}_p \longrightarrow H^2_c(V,\mathcal{A}(p)) \longrightarrow H^2_c(V,\mathcal{A})(p) \longrightarrow 0.$$

Interprete $H^r_c(V,-)$ as $R^r(\pi|V)_!$. Then Theorem 11.1 shows that $\pi_0(H^2_c(U,\mathcal{A}(p)))$ is dual to $H^0(U,T_p\mathcal{A}^t) = T_p(A^t(K))$ (and $H^2_c(U,\mathcal{A}(p))^o$ is dual to $H^1(U,T_p\mathcal{A})^o$). From our assumption, the map $H^2_c(V,\mathcal{A}(p)) \longrightarrow H^2(X,\mathcal{A})(p)$ factors through $\pi_0(H^2_c(U,\mathcal{A}(p)))$, and so we can replace $H^2(X,\mathcal{A})(p)$ in the sequence with $T_p(A^t(K))^*$. Now pass to the direct limit over smaller open sets U.

If A has no constant part, then $A^t(K)$ is finitely generated by the generalized Mordell-Weil theorem [Lang (1983), Chapter 6], and so $T_p(A^t(K)) = 0$.

Remark 11.7. The last theorem is useful in the classification of elliptic surfaces with given generic fibre. See [Cossec and Dolgachev (1986), Chapter 5].

Problem 11.8. Extend as many as possible of the results in [Raynaud (1964/5), II] to the p-part.

Notes: In the case U = X, Theorem 11.1 is in [Artin and Milne (1976)].

Appendix A: Embedding finite group schemes

An *embedding* of one group scheme into a second is a map that is both a homomorphism and a closed immersion.

Theorem A.1. *Let* R *be a local Noetherian ring with perfect residue field* k, *and let* N *be a finite flat group scheme over* Spec R. *Write* m *for the maximal ideal of* R, R_i *for* R/m^{i+1}, *and* N_i *for* $N\otimes_R R_i$. *Then there exists a family of embeddings* $\varphi_i\colon N_i \hookrightarrow \mathcal{A}_i$ *such that* $\mathcal{A}_i$ *is an abelian scheme over* R_i *and* $\varphi_{i+1}\otimes R_i = \varphi_i$ *for all* i. *Consequently there is an embedding of the formal completion* $\hat{N}$ *of* N *into a formal abelian scheme* $\mathcal{A}$ *over* $\mathrm{Spf}(\hat{R})$.

Proof: To deduce the last sentence from the preceding statement, note that the family $(\mathcal{A}_i)$ defines a formal scheme $\mathcal{A}$ over $\mathrm{Spf}(\hat{R})$ and that the φ_i define an embedding of $\hat{N}$ into $\mathcal{A}$ (see [Grothendieck and Dieudonné (1971), §10.6]).

We first prove the theorem in the case that R = k is an algebraically closed field of characteristic exponent p. The only simple finite group schemes over k are $\mathbb{Z}/\ell\mathbb{Z}$ ($\ell\neq p$), $\mathbb{Z}/p\mathbb{Z}$, μ_p, and α_p. The first three of these can be embedded in any nonsupersingular elliptic curve over k, and the last can be embedded in any supersingular elliptic curve. We proceed by induction on the order of N. Consider an exact sequence

$$0 \longrightarrow N' \longrightarrow N \longrightarrow N'' \longrightarrow 0$$

in which N' and N" can be embedded into abelian varieties A' and A". Let e be the class of this extension in $\mathrm{Ext}^1_k(N'',N')$, and let e' be the image of e in $\mathrm{Ext}^1_k(N'',A')$. As $\mathrm{Ext}^2_k(A''/N'',A') = 0$ (see [Milne (1970a), Thm 2]), e lifts to an element $\tilde{e}$ of $\mathrm{Ext}^1_k(A'',A')$, and N embeds into the middle term of any representative of $\tilde{e}$:

$$\begin{array}{ccccccccccc} 0 & \longrightarrow & N' & \longrightarrow & N & \longrightarrow & N'' & \longrightarrow & 0 & & (=e) \\ & & \downarrow & & \downarrow & & \| & & & & \\ 0 & \longrightarrow & A' & \longrightarrow & X & \longrightarrow & N'' & \longrightarrow & 0 & & (=e') \\ & & \| & & \downarrow & & \downarrow & & & & \\ 0 & \longrightarrow & A' & \longrightarrow & A & \longrightarrow & A'' & \longrightarrow & 0 & & (=\tilde{e}). \end{array}$$

We next consider the case that R = k is a perfect field. The first step implies that N embeds into an abelian variety over a finite extension k' of k, $N_{k'} \hookrightarrow A$. Now we can form the restriction of scalars [Demazure and Gabriel (1970), I,§1,6.6] of this map and obtain an embedding $N \hookrightarrow \mathrm{Res}_{k'/k}N_{k'} \hookrightarrow \mathrm{Res}_{k'/k}A$. The fact that k'/k is separable implies that $\mathrm{Res}_{k'/k}A$ is again an abelian variety, because $(\mathrm{Res}_{k'/k}A)\otimes_k k_s = \mathrm{Res}_{k'\otimes_k k_s/k_s} A = A_{k_s}^{[k':k]}$.

To complete the proof, we prove the following statement: let R be a local Artin ring with perfect residue field k = R/m; let I be an ideal in R such that mI = 0, and let $\bar{R} = R/I$; let N be a finite flat group scheme over R, and let $\bar{\varphi}\colon N\otimes_R\bar{R} \hookrightarrow \bar{\mathcal{A}}$ be an embedding of $N\otimes_R\bar{R}$ into an abelian scheme over $\bar{R}$; then $\bar{\varphi}$ lifts to a similar embedding φ over R.

Lemma A.2. *Let* X *be a smooth scheme over* $\bar{R}$, *and let* $\mathcal{L}(X)$ *be the set of isomorphism classes of pairs* (Y,ψ) *where* Y *is a smooth scheme over* R *and* ψ *is an isomorphism* $Y\otimes\bar{R} \xrightarrow{\approx} X$. *Let* $\mathcal{T}_{X_0}$ *be the tangent sheaf on* $X_0 \overset{df}{=} X\otimes_{\bar{R}}k$.

(a) *The obstruction to lifting* X *to* R *is an element* $\alpha \in H^2(X_0,\mathcal{T}_{X_0})\otimes_k I$.

(b) *When nonempty,* $\mathcal{L}(X)$ *is a principal homogeneous space for* $H^1(X_0,\mathcal{T}_{X_0})\otimes_k I$.

(c) *If* X *is an abelian scheme over* $\bar{R}$, *then* $\alpha = 0$.

Proof: See [Grothendieck (1971), III.6.3] and [Oort (1971)].

The lemma shows that there is an $(\mathcal{A},\psi) \in \mathcal{L}(\bar{\mathcal{A}})$. As $\mathcal{A}$ is smooth, the zero section of $\bar{\mathcal{A}}$ over $\bar{R}$ lifts to a section of $\mathcal{A}$ over R. Now the rigidity of abelian schemes [Mumford (1965), 6.15] implies that the group structure on $\bar{\mathcal{A}}$ lifts to a group structure on $\mathcal{A}$, and it follows that $\mathcal{A}$ is an abelian scheme. (See also [Messing (1972), IV.2.8.1].) It remains to show that $\bar{\varphi}$ lifts to an embedding φ of N into $\mathcal{A}$ (after possibly changing the choice of $\mathcal{A}$).

Lemma A.3. *There is an exact sequence*

$$\mathrm{Hom}_R(N,\mathcal{A}) \longrightarrow \mathrm{Hom}_{\bar{R}}(\bar{N},\bar{\mathcal{A}}) \longrightarrow \mathrm{Ext}^1_k(N_0,T_0(\mathcal{A}_0)\otimes_k I)$$

where $T_0(\mathcal{A}_0)$ *is the tangent space at zero to* $\mathcal{A}_0$ *and we have used the same notation for the vector space* $T_0(\mathcal{A}_0)\otimes_k I$ *and the vector group it defines.*

Proof: In disagreement with the rest of the book, we shall write Y_{Fl} for the big flat site on Y (category of all schemes locally of finite-type over Y with the flat topology) and Y_{fl} for the small flat site (category of all schemes locally of finite-type and *flat* over Y with the flat topology). There is a wellknown short exact sequence

$$0 \longrightarrow T_0(\mathcal{A}_0)\otimes_k I \longrightarrow \mathcal{A}(R) \longrightarrow \bar{\mathcal{A}}(\bar{R}) \longrightarrow 0.$$

A similar sequence exists with R replaced by any flat R-algebra, and so, if we write i and $\bar{i}$ for the closed immersions Spec k $\hookrightarrow$ Spec R and Spec $\bar{R} \hookrightarrow$ Spec R, then there is an exact sequence

$$0 \longrightarrow i_*(T_0(\mathcal{A}_0)\otimes_k I) \longrightarrow \mathcal{A} \longrightarrow \bar{i}_*\bar{\mathcal{A}} \longrightarrow 0$$

of sheaves on $(\mathrm{Spec}\ R)_{fl}$. This yields an exact sequence

$$\mathrm{Hom}_{R,fl}(N,\mathcal{A}) \longrightarrow \mathrm{Hom}_{R,fl}(N,\bar{i}_*\bar{\mathcal{A}}) \longrightarrow \mathrm{Ext}^1_{R,fl}(N,i_*(T_0(\mathcal{A}_0)\otimes_k I))$$

where the groups are computed in $(\mathrm{Spec}\ R)_{fl}$. Note that $\mathrm{Hom}_{R,fl}(N,\bar{i}_*\bar{\mathcal{A}}) = \mathrm{Hom}_{\bar{R},fl}(\bar{i}^*N,\bar{\mathcal{A}})$ and that (because N, $\bar{N}$, $\mathcal{A}$ and $\bar{\mathcal{A}}$ are all in the underlying categories of the sites)

$$\mathrm{Hom}_{R,fl}(N,\mathcal{A}) = \mathrm{Hom}_R(N,\mathcal{A}), \qquad \mathrm{Hom}_{\bar{R},fl}(\bar{i}^*N,\bar{\mathcal{A}}) = \mathrm{Hom}_{\bar{R}}(\bar{N},\bar{\mathcal{A}})$$

(the right hand groups are the groups of homomorphisms in the category of group schemes). Let f: $(\mathrm{Spec}\ R)_{Fl} \longrightarrow (\mathrm{Spec}\ R)_{fl}$ be the morphism of sites defined by the identity map. Then f_* is exact and preserves injectives, and so $\mathrm{Ext}^r_{R,Fl}(f^*F,F') = \mathrm{Ext}^r_{R,fl}(F,f_*F')$ for any sheaves F on $(\mathrm{Spec}\ R)_{fl}$ and F' on $(\mathrm{Spec}\ R)_{Fl}$. In our case $f^*N =$ N (see [Milne (1980), II.3.1d]), and so we can replace $\mathrm{Ext}^1_{R,fl}(N,i_*(T_0(\mathcal{A}_0)\otimes_k I))$ in the above sequence with the same group computed in the big flat site on R. Next $R^r i_*(T_0(\mathcal{A}_0)\otimes_k I) = 0$ for $r > 0$, because $T_0(\mathcal{A}_0)\otimes_k I$ is the sheaf defined by a coherent module and so $H^r(V_{Fl},T_0(\mathcal{A}_0)\otimes_k I) = H^r(V_{Zar},T_0(\mathcal{A}_0)\otimes_k I) = 0$ for $r > 0$ when V is an affine k-scheme. Therefore

$$\mathrm{Ext}^r_{R,Fl}(N,i_*(T_0(\mathcal{A}_0)\otimes_k I)) = \mathrm{Ext}^r_{k,Fl}(i^*N,T_0(\mathcal{A}_0)\otimes_k I)$$

for all r. Finally $\mathrm{Ext}^1_{k,Fl}(N_0,T_0(\mathcal{A}_0)\otimes_k I) = \mathrm{Ext}^1_k(N_0,T_0(\mathcal{A}_0)\otimes_k I)$.

We know that $\mathcal{T}_{\mathcal{A}_0}$ is the free sheaf $T_0(\mathcal{A}_0)\otimes_k \mathcal{O}_{\mathcal{A}_0}$. On tensoring the isomorphism $\mathrm{Ext}^1_k(\mathcal{A}_0,\mathbb{G}_a) \xrightarrow{\approx} H^1(\mathcal{A}_0,\mathcal{O}_{\mathcal{A}_0})$ [Serre (1959), VII.17] with $T_0(\mathcal{A}_0)\otimes_k I$ we get an isomorphism,

$$\mathrm{Ext}^1_k(\mathcal{A}_0,T_0(\mathcal{A}_0)\otimes_k I) \xrightarrow{\approx} H^1(\mathcal{A}_0,\mathcal{T}_{\mathcal{A}_0})\otimes_k I.$$

The inclusion $N_0 \hookrightarrow \mathcal{A}_0$ defines a map

$$\mathrm{Ext}^1_k(\mathcal{A}_0,T_0(\mathcal{A}_0)\otimes_k I) \longrightarrow \mathrm{Ext}^1_k(N_0,T_0(\mathcal{A}_0)\otimes_k I),$$

which is surjective because $\mathrm{Ext}^2_k(-,\mathbb{G}_a) = 0$ (see [Oort (1966), p.II.14-2] and (I.0.17)). Consider

$$\begin{array}{ccccc} & & & & H^1(\mathcal{A}_0,\mathcal{T}_{\mathcal{A}_0})\otimes_k I \\ & & & & \downarrow \text{surj} \\ \mathrm{Hom}_R(N,\mathcal{A}) & \longrightarrow & \mathrm{Hom}_{\bar{R}}(\bar{N},\bar{\mathcal{A}}) & \longrightarrow & \mathrm{Ext}^1_k(N_0,T_0(\mathcal{A}_0)\otimes_k I). \end{array}$$

It is clear from this diagram and Lemma A.2b that if $\bar{\varphi}: \bar{N} \longrightarrow \bar{\mathcal{A}}$ does not lift to a map φ from N to $\mathcal{A}$, then a different choice of $\mathcal{A}$ can be made so that $\bar{\varphi}$ does lift. The lifted map φ is automatically an embedding.

Corollary A.4. *In addition to the hypotheses of the theorem, assume that* R *is complete and that* N *has order a power of* p. *Then* N *can be embedded in a* p-*divisible group scheme* H *over* R.

Proof: Take the p-divisible group scheme associated with the formal abelian scheme $\mathcal{A}$.

We next consider the problem of resolving a finite flat group scheme by smooth group schemes. Let N be a finite flat group scheme over a Noetherian scheme S. Then the functor $\mathcal{M}or_S(N,\mathbb{G}_m)$ is representable by $\mathrm{Res}_{N/S}\mathbb{G}_m$, which is a smooth affine group scheme of finite type over S. Note that N^D is (in an obvious way) a closed subgroup of $\mathcal{Mor}_S(N,\mathbb{G}_m)$. Write N^i for $N\times_S\ldots\times_S N$, and let $\mathcal{Mor}_S(N^2,\mathbb{G}_m)_{sym}$ be the kernel of morphism

$$\mathcal{Mor}_S(N^2,\mathbb{G}_m) \longrightarrow \mathcal{Mor}_S(N^2,\mathbb{G}_m)$$

sending f to the function f_{sym}, where $f_{sym}(x,y) = f(y,x)f(x,y)^{-1}$. Finally, let $\mathcal{Z}^2(N,\mathbb{G}_m)_{sym}$ be the kernel of the boundary map

$$d\colon \mathcal{Mor}_S(N^2,\mathbb{G}_m)_{sym} \longrightarrow \mathcal{Mor}_S(N^3,\mathbb{G}_m),$$

$df(x,y,z) = f(y,z)f(xy,z)^{-1}f(x,yz)f(x,y)^{-1}$. The image of the boundary map

$$d\colon \mathcal{Mor}_S(N,\mathbb{G}_m) \longrightarrow \mathcal{Mor}_S(N^2,\mathbb{G}_m),\ df(x,y) = f(xy)f(x)^{-1}f(y)^{-1},$$

is contained in $\mathcal{Z}^2(N,\mathbb{G}_m)_{sym}$.

Theorem A.5. *The sequence*

$$0 \longrightarrow N^D \longrightarrow \mathcal{Mor}_S(N,\mathbb{G}_m) \xrightarrow{\ d\ } \mathcal{Z}^2_S(N,\mathbb{G}_m)_{sym} \longrightarrow 0$$

is an exact sequence of affine group schemes on S. *The final two terms are smooth* over S.

Proof: See [Bégeuri (1980), 2.2.1].

The exact sequence in the theorem is called the *canonical smooth resolution* of N^D.

Theorem A.6. *Let* N *be a finite flat group scheme over a Noetherian scheme* S. *Locally for the Zariski topology on* S, *there is a projective abelian scheme* $\mathcal{A}$ *over* S *and an embedding* $N \hookrightarrow \mathcal{A}$.

Proof: The idea of the proof is to construct (locally) a smooth curve $\pi\colon X \longrightarrow S$ over S and a principal homogeneous space Y for N^D over X. Such a Y defines an element of $R^1\pi_*N^D$, and cup-product with this element defines a map $N = \mathcal{Hom}_S(N^D,\mathbb{G}_m) \longrightarrow R^1\pi_*\mathbb{G}_m = \mathbf{Pic}_{X/S}$ whose image is in the abelian scheme $\mathbf{Pic}^0_{X/S} \subset \mathbf{Pic}_{X/S}$. One shows that the map is a closed immersion. For the details, see [Raynaud (1979)] and [Berthelot, Breen, and Messing (1982), 3.1.1].

Remark A.7. It is possible to construct quotients by finite flat group schemes (see [Dieudonné (1965), p114]). Therefore from (A.1), we get exact sequences

$$0 \longrightarrow N_i \longrightarrow \mathcal{A}_i \longrightarrow \mathcal{B}_i' \longrightarrow 0$$

with $\mathcal{B}_i$ an abelian scheme, and from (A.4), we get an exact sequence

$$0 \longrightarrow N \longrightarrow H \longrightarrow H' \longrightarrow 0$$

with H' a p-divisible group over R. Finally, (A.6) shows that (locally) N fits into an exact sequence

$$0 \longrightarrow N \longrightarrow \mathcal{A} \longrightarrow \mathcal{B} \longrightarrow 0$$

with $\mathcal{A}$ and $\mathcal{B}$ projective abelian schemes.

Remark A.8. Let S be the spectrum of a discrete valuation ring R with field of fractions K, and let N be a quasi-finite flat separated group scheme over S. If the normalization $\tilde{N}$ of N in N_K is flat over R, then N is a subgroup of an abelian scheme $\mathcal{A}$ (because it is an open subgroup of $\tilde{N}$ by Zariski's main theorem, and $\tilde{N}$ is a closed subgroup of an abelian scheme). Conversely, if N is a subgroup of an abelian scheme $\mathcal{A}$ over S, then its normalization is flat (because $N_K \subset (\mathcal{A}_K)_n$ for some n, and the closure of N_K in $\mathcal{A}_n$ is flat, see Appendix B).

The quotient $\mathcal{A}/N$ is represented by an algebraic space [Artin (1969), 7.3], but it is not an abelian scheme unless N is finite because it is not separated (the closure of the zero section is $\tilde{N}/N$).

Notes: Theorem A.1 is due to Oort [Oort (1967)], and Theorem A.6 is due to Raynaud. Lemma A.3 is an unpublished result of Tate; the above proof of it was suggested to me by Messing.

Appendix B: Extending finite group schemes

Let R be a discrete valuation ring with field of fractions K. In [Raynaud (1974), p271], it is asserted that, when K has characteristic p, every finite group scheme over K killed by a power of p extends to a finite group scheme over R (the statement is credited to Artin and Mazur). Our first proposition provides a counterexample to this assertion. Then we investigate some cases where the group does extend. First we recall a wellknown lemma.

Lemma B.1. *Let R be a discrete valuation ring with field of fractions K, and let*

$$0 \longrightarrow N' \longrightarrow N \longrightarrow N'' \longrightarrow 0$$

be an exact sequence of finite group schemes over R. Assume that N extends to a finite flat group scheme $\mathcal{N}$ over R. Then there exists a unique exact sequence

$$0 \longrightarrow \mathcal{N}' \longrightarrow \mathcal{N} \longrightarrow \mathcal{N}'' \longrightarrow 0$$

of finite flat group schemes over R having the original sequence as its generic fibre.

Proof: The group $\mathcal{N}'$ is the closure of N' in $\mathcal{N}$, and $\mathcal{N}''$ is the quotient of $\mathcal{N}$ by $\mathcal{N}'$. (Alternatively, $\mathcal{N}'$ is such that $\Gamma(\mathcal{N}',\mathcal{O}_{\mathcal{N}'})$ is the image of $\Gamma(\mathcal{N},\mathcal{O}_{\mathcal{N}})$ in $\Gamma(N',\mathcal{O}_{N'})$, and $\mathcal{N}''$ is such that $\Gamma(\mathcal{N}'',\mathcal{O}_{\mathcal{N}''}) = \Gamma(\mathcal{N},\mathcal{O}_{\mathcal{N}}) \cap \Gamma(N'',\mathcal{O}_{N''})$.)

Now let R be a discrete valuation ring of characteristic $p \neq 0$. Consider an extension of $\mathbb{Z}/p\mathbb{Z}$ by μ_p over K:

$$0 \longrightarrow \mu_p \longrightarrow N \longrightarrow \mathbb{Z}/p\mathbb{Z} \longrightarrow 0.$$

Such extensions are classified by $\mathrm{Ext}^1_K(\mathbb{Z}/p\mathbb{Z},\mu_p)$, and the following diagram shows that this group is isomorphic to $K^\times/K^{\times p}$:

$$\begin{array}{ccccccc}
 & & & & \mathrm{Hom}_K(\mathbb{Z},\mathbb{G}_m) = K^\times & & \\
 & & & & \downarrow p & & \\
 & & & & \mathrm{Hom}_K(\mathbb{Z},\mathbb{G}_m) = K^\times & & \\
 & & & & \downarrow & & \\
\mathrm{Hom}_K(\mathbb{Z}/p\mathbb{Z},\mathbb{G}_m) & \longrightarrow & \mathrm{Ext}^1_K(\mathbb{Z}/p\mathbb{Z},\mu_p) & \xrightarrow{\approx} & \mathrm{Ext}^1_K(\mathbb{Z}/p\mathbb{Z},\mathbb{G}_m) & \longrightarrow & 0 \\
\| & & & & \downarrow & & \\
0 & & & & \mathrm{Ext}^1_K(\mathbb{Z},\mathbb{G}_m) = 0. & &
\end{array}$$

Let $\alpha(N)$ be the class of the extension in $K^\times/K^{\times p}$, and let $a(N)$ be $\mathrm{ord}(\alpha(N))$ regarded as an element of $\mathbb{Z}/p\mathbb{Z}$.

Proposition B.2. *The finite group scheme* N *extends to a finite flat group scheme over* R *if and only if* $a(N) = 0$. *Therefore, there exists a finite group scheme over* K *killed by* p^2 *that does not extend to a finite flat group scheme over* R.

Proof: Assume that N extends to a finite flat group scheme $\mathcal{N}$ over R, and let

$$0 \longrightarrow \mathcal{N}^0 \longrightarrow \mathcal{N} \longrightarrow \mathcal{N}/\mathcal{N}^0 \longrightarrow 0,$$

be the extension of the sequence given by Lemma B.1. The isomorphism $\mathbb{Z}/p\mathbb{Z} \xrightarrow{\approx} (\mathcal{N}/\mathcal{N}^0)_K$ extends to a map $\mathbb{Z}/p\mathbb{Z} \longrightarrow \mathcal{N}/\mathcal{N}^0$ over R: simply map each section of $\mathbb{Z}/p\mathbb{Z}$ over R to the closure of its image under the first map. Similarly, the Cartier dual of the isomorphism $\mathcal{N}^0_K \xrightarrow{\approx} \mu_p$ extends to a map $\mathbb{Z}/p\mathbb{Z} \longrightarrow (\mathcal{N}^0_K)^D$, and the dual of this map is a map

$\mathcal{N}^0 \longrightarrow \mu_p$ whose generic fibre is the original isomorphism. Now after pulling back by $\mathbb{Z}/p\mathbb{Z} \longrightarrow \mathcal{N}/\mathcal{N}^0$ and pushing out by $\mathcal{N}^0 \longrightarrow \mu_p$, we obtain an extension of $\mathbb{Z}/p\mathbb{Z}$ by μ_p over R whose generic fibre is exactly the original extension. Thus we see that N extends over R if and only if the original extension of $\mathbb{Z}/p\mathbb{Z}$ by μ_p extends over R. The result is now obvious from the commutative diagram,

$$\begin{array}{ccccccc} \mathrm{Ext}^1_K(\mathbb{Z}/p\mathbb{Z},\mu_p) & \xrightarrow{\approx} & \mathrm{Ext}^1_K(\mathbb{Z}/p\mathbb{Z},\mathbb{G}_m) & \xrightarrow{\approx} & H^1(K,\mu_p) & \xrightarrow{\approx} & K^\times/K^{\times p} \\ \uparrow & & \uparrow & & \uparrow & & \uparrow \\ \mathrm{Ext}^1_R(\mathbb{Z}/p\mathbb{Z},\mu_p) & \xrightarrow{\approx} & \mathrm{Ext}^1_R(\mathbb{Z}/p\mathbb{Z},\mathbb{G}_m) & \xrightarrow{\approx} & H^1(R,\mu_p) & \xrightarrow{\approx} & R^\times/R^{\times p} \end{array}$$

because $R^\times/R^{\times p}$ is the kernel of $K^\times/K^{\times p} \xrightarrow{\mathrm{ord}} \mathbb{Z}/p\mathbb{Z}$.

Remark B.3. The same argument shows that an extension N of $\mathbb{Z}/p\mathbb{Z}$ by $\mathbb{Z}/p\mathbb{Z}$ need not extend to an extension of $\mathbb{Z}/p\mathbb{Z}$ by $\mathbb{Z}/p\mathbb{Z}$ over R, because $\mathrm{Ext}^1_K(\mathbb{Z}/p\mathbb{Z},\mathbb{Z}/p\mathbb{Z}) = H^1(K,\mathbb{Z}/p\mathbb{Z}) \approx K/\wp K$ and $\mathrm{Ext}^1_R(\mathbb{Z}/p\mathbb{Z},\mathbb{Z}/p\mathbb{Z}) = H^1(R,\mathbb{Z}/p\mathbb{Z}) \approx R/\wp R$. However, N does extend to an extension of $\mathbb{Z}/p\mathbb{Z}$ by some finite flat group scheme over R. Indeed, in (7.5) we note that if $\mathcal{N}' \approx \mathcal{N}_{a,0}$ in the Oort-Tate classification (0.9) with $a = t^{c(p-1)}$, then $H^1(R,\mathcal{N}')$ is the image of $t^{-cp}R$ in $K/\wp K$, and so if c is chosen sufficiently large, the class of the extension in $\mathrm{Ext}^1_K(\mathbb{Z}/p\mathbb{Z},\mathbb{Z}/p\mathbb{Z})$ will lie in $\mathrm{Ext}^1_R(\mathcal{N}',\mathbb{Z}/p\mathbb{Z})$.

Proposition B.4. *Let X be a regular quasi-projective scheme of characteristic $p \neq 0$ over a ring R, and let K be the field of rational functions on X. Any finite flat group scheme N of height one over K extends to a finite flat group scheme of height one over X.*

Proof: We first prove this in the case that N has order p. Then $N = N_{0,b}$ for some $b \in K$. We have to show that there exists an invertible sheaf $\mathcal{L}$ on X, a trivialization $\mathcal{L}_K \xrightarrow{\approx} K$, and a global section β of $\mathcal{L}^{\otimes 1-p}$ corresponding to b under the trivialization. Let D be a Weil divisor such that $D \leq 0$ and $(b) \geq D$, and let $\mathcal{L} = \mathcal{O}(D)$. Then under the usual identification of $\mathcal{O}(D)_K$ with K,

$$\Gamma(X,\mathcal{L}^{\otimes 1-p}) = \Gamma(X,\mathcal{O}((1-p)D) = \{g \in K \mid (g) \geq (p-1)D\}.$$

Clearly $b \in \Gamma(X,\mathcal{L}^{\otimes 1-p})$.

We now consider the general case. Recall [Demazure and Gabriel (1970), II, §7] that a (commutative) p-Lie algebra $\mathcal{V}$ on a scheme Y of characteristic $p \neq 0$ is a coherent sheaf of $\mathcal{O}_Y$-modules together with a map $\varphi\colon \mathcal{V} \longrightarrow \mathcal{V}$ such that $\varphi(x+y) = \varphi(x) + \varphi(y)$ and $\varphi(ax) = a^p\varphi(x)$. With each locally free p-Lie algebra $\mathcal{V}$ there is a canonically associated finite flat group scheme $N = G(\mathcal{V})$ of height ≤ 1. Moreover, when Y is the spectrum of a field, every finite group scheme N is of the form $G(\mathcal{V})$ for some p-Lie algebra $\mathcal{V}$. Note that to give φ is the same as to give an $\mathcal{O}_Y$-linear map $\mathcal{V} \longrightarrow \mathcal{V}^{(p)}$.

Thus let (V,φ) be the p-Lie algebra associated with N over K. We have to show that (V,φ) extends to a p-Lie algebra over X. Extend V in some trivial way to a locally free sheaf $\mathcal{V}$ on X, and regard φ as a linear map $V \longrightarrow V^{(p)}$. Then φ is an element of $\mathrm{Hom}_K(V,V^{(p)})$ and we would like to extend it to a section of $\mathcal{H}om_{\mathcal{O}_X}(\mathcal{V},\mathcal{V}^{(p)})$. This will not be possible in general, unless we first twist by an ample invertible sheaf. Let $\mathcal{L}$ be such a sheaf on X, and write $\mathcal{V}(r)$ for $\mathcal{V}\otimes\mathcal{L}^{\otimes r}$. Then $\mathcal{H}om_{\mathcal{O}_X}(\mathcal{V}(r),\mathcal{V}(r)^{(p)}) = \mathcal{H}om_{\mathcal{O}_X}(\mathcal{V}(r),\mathcal{V}^{(p)}(pr)) = \mathcal{H}om_{\mathcal{O}_X}(\mathcal{V},\mathcal{V}^{(p)})(pr-p)$, and so for a sufficiently high r, $\mathrm{Hom}_{\mathcal{O}_X}(\mathcal{V}(r),\mathcal{V}(r)^{(p)})$ will be generated by its global sections [Hartshorne (1977), II.7]. Therefore, we can write $\varphi = \sum \alpha_i\varphi_i$ with $\alpha_i \in K$ and $\varphi_i \in \mathrm{Hom}_{\mathcal{O}_X}(\mathcal{V}(r),\mathcal{V}(r)^{(p)})$. Now choose a divisor D such that $(\alpha_i) \geq D$ for all i. Then φ is a global section of

$$\mathcal{H}om_{\mathcal{O}_X}(\mathcal{V}(r)\otimes\mathcal{O}_X(D),\mathcal{V}(r)\otimes\mathcal{O}_X(D)^{(p)}) = \mathcal{H}om_{\mathcal{O}_X}(\mathcal{V}(r),\mathcal{V}(r)^{(p)})\otimes\mathcal{O}_X((p-1)D).$$

Corollary B.5. *Let X be a regular quasi-projective scheme of characteristic* $p \neq 0$ *over a ring R, and let K be the field of rational functions on X. Any finite flat group scheme N whose Cartier dual is of height one over K extends to a finite flat group scheme over X.*

Proof: Apply the proposition to N^D, and take the Cartier dual of the resulting finite flat group scheme.

Remark B.6. It is also possible to prove (B.5) directly from (5.5).

Notes: The counterexample (B.1) was found by the author in 1977.

Appendix C: Biextensions and Néron models

Throughout, X will be a locally Noetherian scheme endowed with either the smooth or the flat topology.

Biextensions

Let A, B, and G be group schemes over X (commutative and of finite type as always). A *biextension* of (B,A) by G is a scheme W with a surjective morphism $\pi\colon W \longrightarrow B\times_X A$ endowed with the following structure:

(a) an action $W\times_{B\times A}G_{B\times A} \longrightarrow W$ of $G_{B\times A}$ on W making W into a $G_{B\times A}$-torsor;

(b) a B-morphism $m_B\colon W\times_B W \longrightarrow W$ and a section e_B of W over B making W into a commutative group scheme over B; and

(c) an A-morphism $m_A\colon W\times_A W \longrightarrow W$ and a section e_A of W over A making W into a commutative group scheme over A.

These structures are to satisfy the following conditions:

(i) if $G_B \longrightarrow W$ is the map $g \mapsto e_B\cdot g$, then

$$0 \longrightarrow G_B \longrightarrow W \xrightarrow{\pi} A_B \longrightarrow 0$$

is an exact sequence of group schemes over B;

(ii) if $G_A \longrightarrow W$ is the map $g \mapsto e_A\cdot g$, then

$$0 \longrightarrow G_A \longrightarrow W \xrightarrow{\pi} B_A \longrightarrow 0$$

is an exact sequence of group schemes over A;

(iii) the following diagram commutes

$$\begin{array}{ccccc} (W\times_A W)\times_{B\times B}(W\times_A W) & \xrightarrow{m_A\times m_A} & W\times_B W & & \\ \| & & & \searrow^{m_B} & \\ & & & & W \\ & & & \nearrow^{m_A} & \\ (W\times_B W)\times_{A\times A}(W\times_B W) & \xrightarrow{m_B\times m_B} & W\times_A W & & \end{array}.$$

(The "equality" at left is $(w_1,w_2;w_3,w_4) \longleftrightarrow (w_1,w_3;w_2,w_4)$.) See [Grothendieck (1972), VII].

Example C.1. Let A and B be abelian varieties of the same dimension over a field. We call an invertible sheaf $\mathcal{P}$ on B×A a *Poincaré sheaf* if its restrictions to {0}×A and B×{0} are both trivial and if $\chi(B\times A,\mathcal{P}) = \pm 1$. It is known [Mumford (1970), §13, p131] that then the map of functors $b \mapsto (b\times 1)^{*}\mathcal{P}\colon B(T) \longrightarrow \mathrm{Pic}(A_T)$ identifies B with the dual abelian variety A^t of A. Moreover, for any abelian variety A over a field, there exists an essentially unique pair $(B,\mathcal{P})$ with $\mathcal{P}$ a Poincaré sheaf on B×A (ibid. §8, §10-12).

Let $\mathcal{P}$ be a Poincaré sheaf on B×A. With $\mathcal{P}$, we can associate a $\mathbb{G}_m$-torsor $W = \mathcal{I}\!som_{B\times A}(\mathcal{O}_{B\times A},\mathcal{P})$ (less formally, W is the line bundle associated with $\mathcal{P}$ with the zero section removed). For each point

$b \in B$, $\mathcal{P}_b$ is a line bundle on A, and W_b has a canonical structure of a group scheme over A such that

$$0 \longrightarrow \mathbb{G}_m \longrightarrow W_b \longrightarrow A \longrightarrow 0$$

is an exact sequence of algebraic groups [Serre (1959), VII, §3]. This construction can be carried out universally (on B), and gives a group structure to W regarded as an A-scheme which is such that

$$0 \longrightarrow \mathbb{G}_{mB} \longrightarrow W \longrightarrow A_B \longrightarrow 0$$

is an exact sequence of group schemes over B. By symmetry, we get a group structure on W regarded as a group scheme over A, and these two structures form a biextension of (B,A) by $\mathbb{G}_m$. Any biextension arising in this way from a Poincaré sheaf will be called a *Poincaré biextension*.

When A, B, and G are sheaves on X_{fl} (or X_{sm}), it is possible to modify the above definition in an obvious fashion to obtain the notion of a biextension of (B,A) by G: it is a sheaf of sets W with a surjective morphism $W \longrightarrow B\times A$ having the structure of a $G_{B\times A}$-torsor and partial group structures satisfying the conditions (i), (ii), and (iii). When A, B, and G are group schemes, we write $\mathrm{Biext}_X(B,A;G)$ for the set of biextensions of (B,A) by G, and when A, B, and G are sheaves, we write $\mathrm{Biext}_{X_{fl}}(B,A;G)$ (or $\mathrm{Biext}_{X_{sm}}(B,A;G)$) for the similar set of sheaves. Clearly, there is a map

$$\mathrm{Biext}_X(B,A;G) \longrightarrow \mathrm{Biext}_{X_{fl}}(B,A;G)$$

and also $\mathrm{Biext}_X(B,A;G) \longrightarrow \mathrm{Biext}_{X_{sm}}(B,A;G)$ when G is smooth over X).

Proposition C.2. *Let A, B, and G be group schemes over X. If G is flat and affine over X, then the map*

$$\mathrm{Biext}_X(B,A;G) \longrightarrow \mathrm{Biext}_{X_{fl}}(B,A;G)$$

is bijective, and if G is smooth and affine over X, then

$$\mathrm{Biext}_X(B,A;G) \longrightarrow \mathrm{Biext}_{X_{sm}}(B,A;G)$$

is bijective.

Proof: The essential point is that, in each case, torsors in the category of sheaves are representable, and torsors in the category of schemes are locally trivial for the respective topologies (see [Milne (1980), III.4.2 and 4.3]).

Consider a biextension (of schemes) W of (B,A) by G. Given an X-scheme T and a T-valued point t of B, we can pull-back the sequence in (i) and so obtain an extension

$$0 \longrightarrow G_T \longrightarrow W(t) \longrightarrow A_T \longrightarrow 0.$$

This gives us a map $B(T) \longrightarrow \mathrm{Ext}^1_T(A_T,G_T)$, which can be shown to be a group homomorphism. In this manner, a biextension of (B,A) by G defines homomorphisms of sheaves $B \longrightarrow \mathcal{E}xt^1_{X_{fl}}(A,G)$ and $A \longrightarrow \mathcal{E}xt^1_{X_{fl}}(B,G)$. A biextension of sheaves determines similar maps. This has a pleasant restatement in terms of derived categories.

Proposition C.3. *There is a canonical isomorphism*

$$\mathrm{Biext}_{X_{fl}}(B,A;G) \xrightarrow{\approx} \mathrm{Hom}_{X_{fl}}(B\otimes^L A, G[1])$$

(and similarly for the smooth topology).

Proof: See [Grothendieck (1972), VII.3.6.5].

Given a biextension of (B,A) by G, we can define pairings

$$H^r(X,B) \times H^s(X,A) \longrightarrow H^{r+s+1}(X,G)$$

in three different ways: directly from the map $B \otimes^{L} A \longrightarrow G[1]$, by using the map $B \longrightarrow \mathcal{E}xt^{1}(A,G)$, or by using the map $A \longrightarrow \mathcal{E}xt^{1}(B,G)$.

Proposition C.4. *The three pairings are equal (up to sign).*

$$\mathrm{Biext}_{X}(B,A;G) \longrightarrow \mathrm{Biext}_{X_{sm}}(B,A;G)$$

Proof: See Theorem 0.15.

Let W be a biextension of (B,A) by G. For any integer n, the map $B \otimes^{L} A \longrightarrow G[1]$ defines in a canonical way a map $B_n \otimes A_n \longrightarrow G$ [Grothendieck (1972), VIII.2]. Up to sign, the following diagram commutes

$$\begin{array}{ccccccc} 0 \longrightarrow & B_n & \longrightarrow & B & \xrightarrow{n} & B \\ & \downarrow & & \downarrow & & \downarrow \\ & \mathcal{H}om(A_n,G) & \longrightarrow & \mathcal{E}xt^{1}(A,G) & \xrightarrow{n} & \mathcal{E}xt^{1}(A,G) \end{array}$$

(and similarly with A and B interchanged). The pairing $B_n \times A_n \longrightarrow G$ defines pairings of cohomology groups.

Corollary C.5. *The diagram*

$$\begin{array}{ccccc} H^{r}(X,B) & \times & H^{s}(X,A) & \longrightarrow & H^{r+s+1}(X,G) \\ \downarrow & & \uparrow & & \| \\ H^{r+1}(X,B_n) & \times & H^{s}(X,A_n) & \longrightarrow & H^{r+s+1}(X,G) \end{array}$$

commutes.

Proof: This is obvious from the definitions.

When W is a Poincaré biextension on B×A, the pairing $B_n \times A_n \longrightarrow \mathbb{G}_m$ identifies each group with the Cartier dual of the other. For n prime to the characteristic, it agrees with Weil's e_n-pairing.

Néron models

From now on X is a Noetherian normal integral scheme of dimension one with perfect residue fields. The fundamental theorem of Néron [Néron (1964)] on the existence of canonical models can be stated as follows.

Theorem C.6. *Let* $g: \eta \longrightarrow X$ *be the inclusion of the generic point of* X *into* X. *For any abelian variety* A *over* η, g_*A *is represented on* X_{sm} *by a smooth group scheme* $\mathcal{A}$.

Proof: For a modern account of the proof, see [Artin (1986)].

The group scheme $\mathcal{A}$ is called the *Néron model* of A. It is separated and of finite type, and $\mathcal{A}_\eta = A$. It is obviously uniquely determined up to a unique isomorphism. Its formation commutes with étale maps $X' \longrightarrow X$ and with Henselization and completion. Let $\mathcal{A}^o$ be the open subscheme of $\mathcal{A}$ having connected fibres. Then there is an exact sequence of sheaves on X_{sm}

$$0 \longrightarrow \mathcal{A}^o \longrightarrow \mathcal{A} \longrightarrow \Phi \longrightarrow 0$$

in which Φ is a finite sum $\oplus\, i_{v*}\Phi_v$ with a Φ_v finite sheaf on v.

Assume now that X is the spectrum of a discrete valuation ring R, and write K and k for the field of fractions of R and its (perfect) residue field. As usual, i and j denote the inclusions of the closed and open points of X into X.

If the identity component $\mathcal{A}_0^o$ of the closed fibre $\mathcal{A}_0$ of $\mathcal{A}$ is an extension of an abelian variety by a torus, then A is said to have *semistable reduction*. In this case, the formation of $\mathcal{A}^o$ commutes with all finite field extensions $K \longrightarrow L$.

Theorem C.7. *There exists a finite separable extension* L *of* K *such that* A_L *has semistable reduction.*

Proof: There are several proofs; see for example [Grothendieck (1972), IX.3] or [Artin and Winters (1971)].

Let φ: A $\longrightarrow$ B be an isogeny of abelian varieties over K. From the definition of $\mathcal{B}$, we see that φ extends uniquely to an homomorphism φ: $\mathcal{A} \longrightarrow \mathcal{B}$. Write φ^o for the restriction of φ to $\mathcal{A}^o$, and let N = Ker(φ^o); it is group scheme over X.

Proposition C.8. *Let* φ: $\mathcal{A} \longrightarrow \mathcal{B}$ *be the map defined by an isogeny* A $\longrightarrow$ B. *The following conditions are equivalent:*

(a) φ: $\mathcal{A} \longrightarrow \mathcal{B}$ *is flat;*

(b) N *is flat over* X;

(c) N *is quasi-finite over* X;

(d) $(\varphi^o)\otimes_R k$ *is surjective.*

When these conditions are realized, then the following sequence is exact on X_{fl}:

$$0 \longrightarrow N \longrightarrow \mathcal{A}^o \xrightarrow{\varphi^o} \mathcal{B}^o \longrightarrow 0.$$

Proof: The same arguments as those in [Grothendieck (1972), IX.2.2.1] suffice to prove this result.

Corollary C.9. *Let* A *be an abelian variety over* K, *and let* n *be an integer. The following conditions are equivalent:*

(a) n: $\mathcal{A} \longrightarrow \mathcal{A}$ *is flat;*

(b) n: $\mathcal{A}^o \longrightarrow \mathcal{A}^o$ *is surjective;*

(c) $\mathcal{A}^o_n$ *is quasi-finite;*

(d) n *is prime to the characteristic of* k *or* A *has semistable re-*

duction.

Proof: It is easy to see from the structure of $\mathcal{A}^o\otimes_R k$ that condition (d) of the corollary is equivalent to condition (d) of the proposition. The corollary therefore follows directly from the proposition.

Biextensions of Néron models

From now on, we endow X with the smooth topology. Also we continue to assume that X is the spectrum of a discrete valuation ring, and we write x for its closed point. Let W be a Poincaré biextension on B×A, and write $i_*\Phi'$ and $i_*\Phi$ for $\mathcal{B}/\mathcal{B}^o$ and $\mathcal{A}/\mathcal{A}^o$ respectively. According to [Grothendieck (1972), VIII], there is a canonical pairing of $Gal(k_s/k)$-modules

$$\Phi' \times \Phi \longrightarrow \mathbb{Q}/\mathbb{Z}$$

which represents the obstruction to extending W to a biextension of $(\mathcal{B},\mathcal{A})$ by $\mathbb{G}_m$. We review this theory, but first we need a lemma.

If Γ is a submodule of Φ, then we write $\mathcal{A}^\Gamma$ for the inverse image of $i_*\Gamma$ in $\mathcal{A}$. Thus $\mathcal{A}$ has the same generic fibre as $\mathcal{A}$ and $\mathcal{A}_x/\mathcal{A}^o_x = \Gamma$.

Lemma C.10. *For any submodule $\Gamma \subset \Phi$, there is a canonical isomorphism*

$$j_*\mathcal{E}xt^1_{K_{sm}}(A,\mathbb{G}_m) \approx \mathcal{E}xt^1_{X_{sm}}(\mathcal{A}^\Gamma, j_*\mathbb{G}_{mK});$$

therefore $\mathcal{B} \approx \mathcal{E}xt^1_{X_{sm}}(\mathcal{A}^\Gamma, j_*\mathbb{G}_{mK})$.

Proof: We first show that $R^1j_*\mathbb{G}_m$ (computed for the smooth topology) is zero. For each Y smooth over X, $R^1j_*\mathbb{G}_m|Y_{et}$ is the sheaf (for the étale topology) associated with the presheaf $U \mapsto Pic(U_K)$. We shall show that in fact the sheaf associated with $U \mapsto Pic(U_K)$ for the Zariski topology is zero.

Let $y \in Y$, and let π be a uniformizing parameter for R. We have to show that $\mathrm{Pic}(\mathcal{O}_{Y,y}[\pi^{-1}]) = 0$. Let $Y' = \mathrm{Spec}\,\mathcal{O}_{Y,y}$ and write i' and j' for the inclusions $Z' \hookrightarrow Y'$ and $U' \hookrightarrow Y'$ corresponding to the maps $\mathcal{O}_{Y,y} \twoheadrightarrow \mathcal{O}_{Y,y}/(\pi)$ and $\mathcal{O}_{Y,y} \hookrightarrow \mathcal{O}_{Y,y}[\pi^{-1}]$. If $Z' = \emptyset$, then $\mathcal{O}_{Y,y}[\pi^{-1}] = \mathcal{O}_{Y,y}$, and the assertion is obvious. In the contrary case, Z' is a prime divisor on the regular scheme Y' (because Y is smooth over X), and so there is an exact sequence

$$0 \longrightarrow \mathbb{G}_m \longrightarrow j'_*\mathbb{G}_m \longrightarrow i'_*\mathbb{Z} \longrightarrow 0.$$

The cohomology sequence of this is

$$\cdots \longrightarrow \mathrm{Pic}(Y') \longrightarrow \mathrm{Pic}(U') \longrightarrow H^1(Y', i'_*\mathbb{Z}) \longrightarrow \cdots .$$

But $\mathrm{Pic}(Y') = 0$ because Y' is the spectrum of a local ring, and $H^1(Y', i'_*\mathbb{Z}) = H^1(Z', \mathbb{Z})$, which is zero because Z' is normal. Therefore $\mathrm{Pic}(U') = 0$.

As $j^*\mathcal{A}^\Gamma = A$, there is a canonical isomorphism of functors

$$j_*\mathcal{H}om_K(A,-) \approx \mathcal{H}om_X(\mathcal{A}^\Gamma, j_*-)$$

(see [Milne (1980), II.3.22]). We form the first right derived functor of each side and evaluate it at $\mathbb{G}_m$. Because $\mathcal{H}om_K(A,\mathbb{G}_m) = 0$, on the left we get $j_*\mathcal{E}xt^1_K(A,\mathbb{G}_m)$, and because $R^1j_*\mathbb{G}_m = 0$, on the right we get $\mathcal{E}xt^1_X(\mathcal{A}^\Gamma, j_*\mathbb{G}_m)$, which proves the lemma.

Let $B \xrightarrow{\approx} \mathcal{E}xt^1_K(A,\mathbb{G}_m)$ be the map defined by W. On applying j_*, we get a map $\mathcal{B} \xrightarrow{\approx} j_*\mathcal{E}xt^1_K(A,\mathbb{G}_m) \xrightarrow{\approx} \mathcal{E}xt^1_X(\mathcal{A}, j_*\mathbb{G}_{mK})$. From the exact sequence

$$0 \longrightarrow \mathbb{G}_m \longrightarrow j_*\mathbb{G}_m \longrightarrow i_*\mathbb{Z} \longrightarrow 0$$

we get an exact sequence

$$\mathcal{H}om_X(\mathcal{A}, i_*\mathbb{Z}) \longrightarrow \mathcal{E}xt^1_X(\mathcal{A},\mathbb{G}_m) \longrightarrow \mathcal{E}xt^1_X(\mathcal{A}, j_*\mathbb{G}_m) \longrightarrow \mathcal{E}xt^1_X(\mathcal{A}, i_*\mathbb{Z}).$$

But $\mathcal{E}xt^r_X(\mathcal{A}, i_*\mathbb{Z}) = i_*\mathcal{E}xt^r_X(i^*\mathcal{A},\mathbb{Z})$, and $i^*\mathcal{A} = \mathcal{A}_k$ because $\mathcal{A}$ is in the underlying category of X_{sm} (see [Milne (1980), II.3.1d]). Therefore $\mathcal{H}om_X(\mathcal{A}, i_*\mathbb{Z}) = i_*\mathcal{H}om_x(\mathcal{A}_x, \mathbb{Z}) = 0$ and $\mathcal{E}xt^1_X(\mathcal{A}, i_*\mathbb{Z}) = i_*\mathcal{E}xt^1_x(\mathcal{A}_x, \mathbb{Z}) = i_*\mathcal{H}om_x(\mathcal{A}_x, \mathbb{Q}/\mathbb{Z}) = i_*\mathcal{H}om_x(\Phi, \mathbb{Q}/\mathbb{Z})$. This gives us the lower row of the diagram below.

Lemma C.11. *There is a unique map* $\Phi' \longrightarrow \mathcal{H}om_x(\Phi, \mathbb{Q}/\mathbb{Z})$ *making*

$$\begin{array}{ccccccccc} 0 & \longrightarrow & \mathcal{B}^o & \longrightarrow & \mathcal{B} & \longrightarrow & i_*\Phi' & \longrightarrow & 0 \\ & & \downarrow & & \downarrow \approx & & \downarrow & & \\ 0 & \longrightarrow & \mathcal{E}xt^1_X(\mathcal{A}, \mathbb{G}_m) & \longrightarrow & \mathcal{E}xt^1_X(\mathcal{A}, j_*\mathbb{G}_m) & \longrightarrow & i_*\mathcal{H}om_x(\Phi, \mathbb{Q}/\mathbb{Z}) & \longrightarrow & 0 \end{array}$$

commute.

Proof: Obviously the composite of the maps

$$\mathcal{B} \longrightarrow \mathcal{E}xt^1_X(\mathcal{A}, j_*\mathbb{G}_m) \longrightarrow i_*\mathcal{H}om_x(\Phi, \mathbb{Q}/\mathbb{Z})$$

factors through $\mathcal{B}/\mathcal{B}^o = i_*\Phi'$.

To give a map of sheaves $\Phi' \longrightarrow \mathcal{H}om_x(\Phi, \mathbb{Q}/\mathbb{Z})$ is the same as to give a map of $\mathrm{Gal}(k_s/k)$-modules $\Phi' \longrightarrow \mathrm{Hom}(\Phi, \mathbb{Q}/\mathbb{Z})$, or to give an equivariant pairing $\Phi' \times \Phi \longrightarrow \mathbb{Q}/\mathbb{Z}$. We shall refer to the pairing defined by the map in the lemma as the *canonical pairing*.

Proposition C.12. *The biextension W of (B,A) by* $\mathbb{G}_m$ *extends to a biextension of* $(\mathcal{B}^{\Gamma'}, \mathcal{A}^{\Gamma})$ *by* $\mathbb{G}_m$ *if and only if* Γ' *and* Γ *annihilate each other in the canonical pairing between* Φ' *and* Φ. *The extension, if it exists, is unique.*

Proof: For a detailed proof, see [Grothendieck (1972), VIII.7.1b]. We merely note that it is obvious from the following diagram (extracted from C.11)

$$\begin{array}{ccccccccc} 0 \longrightarrow & \mathcal{B}^{o} & \longrightarrow & \mathcal{B}^{\Gamma'} & \longrightarrow & i_{*}\Gamma' & \longrightarrow 0 \\ & \downarrow & & \downarrow \text{inj} & & \downarrow & \\ 0 \longrightarrow & \mathcal{E}xt^{1}_{X}(\mathcal{A}^{\Gamma},\mathbb{G}_{m}) & \longrightarrow & \mathcal{E}xt^{1}_{X}(\mathcal{A}^{\Gamma},j_{*}\mathbb{G}_{m}) & \longrightarrow & i_{*}\mathcal{H}om_{X}(\Gamma,\mathbb{Q}/\mathbb{Z}) & \longrightarrow 0 \end{array}$$

that $\mathcal{B}^{\Gamma'} \longrightarrow \mathcal{E}xt^{1}_{X}(\mathcal{A}^{\Gamma},j_{*}\mathbb{G}_{m})$ factors through $\mathcal{E}xt^{1}_{X}(\mathcal{A}^{\Gamma},\mathbb{G}_{m})$ if and only if Γ' maps to zero in $\mathrm{Hom}(\Gamma,\mathbb{Q}/\mathbb{Z})$.

Conjecture C.13. *The canonical pairing $\Phi' \times \Phi \longrightarrow \mathbb{Q}/\mathbb{Z}$ is nondegenerate.*

The conjecture is due to Grothendieck (ibid., IX.1.3). We shall see in the main body of the chapter that it is a consequence of various duality theorems for abelian varieties.

Proposition C.14. *If Γ' annihilates Γ, then there is a commutative diagram*

$$\begin{array}{ccccccccc} 0 \longrightarrow & \mathcal{B}^{\Gamma'} & \longrightarrow & \mathcal{B} & \longrightarrow & i_{*}(\Phi'/\Gamma') & \longrightarrow 0 \\ & \downarrow & & \downarrow \approx & & \downarrow & \\ 0 \longrightarrow & \mathcal{E}xt^{1}_{X}(\mathcal{A}^{\Gamma},\mathbb{G}_{m}) & \longrightarrow & \mathcal{E}xt^{1}_{X}(\mathcal{A}^{\Gamma},j_{*}\mathbb{G}_{m}) & \longrightarrow & i_{*}\mathcal{H}om_{X}(\Gamma,\mathbb{Q}/\mathbb{Z}) & \longrightarrow 0. \end{array}$$

If conjecture (C.13) holds and Γ' is the exact annihilator of Γ, or if $\Gamma = 0$ and $\Gamma' = \Phi'$, then all the vertical maps are isomorphisms.

Proof: The diagram can be constructed the same way as the diagram in (C.11). The final statement is obvious.

The Raynaud group

Assume now that A has semistable reduction, and write Φ_0 for $\mathcal{A}_x/\mathcal{A}^{o}_x$. Over k there is an exact sequence

$$0 \longrightarrow T \longrightarrow \mathcal{A}^{o}_{x} \longrightarrow B \longrightarrow 0$$

with B an abelian variety over k. The next theorem shows that this sequence has a canonical lifting to R.

For any group scheme G over X, we write $\hat{G}$ for the formal completion of G along the closed point x of X [Hartshorne (1977), II.9].

Theorem C.15. *There is a smooth group scheme* $\mathcal{A}^{\#}$ *over* R *and canonical isomorphisms* $\hat{\mathcal{A}} \xrightarrow{\approx} (\mathcal{A}^{\#})\hat{}$ *and* $\hat{\mathcal{A}}^{o} \xrightarrow{\approx} (\mathcal{A}^{\#o})\hat{}$.

(a) *There is an exact sequence over* R

$$0 \longrightarrow \mathcal{T} \longrightarrow \mathcal{A}^{\#o} \longrightarrow \mathcal{B} \longrightarrow 0$$

with $\mathcal{T}$ *a torus and* $\mathcal{B}$ *an abelian scheme; the reduction of this sequence modulo the maximal ideal of* R *is the above sequence.*

(b) *Let* $\Phi = \mathcal{A}^{\#}/\mathcal{A}^{\#o}$; *then* Φ *is a finite étale group scheme over* R *whose special fibre is* Φ_0.

(c) *Let* $\mathcal{N} = (\mathcal{A}^{\#o})_p$; *then* $\mathcal{N}$ *is the maximal finite flat subgroup scheme of the quasi-finite flat group scheme* $\mathcal{A}^{o}_p$, *and there is a filtration*

$$A_p = (\mathcal{A}^{o}_p)_K \supset N \supset T_p \supset 0$$

with $N = \mathcal{N}_K$, $T_p = (\mathcal{T}_p)_K$, *and* $N/T_p = (\mathcal{B}_p)_K$.

(d) *Let* A^t *be the dual abelian variety to* A, *and denote the objects corresponding to it with a prime. The nondegenerate pairing of finite group schemes over* K

$$A'_p \times A_p \longrightarrow \mathbb{G}_m$$

induces nondegenerate pairings

$$N'/T'_p \times N/T_p \longrightarrow \mathbb{G}_m$$

$$A'_p/T'_p \times N \longrightarrow \mathbb{G}_m.$$

(e) *Assume* R *is Henselian. If* $A(K)_p = A(K_a)_p$, *then* Φ_p *has order*

p^{μ}, *where* μ *is the dimension of the maximal subtorus of* $\mathcal{A}_x$.

Proof: For (a), (b), and (c) see [Grothendieck (1972), IX.7].

(d) The restriction of the pairing on $A'_p \times A_p$ to $N' \times N$ extends to a pairing $\mathcal{N}' \times \mathcal{N} \longrightarrow \mathbb{G}_m$ induced by the biextension of $(\mathcal{A}^{to},\mathcal{A}^{o})$ by $\mathbb{G}_m$. This pairing is trivial on $\mathcal{T}'_p$ and $\mathcal{T}'_p$, and the quotient pairing on $\mathcal{B}'_p \times \mathcal{B}_p$ is that defined by the canonical extension of a Poincaré bi-extension of $(\mathcal{B}'_K,\mathcal{B}_K)$ by $\mathbb{G}_m$. This shows that T'_p and T_p are the left and right kernels in the pairing $N' \times N \longrightarrow \mathbb{G}_m$. The pairing $A'_p/T'_p \times N \longrightarrow \mathbb{G}_m$ is obviously right nondegenerate. But A'_p/T'_p has order $p^{2d-\mu}$ where d is the common dimension of A and A^t and μ is the common dimension of $\mathcal{T}$ and $\mathcal{T}'$, and N has order $p^{\mu+2\alpha}$ where α is the common dimensions of $\mathcal{B}$ and $\mathcal{B}'$. As $d = \mu+\alpha$, this proves that the pairing is also left nondegenerate.

(e) From the diagram

$$\begin{array}{ccccccccc} 0 & \longrightarrow & \mathcal{A}^o(R) & \longrightarrow & A(K) & \longrightarrow & \Phi(k) & \longrightarrow & 0 \\ & & \downarrow p & & \downarrow p & & \downarrow p & & \\ 0 & \longrightarrow & \mathcal{A}^o(R) & \longrightarrow & A(K) & \longrightarrow & \Phi(k) & \longrightarrow & 0 \end{array}$$

we obtain an exact sequence

$$0 \longrightarrow \mathcal{A}^o(R)_p \longrightarrow A(K)_p \longrightarrow \Phi(k)_p \longrightarrow \mathcal{A}^o(R)^{(p)}.$$

Let $a \in \Phi(k)_p$. There will exist a finite flat local extension R' of R such that a maps to zero in $\mathcal{A}^o(R')^{(p)}$ (because $p\colon \mathcal{A}^o \longrightarrow \mathcal{A}^o$ is a finite flat map), and so the image of a in $\Phi(k')$ lifts to $A(K')_p$. By assumption, $A(K)_p \xrightarrow{\approx} A(K')_p$, and so a lifts to $A(K)_p$. This shows that

$$0 \longrightarrow \mathcal{A}^o(R)_p \longrightarrow A(K)_p \longrightarrow \Phi(k)_p \longrightarrow 0$$

is exact, and the result follows by counting.

The group $\mathcal{A}^{\#}$ is called the *Raynaud group scheme*.

Néron models and Jacobians

Let X again be any Noetherian normal integral scheme of dimension one, and let π: Y $\longrightarrow$ X be a flat proper morphism of finite-type. Recall that $\mathbf{Pic}_{Y/S}$ is defined to be the sheaf on X_{Et} associated with the presheaf $X' \mapsto \text{Pic}(Y \times_X X')$. Write $\mathbf{P} = \mathbf{Pic}_{Y/X}$. When $\mathbf{P}$ is representable by an algebraic space, then $\mathbf{P}^{\tau} = \mathbf{Pic}^{\tau}_{Y/X}$ is defined to be the subsheaf of $\mathbf{P}$ such that, for all X-schemes X', $\mathbf{P}(X')$ consists of the sections ξ whose image in $(\mathbf{P}_x/\mathbf{P}^o_x)_{(X')}$ is torsion for all $x \in X$.

Assume now that Y is regular, that the fibres of Y over X are all pure of dimension 1, and that Y_η is a smooth and geometrically connected. Moreover, assume that $\mathcal{O}_X \xrightarrow{\approx} \pi_* \mathcal{O}_Y$ universally (that is, as sheaves on X_{fl}). For each closed point x of X, define d_x to be the greatest common divisor of the multiplicities of the irreducible components of Y_x (the multiplicity of $Y_i \subset Y_x$ is the length of $\mathcal{O}_{Y_i, y_i}$ where y_i is the generic point of Y_i).

Theorem C.16. (a) *The functor* $\mathbf{P}$ *is representable by an algebraic space locally of finite type over* X.

(b) *If* $d_x = 1$ *for all* x, *then* $\mathbf{P}^{\tau}$ *is representable by a separated group scheme over* X.

(c) *Assume that the residue fields of* X *are perfect. Under the hypothesis of* (b), $\mathbf{P}^{\tau}$ *is the Néron model of the Jacobian of* Y_η.

Proof: (a) Our assumption that $\mathcal{O}_S \xrightarrow{\approx} \pi_* \mathcal{O}_X$ universally says that π is cohomologically flat in dimension zero. Therefore the statement is a special case of a theorem of Artin [Artin (1969b)].

(b) This is a special case of [Raynaud (1970), 6.4.5].

(c) This is a special case of [Raynaud (1970), 8.1.4].

Remark C.17. The hypotheses in the theorem are probably too stringent.

The autoduality of the Jacobian

Let C be a smooth complete curve over a field k. Then there is a canonical biextension of (J,J) by $\mathbb{G}_m$, and the two maps $J \longrightarrow \mathcal{E}xt^1_k(J,\mathbb{G}_m)$ are isomorphisms (and differ only be a minus sign) (see [Milne (1986c), §6] or [Moret-Bailly (1985)]). It is this biextension which we wish to extend to certain families of curves.

Let $\pi\colon Y \longrightarrow X$ be a flat projective morphism with fibres pure of dimension one and with smooth generic fibre; assume that Y is regular and that Y has a section s over X. Endow both Y and X with the smooth topology. Then $R^1\pi_*\mathbb{G}_m$ is representable by a smooth group scheme $\mathbf{Pic}_{Y/X}$ over X. Write P^0 for the kernel of the degree map on the generic fibre; thus $P^0(X)$ is the set of isomorphism classes of invertible sheaves on Y whose restriction to s(X) is trivial and whose restriction to Y_η has degree zero. Note that $\mathbf{Pic}_{Y/X} = P^0 \oplus \mathbb{Z}$.

Theorem C.18. *There exists a biextension of* (P^0,P^0) *by* $\mathbb{G}_m$ *whose restriction to the generic fibre is the canonical biextension.*

Proof: In the case that **P** has connected fibres, this follows from the result [Grothendieck (1972), VIII.7.1b] that for any two group schemes B and A over X with connected fibres, and any nonempty open subset U of X, the restriction functor

$$\mathrm{Biext}_X(A,B;\mathbb{G}_m) \longrightarrow \mathrm{Biext}_U(A,B;\mathbb{G}_m)$$

is a bijection. See also [Moret-Bailly (1985), 2.8.2]. For the general case, we refer the reader to [Artin (1967)].

Connjecture C.19. *Assume that* Y *is the minimal model of its generic fibre. Then the maps* $P^0 \longrightarrow \mathcal{E}xt^1_X(P^0, \mathbb{G}_m)$ *(of sheaves for the smooth topology) induced by the biextension in* (C.18) *are isomorphisms.*

A proof of this conjecture has been announced by Artin and Mazur [Artin (1967)], at least in some cases. We shall refer to this as the *autoduality hypothesis*.

Notes: The concept of a biextension was introduced by Mumford [Mumford (1969)], and was developed by Grothendieck in [Grothendieck (1972)]. Apart from Néron's Theorem C.6, the theorems of Artin and Raynaud (C.16), most of the results are due to Grothendieck. The exposition is partly based on [McCallum (1986)].

... and so there ain't nothing more to write about, and I am rotten glad of it, because if I'd knowed what a trouble it was to make a book I wouldn't a tackled it and ain't agoing to no more. But I reckon I got to light out for the Territory ahead of the rest, because Aunt Sally she's going to sivilize me and I can't stand it. I been there before.

H. Finn

BIBLIOGRAPHY

[The date is the date of publication, except for a few seminar talks. Recent preprints have been assigned the date 1986.]

Artin, E. and Tate, J. 1961
Class Field Theory, Harvard University. (Reprinted by Benjamin, New York, 1968.)

Artin, M. 1967
Autoduality of the Jacobian (Notes by B. Wyman), Bowdoin College (mimeographed notes).

Artin, M. 1969a
The implicit function theorem in algebraic geometry, In *Algebraic Geometry*, (Internat. Coll. Bombay, 1968), Oxford U. Press, London, 21-71.

Artin, M. 1969b
Algebraization of formal moduli I, In *Global Analysis*, Tokyo, 21-71.

Artin, M. 1986
Néron models, In *Arithmetic Geometry* (Cornell and Silverman, Ed.), Springer, Heidelberg.

Artin, M., Grothendieck, A., and Verdier, J.-L. 1972/73
Théorie des Topos et Cohomologie Etale des Schémas, Lecture Notes in Math. 269, 270, 305, Springer, Heidelberg, 1972-73.

Artin, M. and Milne, J. 1976
Duality in the flat cohomology of curves, Invent. Math. **35** 111-129.

Artin, M. and Verdier, J.-L. 1964
Seminar on étale cohomology of number fields, AMS Woods Hole Summer Institute 1964, notes produced in a limited edition only, 5pp. (A translation has been published in Uspecki Mat. Nauk. **20** (1965), 13-18.)

Artin, M. and Winters, G. 1971
Degenerate fibres and stable reduction of curves, Topology **10** 373-383.

Atiyah, M. and MacDonald, I. 1969
Introduction to Commutative Algebra, Addison-Wesley, Reading.

Bashmakov, M. 1972
The cohomology of abelian varieties over a number field, Russian Math. Surveys **27** no.6, 25-71.

Bégeuri, L. 1980
Dualité sur un corps local à corps résiduel algébriquement clos, Mem. Soc. Math. Fr., **108**, fasc. 4, 121pp.

Berthelot, P. 1981
Le théorème de dualité plate pour les surfaces (d'apres J. S. Milne). In *Surfaces Algébriques* (Orsay, 1976 - 78) Lecture Notes in Math., 868, Springer, Heidelberg 1981, pp. 203 - 237.

Berthelot, P., Breen, L, and Messing, W. 1982
Théorie de Dieudonné Cristalline II, Lecture Notes in Math., 930, Springer, Heidelberg.

Bester, M. 1978
Local flat duality of abelian varieties, Math. Ann. **235**, 149-174.

Bloch, S. 1986

Algebraic cycles and higher K-theory, preprint.
Bloch, S. and Kato, K. 1986
p-adic etale cohomology, preprint.
Bogomolov, F. 1981
Points of finite order on an abelian variety, Math. USSR Izv. **17**, 55 -72.
Borel, A. 1979
Automorphic L-functions, In *Automorphic Forms, Representations, and L-functions*. Proceedings of Symposia in Pure Math. 33, part 2, 27-61.
Borel, A. and Harder, G. 1978
Existence of discrete cocompact subgroups of reductive groups over local fields, J. reine angew. Math. **298**, 53-64.
Bourbaki, N. Alg
Algèbre, Eléments de Math. 4,6,7,11,14,23,24, Hermann, Paris, 1947-59.
Bourbaki, N. Tpgy.
General Topology, Part 1, Addison-Wesley, Reading, 1966.
Breen, L. 1969a
Extensions of abelian sheaves and Eilenberg-MacLane algebras, Inventiones math. **9**, 15-44.
Breen, L. 1969b
On a nontrivial higher extension of representable abelian sheaves, Bull. A.M.S. **75**, 1249-1253.
Breen, L. 1975
Un théorème d'annulation pour certains Ext^i de faisceaux abeliens. Ann. Sci. Ecole Norm. Sup. (4), **8**, 339 - 352.
Breen, L. 1981
Extensions du groupe additif sur le site parfait. In *Surfaces Algébriques* (Orsay, 1976 - 78), Lecture Notes in Math. 868, Springer, Heidelberg 1981, 238 - 262,
Browder, W. 1977
Algebraic K-theory with coefficients $\mathbb{Z}/p$. In *Geometric Applications of Homotopy Theory* I, Lecture Notes in Math. 657, Springer, Heidelberg, pp 40-84.
Brumer, A. 1966
Pseudocompact algebras, profinite groups and class formations, J. Algebra **4**, 442-470.
Cartan, H. and Eilenberg, S. 1956
Homological Algebra, Princeton U. Press, Princeton, 1956.
Cassels, J. 1962
Proof of the Hauptvermutung, J. Reine Angew. Math. **211**, 95-112.
Cassels, J. 1964
The dual exact sequence, J. Reine Angew. Math. **216**, 150-158.
Cassels, J. 1965
On conjectures of Birch and Swinnerton-Dyer, J. Reine Angew. Math. **217**, 180-199.
Cossec, F. and Dolgachev, I. 1986
Enriques Surfaces. In preparation.
Deligne, P. 1966
Cohomologie à support propre et construction du foncteur $f^!$, 404-421; in Hartshorne 1966.
Deligne, P. 1977
Théorèmes de finitude en cohomologie ℓ-adique. In *Cohomologie Etale* (SGA 4½). Lecture Notes in Math. 569. Springer, Heidleberg, 233-261.
Demazure, M. and Gabriel, P. 1970

Groupes Algébriques, Tome 1, Géométrie algébriques - généralités - groupes commutatifs, Masson, Paris.

Deninger, C. 1984
On Artin-Verdier duality for function fields, Math. Z. **188**, 91-100.

Deninger, C. 1986a
An extension of Artin-Verdier duality to non-torsion sheaves, J. Reine Angew. Math. **366**, 18-31.

Deninger, C. 1986b
ℓ-adic Lefschetz numbers of arithmetic schemes, J. Reine Angew. Math., to appear.

Deninger, C. 1986c
Duality in the etale cohomology of one-dimensional proper schemes and generalizations, preprint.

Deninger, C. and Wingberg, K. 1986
Artin-Verdier duality for n-dimensional local fields involving higher dimensional algebraic K-sheaves, J. Pure Appl. Algebra, to appear.

Dieudonné, J. 1964
Fondements de la Géométrie Algébriques Moderne, La Presses de l'Université de Montréal.

Etesse, J.-Y. 1982a
Cohomologie du complexe de de Rham-Witt a coefficients dans un F-cristal unite et dualite plate pour les surfaces. C. R. Acad. Sci. Paris Ser. I Math., **294**, no. 18, 621-624.

Etesse, J.-Y. 1982b
Complexe de de Rham-Witt a coefficients dans un cristal. C. R. Acad. Sci. Paris Ser. I Math., **294**, no. 17, 581-584.

Fulton, W. 1984
Intersection Theory, Springer, Heidelberg.

Gabber, O. 1983
K-theory of Henselian local rings and Henselian pairs, Handwritten notes, 1983.

Gamst, J. and Hoechsmann, K. 1970
Products in sheaf cohomology, Tohoku Math. J. **22**, 143-162.

Gordon, W. 1979
Linking the conjectures of Artin-Tate and Birch-Swinnerton-Dyer, Compos. Math. **38**, 163-199.

Greenberg, M. 1961
Schemata over local rings, Annals of Math. **73**, 624-648.

Greenberg, M. 1966
Rational points in Henselian discrete valuation rings, Publ. Math. I.H.E.S, **31**, 59-64.

Grothendieck, A. 1971
Revêtements Etales et Groupe Fondamental. Séminaire de Géométrie Algébrique du Bois Marie 1960-1961 (SGA 1). Lecture Notes in Math. 224, Springer, Heidelberg 1971.

Grothendieck, A. 1972
Groupes de Monodromie en Géométrie Algébrique. I. Séminaire de Géométrie Algébrique du Bois-Marie 1967-1969 (SGA 7 I). Lecture Notes in Math. 288. Springer, Heidelberg 1972.

Grothendieck, A. and Dieudonné, J. 1971
Eléments de Géométrie Algébrique, I, Springer, Heidelberg.

Haberland, K. 1978
Galois Cohomology of Algebraic Number Fields, VEB Deutscher Verlag der Wissenschaften, Berlin.

Harder, G. 1965/66
Über die Galoiskohomologie halbeinfacher Matrizengruppen, I,

Math. Zeit. **90** (1965), 404-428, II, Ibid., **92** (1966), 396-415.
Harder, G. 1967/68
Bericht über neuere Resultate der Galoiskohomologie halbeinfacher Matrizengruppen, Jahresbericht d. D.M.V. **70** (1967/68), 182-216.
Harris, M. 1979
p-adic representations arising from descent on abelian varieties, Composit. Math. **39**, 177-245.
Hartshorne, R. 1966
Residues and Duality, Lecture Notes in Math, **20**, Springer, Heidelberg.
Hartshorne, R 1972
Cohomology with compact supports for coherent sheaves on an algebraic variety. Math. Ann., **195**, 199 - 207.
Hartshorne, R, 1977
Algebraic Geometry. Graduate Texts in Mathematics, No. 52. Springer, Heidelberg.
Hazewinkel, M. 1969
Abelian extensions of local fields, Thesis, Amsterdam.
Herstein, I. 1968
Noncommutative Rings, Carus Math. Monograph 15, MAA.
Hewitt, E. and Ross, K. 1963
Abstract Harmonic Analysis I, Springer, Heidelberg.
Hilton, P. and Stammbach, U. 1970
A Course in Homological Algebra, Springer, Heidelberg.
Illusie, L. 1979
Complexe de de Rham-Witt et cohomologie cristalline, Ann. Sci. Ecole Norm. Sup. (4), **12**, 501 - 661.
Jannsen, U. 1982
Galoismoduln mit Hasse-Prinzip, J. Reine Angew. Math. **337**, 154-158.
Jarden, M. 1969
Rational points on algebraic varieties over large number fields, Bull. AMS, **75**, 603-606.
Jarden, M. 1972
Elementary statements over large algebraic fields, Trans. AMS, **164**, 67-91.
Jensen, C. 1972
Les Foncteurs Dérivés de $\varprojlim$ *et leurs Applications en Théorie des Modules*, Lecture Notes in Math., 254, Springer, Heidelberg.
Kato, Katzuya 1979
A generalization of local class field theory by using K-groups, I, J. Fac. Science, Univ. Tokyo, Sec IA, **26**, 303-376.
Kato, Kazuya 1985/86
Duality theories for the p-primary etale cohomology, I, Algebraic and Topological Theories, 127-148; II, preprint.
Katz, N. 1970
Nilpotent connections and the monodromy theorem: Applications of a result of Turritin. Inst. Hautes Etudes Sci. Publ. Math., **39**, 175 - 232.
Katz, N. and Lang, S. 1981
Finiteness theorems in geometric classfield theory. (With an appendix by Kenneth A. Ribet.) Enseign. Math. (2), **27**, 285-319.
Kawada, Y. 1960
Class formations. VI. Restriction to a subfamily, J. Fac. Sci. Univ. Tokyo Sect. I **8**, 229-262.
Kawada, Y., 1971
Class formations, Proc. Symp. Pure Math., Vol XX, 96-114, AMS, Providence.

Kawada, Y. and Tate, J. 1955
On the Galois cohomology of unramified extensions of function fields in one variable, Amer. J. Math. **77**, 197-217.
Kazarnanovski, B. 1972
Proof of a theorem of Tate, Russ. Math. Surveys **27**, 56-68.
Klingen, N. 1983
Einbettungsprobleme mit proendlichen Kern und ihre Lösbarkeit bei Beschrankung der Verzweigung, J. No. Theory **16**, 106-118.
Kneser, M. 1965
Galois-Kohomologie halbeinfacher algebraischer Gruppen über p-adischen Körpern, I. Math. Zeit. **88** (1965), 40-47; II. ibid. **89** (1965), 250-272.
Kneser, M. 1966
The Hasse principle for H^1 of simply connected groups, Proc. Sympos. Pure Math., A.M.S. **9**, 159-163.
Kneser, M. 1969
Lectures on Galois Cohomology of Classical Groups, Lectures on Math. 47, Tata, Bombay.
Kottwitz, R. 1984
Stable trace formula: cuspidal tempered terms, Duke Math. J. **51**, 611-650.
Kramer, K. 1977
Two-descent for elliptic curves in characteristic two, Trans. Amer. Math. Soc. **232**, 279-295.
Labesse, J.P. 1984
Cohomologie, L-groups et fonctorialité, Compositio Math. **55**, 163-184.
Lang, S. 1958
Introduction to Algebraic Geometry, Interscience, New York.
Lang, S. 1959
Abelian Varieties, Interscience, New York.
Lang, S. 1966
Rapport sur la Cohomologie des Groupes, Benjamin, New York.
Lang, S. 1970
Algebraic Number Theory, Addison-Wesley, Reading.
Lang, S. 1983
Fundamentals of Diophantine Geometry, Springer, Heidelberg.
Lang, S. and Tate, J. 1958
Principal homogeneous spaces over abelian varieties, Amer. J. Math. **80**, 659-684.
Langlands, R. 1968
Representations of algebraic groups, Mimeographed notes, Yale, 31pp.
Langlands, R. 1983
Les Débuts d'une Formule des Traces Stable, Publ. Math. Univ. Paris VII, 13.
Lichtenbaum, S. 1984
Values of zeta-functions at non-negative values, In *Number Theory*, Noordwijkerhout 1983, Lecture Notes in Math. 1068, Springer, Heidelberg, 1984.
Lichtenbaum, S. 1986
The construction of weight two arithmetic cohomology, preprint.
Litvak, G. 1980
Nondegeneracy of the Tate pairing in Galois cohomology of finite modules over general local fields of zero characteristic, Ukrain. Math. J. **32**, 392-394.
McCallum, W. 1986
Duality theorems for Néron models, Duke Math. J., to appear.

Manin, Ju. 1971
Cyclotomic fields and modular curves, Russ. Math. Surveys, **26**, no. 6, 7-78.
Mattuck, A. 1955
Abelian varieties over p-adic ground fields, Ann. of Math. **62**, 92-119.
Mazur, B. 1970a
Local flat duality, Amer. J. Math. **92**, 343-361.
Mazur, B. 1970b
Finite flat structures, Proc. Symp. Pure Math. **17**, A.M.S., Providence, 219-225.
Mazur, B. 1972
Rational points of abelian varieties with values in towers of number fields, Invent. Math. **18**, 183-266.
Mazur, B. 1973
Notes on étale cohomology of number fields, Ann. Sci. Ecole Norm. Sup. **6**, 521-552.
Mazur, B. and Roberts, L. 1970
Local Euler characteristics, Invent. Math. **9**, 201-234.
Merkur'rev, A. and Suslin, A. 1982
K-cohomology of Severi-Brauer varieties and the norm residue homomorphism, Izv. Akad. Nauk. SSSR Ser. Mat., **46**, 1011-1046, 1135-1136.
Messing, W. 1972
The Crystals Associated to Barsotti-Tate Groups: with Applications to Abelian Schemes, Lecture Notes in Math. 264, Springer, Heidelberg.
Milne, J. 1967
The conjectures of Birch and Swinnerton-Dyer for constant abelian varieties over function fields, Thesis, Harvard University.
Milne, J. 1968
The Tate-Safarevic group of a constant abelian variety, Invent. Math. **5**, 63-84.
Milne, J. 1970a
The homological dimension of commutative group schemes over a perfect field, J. of Algebra, **16**, 436-441.
Milne, J. 1970b
Elements of order p in the Tate-Safarevic group, Bull. London Math. Soc. **2**, 293-296.
Milne, J. 1970/72
Weil-Chatelet groups over local fields, Ann. Sci. Ecole Norm. Sup. **3** (1970), 273-284; addendum, ibid., **5** (1972), 261-264.
Milne, J. 1972
On the arithmetic of abelian varieties, Invent. math. **17**, 177-190.
Milne, J. 1973
On a theorem of Mazur and Roberts, Amer. J. Math. **95**, 80-86.
Milne, J. 1975
On a conjecture of Artin and Tate, Annals of Math. **102**, 517-533.
Milne, J. 1976
Duality in the flat cohomology of a surface, Ann. scient. ENS, **9**, 171-202.
Milne, J. 1980
Etale Cohomology, Princeton Univ. Press, Princeton.
Milne, J. 1981
Comparison of the Brauer group with the Tate-Safarevic group, J. Fac. Science, Univ. Tokyo, Sec. IA, **28**, 735-743.

Milne, J. 1986a
Values of zeta functions of varieties over finite fields, Amer. J. Math., **108**, 297-360.

Milne, J. 1986b
Abelian varieties, In *Arithmetic Geometry* (Cornell, G. and Silverman, J. Eds.), Springer, Heidelberg.

Milne, J. 1986c
Jacobian varieties, *Arithmetic Geometry* (Cornell and Silverman, Ed.), Springer, Heidelberg.

Mitchell, B. 1965
Theory of Categories, Academic Press, New York.

Moret-Bailly, L. 1985
Métriques permises, In *Séminaire sur les Pinceaux Arithmétiques: La Conjecture de Mordell*, (Szpiro ed.), Astérisque **127**, 29-87.

Mumford, D. 1965
Geometric Invariant Theory, Springer, Heidelberg.

Mumford, D. 1969
Biextensions of formal groups, In *Algebraic Geometry*, (Internat. Coll. Bombay, 1968), Oxford University Press, Oxford, 1969.

Mumford, D. 1970
Abelian Varieties, Oxford University Press, Oxford.

Narkiewicz, W. 1974
Elementary and Analytic Theory of Algebraic Numbers, PWN -- Polish Scientific Pub., Warsaw.

Néron, A. 1964
Modèles minimaux des variétés abéliennes sur les corps locaux et globaux, Publ. Math. I.H.E.S. **21**, 1-128.

Neumann, O. 1977
Proper solutions of the imbedding problem with restricted ramification, Acta Arith. **33**, 49-52.

Oersterlé, J. 1982/83
Compatibilité de la suite exacte de Poitou-Tate aux mesures de Haar, Séminaire de Théorie des Nombres de Bordeaux, 1982/83, Ex 19.

Oesterlé, J. 1984
Nombres de Tamagawa et groupes unipotents en caractristique p, Inventiones math., **78**, 13-88.

Ogg, A. 1962
Cohomology of abelian varieties over function fields, Ann. of Math., **76**, 284-292.

Ono, Takahashi 1961
Arithmetic of algebraic tori, Annals of Math., **74**, 101-139.

Ono, Takahashi 1963
On the Tamagawa number of algebraic tori, Annals of Math., **78**, 47-73.

Oort, F. 1966
Commutative Group Schemes, Lecture Notes in Math. 15, Springer, Heidelberg.

Oort, F. 1967
Embeddings of finite group schemes into abelian schemes Mimeographed notes, Bowdoin College, 9pp..

Oort, F. 1971
Finite group schemes, local moduli for abelian varieties, and lifting problems. Compositio Math., **23**, 265 - 296.

Oort, F. and Tate, J. 1970
Group schemes of prime order, Ann. scient. E.N.S. **3**, 1-21.

Poitou, G. 1966
Remarques sur l'homologie des groupes profini, In *Les Tendances*

Géometrie en Algèbre et Théorie des Nombres, CNRS, Paris, 201-213.

Poitou, G. 1967
Cohomologie Galoisienne des Modules Finis, Dunod, Paris.

Raynaud, M. 1964/65
Caractéristique d'Euler-Poincaré d'un faisceau et cohomologie des variétés abéliennes (d'après Ogg-Shafarévitch et Grothendieck), Séminaire Bourbaki 1964/65, 286. (Published in: *Dix Exposés sur la Cohomologie des Schémas*, North-Holland, Amsterdam, 1968, 12-30.)

Raynaud, M. 1966
Modèles d'Néron, C.R. Acad. Sci. Paris **262**, 413-416.

Raynaud, M. 1970
Anneaux Locaux Henséliens, Lecture Notes in Math. 169.

Raynaud, M. 1971
Specialization du foncteur de Picard, Publ. Math. I.H.E.S. **38**, 27-76.

Raynaud, M. 1974
Schémas en groupes de type (p,...,p). Bull. Soc. Math. France, **102**, 241 - 280.

Raynaud, M. 1979
"p-torsion" du schema de Picard. In *Journées de Géométrie Algébrique de Rennes* (Rennes, 1978), Vol. II, Astérisque 64 (1979), pp. 87-148.

Rim, D. and Whaples, G. 1966
Global norm residue symbol map over quasi-finite field, Nagoya Math J. **27**, 323-329.

Roberts, L. 1973
The flat cohomology of group schemes of rank p, Amer. J. Math. **95**, 688-702.

Roos, J.-E. 1961
Sur les foncteurs dérivés de $\varprojlim$. Applications. Compte Rendu Acad. Sci. Paris **252**, 3702-3704. Erratum ibid. **254** p1722.

Rubin, K. 1985
Elliptic curves and $\mathbb{Z}_p$-extensions, Compos. Math. **56**, 237-250.

Sansuc, J.-J. 1981
Groupe de Brauer et arithmétique des groupes algébriques linéaires sur un corps de nombres, J. reine angew. Math. **327**, 12-80.

Serre, J.-P. 1959
Groupes Algébriques et Corps de Classes, Hermann, Paris.

Serre, J.-P. 1960
Groupes Proalgébriques, Publ. Math. IHES, **7**, 341-403.

Serre, J.-P. 1961
Sur les corps locaux à corps résiduel algébriquement clos, Bull. Soc. Math. France **89**, 105-154.

Serre, J.-P. 1962
Corps Locaux, Hermann, Paris.

Serre, J.-P. 1964
Cohomologie Galoisienne, Lecture Notes in Math. 5, Springer, Heidelberg.

Serre, J-P. 1964/71
Sur les groupes de congruence des variétés abéliennes, I. Izv. Akad. Nauk SSSR, Ser. Mat.**28** (1964), 3-20; II. ibid., **35** (1971), 731-737.

Serre, J.-P. 1967
Local class field theory, In *Algebraic Number Theory*, (Ed.

Cassels J. and Fröhlich, A.), Academic Press, London and New York, 129-161.
Serre, J.-P. 1967b
Représentations Linéaires des Groupes Finis, (Second edition) Hermann, Paris.
Serre, J.-P. 1977
Modular forms of weight one and Galois representations, In *Algebraic Number Fields: L-Functions and Galois Properties*, Academic Press, London, 193-268.
Shafarevich, I. 1954
Construction of fields of algebraic numbers with given solvable Galois group, Izv. Akad. Nauk SSSR, Ser. Mat. **18**, 525-578.
Shafarevich, I. 1962
Principal homogeneous spaces defined over a function field, A.M.S. Transl., Series 2, **36**, 85-114.
Shatz, S. 1964
Cohomology of artinian group schemes over local fields, Ann. of Math., **79**, 411-449.
Shatz, S. 1966
The cohomological dimension of certain Grothendieck topologies, Ann. of Math., **83**, 572-595.
Shatz, S. 1967
The cohomology of certain elliptic curves over local and quasi-local fields, Illinois J. Math. **11**, 234-241.
Shatz, S. 1972
Profinite groups, Arithmetic, and Geometry, Annals of Math. Studies 67, Princeton U. P., Princeton.
Shatz, S. 1986
Group schemes, formal groups, and p-divisible groups, In *Arithmetic Geometry*, (Cornell and Silverman, Ed.) Springer, Heidelberg.
Shelstad, D. 1986
Twisted endoscopic groups in the abelian case, preprint.
Suslin, A. 1983a
Torsion in K_2 of fields, preprint, I.H.E.S.
Suslin, A. 1983b
On the K-theory of algebraically closed fields, Invent. math., **73**, 241-245.
Suslin, A. 1984
On the K-theory of local fields, J. Pure Appl. Algebra, **34**, 301-318.
Swan, R. 1960
Induced representations and projective modules, Ann of Math., **77**. 267-291.
Szpiro, L. 1985
Degrés, intersections, hauteurs, In *Séminaire sur les Pinceaux Arithmétiques: La Conjecture de Mordell*, (Szpiro ed.), Astérisque 127, 11-28.
Takahashi, Toyofumi 1969
Galois cohomology of finitely generated modules, Tohoku Math. J. **11**, 102-111.
Tate, J. 1957/58
WC-groups over p-adic fields, Séminaire Bourbaki, Exposé 156, 13pp.
Tate, J. 1962
Duality theorems in Galois cohomology over number fields, *Proc. Intern. Congress Math.*, Stockholm, pp234-241.
Tate, J. 1965/66

On the conjectures of Birch and Swinnerton-Dyer and a geometric analog, Séminaire Bourbaki, 306. (Reprinted in *Dix Exposés sur la Cohomologie des Schémas*, North-Holland, Amsterdam, 1968.)

Tate, J. 1966
Letter to T. Springer, dated 13 Jan 1966, 6pp.

Tate, J. 1967a
Global class field theory, In *Algebraic Number Theory* (Cassels, J and Frohlich, A., ed.) Academic Press, London and New York, 163-203.

Tate, J. 1967b
p-divisible groups, *Proc. of a Conf. on Local Fields* (at Driebergen), Springer, Heidelberg, 1967, 158-183.

Tate, J. 1979
Number theoretic background. In *Automorphic Forms, Representations, and L-functions*, Part 2, Proc. Symp. Pure Math. 33, AMS, 3-26.

Thomason, R. 1984a
Absolute cohomological purity. Bull. Soc. Math. France, **112**, 397-406.

Uchida, K. 1969
On Tate duality theorems on Galois cohomology, Tohoku Math. J. **11**, 92-101.

Vvedens'kii, O. 1973
On local "class fields" of elliptic curves, Math. USSR Izvestija, **7**, 19-84.

Vvedens'kii, O. 1976
On quasi-local "class fields" of elliptic curves, I, Math. USSR Izvestija, **10**, 913-936.

Vvedens'kii, O. 1979a
The frequency of the appearance of the Artin-Milne effect in elliptic curves, Dokl. Akad. Nauk. SSSR **245**, 780-781.

Vvedens'kii, O. 1979b
The Artin effect in elliptic curves, I, Izv. Akad. Nauk. SSSR, Ser. mat. **43**, 1042-1053, 1197.

Vvedens'kii, O. 1980/81
The Artin effect in abelian varieties, I , Dokl. Akad. Nauk. SSSR **252** (1980), 20; II, Izv. Akad. Nauk SSSR **45** (1981), 23-46, 239.

Vvedens'kii, O. and Krupjak, I. 1976
A remark on the nondegeneracy of the Tate pairing in the Galois cohomology of finite modules over a general local field, Ukrain. Mat. Z. **28**, 526-528.

Wake, R. 1986
Phantom points on abelian varieties, preprint.

Weil, A. 1961
Adeles and Algebraic Groups, Mimeographed notes, IAS, 1961. (Reprinted as Progress in Math. 23, Birkhauser, Boston, 1982.)

Weil, A. 1967
Basic Number Theory, Springer, Heidelberg.

Weiss, E. 1969
Cohomology of Groups, Academic Press, New York.

Whaples, G. 1952/54
Generalized local class field theory, I, Duke Math. J. **19** (1952), 505-517; II, ibid. **21** (1954), 247-255; III, ibid. **21** (1954), 575-581; IV, ibid. **21** (1954), 583-586.

Zink, Th. 1978
Etale cohomology and duality in number fields, Appendix to Haberland 1978, p127-145.

INDEX